AF450148

Pests of Plantation Crops

The Editors

Dr. Pallem Chowdappa received M.Sc. in 1980 from Sri Venkateswara University, Tirupathi, Ph.D in 1985 from Mangalore University, Mangalore, Karnataka and post doctoral research at CABI Bioscience, U.K. He joined as Scientist-SI in 1985 at ICAR-Central Plantation Crops Research Institute, Kasaragod, Kerala and was elevated to Principal Scientist in 2006 at Indian Institute of Horticultural Research, Bangalore. Dr. Chowdappa served as Scientist-in-Charge, Central Plantation Crops Research Institute Research Centre, Hirehalli and Head, Central Horticultural Experimental Station, Hirehalli from December, 2000 till April, 2006. He became Director, Central Plantation Crops Research Institute, Kasaragod in September, 2014. Dr. Chowdappa is specialized in molecular plant pathology and has over 30 years of research experience in molecular characterization and management of *Alternaria, Colletotrichum* and *Phytophthora* associated with diseases of horticultural crops. He attended international training program on 'Oomycetes bioinformatics' at Virginia Tech, USA in 2014. Dr. Chowdappa was awarded DFID fellowship for Post-Doctoral research at CABI Bioscience, UK in 1998. Dr. Chowdappa has published more than 120 research papers in leading national and international journals, 12 books, 35 technical bulletins, 42 book chapters and 65 experimental manuals. He is a fellow of Scientific Academia and has won several awards of repute. He is also president of many scientific societies in India.

Dr. Chandrika Mohan has completed her M. Sc and M. Phil from Calicut University, Kerala and Ph.D. from Kerala University. She has more than 28 years of research experience in coconut pest management and is currently working as Principal Scientist (Agrl. Entomology) at ICAR-Central Plantation Crops Research Institute, Regional Station, Kayamkulam. She has contributed towards the development and validation of biocontrol strategies for sustainable pest management in coconut palm. She was associated with many projects funded by NATP, ICAR-AP cess fund, ICAR- out reach network projects, DBT and APCC on various IPM aspects of coconut. She has published more than 200 technical articles including more than 50 peer reviewed research articles in National and International Journals. She has received several awards including Fr. Gabriel Memorial Gold medal. She has functioned as focal point expert from India in International Workshops organized by SAARC and FAO. She is a fellow of Association of Advancement of Pest Management in Horticultural Ecosystems and life member in six scientific societies

Dr. A. Josephrajkumar is currently working as Principal Scientist in the discipline of Agricultural Entomology at ICAR-Central Plantation Crops Research Institute, Regional Station, Kayamkulam. He has undergone his undergraduate and postgraduate programme from Tamil Nadu Agricultural University, Coimbatore and completed his doctorate from Indian Agricultural Research Institute, Pusa, New Delhi. He has undergone a DBT-Post doctoral fellowship at SPIC Science Foundation, Guindy, Chennai on "Gut proteinases of cardamom shoot and capsule borer, *Conogethes punctiferalis*". He had worked as Assistant Professor at Cardamom Research Station, Kerala Agricultural University, Pamapdumpara during 1999-2007, refined IPM strategies for cardamom and black pepper and was instrumental in the release of bold capsuled cardamom variety PV-2. Subsequently, he has joined ICAR-CPCRI, Regional Station, Kayamkulam during 2007 and has been working on the IPM of coconut and vector entomology of root (wilt) disease. He was the Principal Investigator of ICAR-Consortium Research Platform on Borers targeting on red palm weevil by disruption of gut proteinases and endosymbionts. He has published more than 40 research papers in peer reviewed journals and has presented several papers in National and International symposia bestowed with oral presentation awards. He has received several National Awards including ICAR-Jawaharlal Nehru Award for best doctorate thesis, ICAR-Lal Bahadur Shastri Young Scientist Award and AZRA-Young Scientist award. He is also the Fellow of ICAR-Association for Advancement of Pest management in Horticultural Ecosystem, Bengaluru as well as IACR-Society for Plant Protection Sciences, New Delhi.

Pests of Plantation Crops

— Editors —

P. Chowdappa

Chandrika Mohan

A. Josephrajkumar

2018

Daya Publishing House®

A Division of

Astral International Pvt. Ltd.

New Delhi – 110 002

Publisher's Note:

Every possible effort has been made to ensure that the information contained in this book is accurate at the time of going to press, and the publisher and author cannot accept responsibility for any errors or omissions, however caused. No responsibility for loss or damage occasioned to any person acting, or refraining from action, as a result of the material in this publication can be accepted by the editor, the publisher or the author. The Publisher is not associated with any product or vendor mentioned in the book. The contents of this work are intended to further general scientific research, understanding and discussion only. Readers should consult with a specialist where appropriate.

Every effort has been made to trace the owners of copyright material used in this book, if any. The author and the publisher will be grateful for any omission brought to their notice for acknowledgement in the future editions of the book.

All Rights reserved under International Copyright Conventions. No part of this publication may be reproduced, stored in a retrieval system, or transmitted in any form or by any means, electronic, mechanical, photocopying, recording or otherwise without the prior written consent of the publisher and the copyright owner.

Cataloging in Publication Data--DK
Courtesy: D.K. Agencies (P) Ltd. <docinfo@dkagencies.com>

Pests of plantation crops / editors, P. Chowdappa, Chandrika Mohan, A. Josephrajkumar.
pages cm
ISBN 9789351249665 (Int. Edition)

1. Tropical crops--Diseases and pests--Integrated control.
I. Chowdappa, P. (Pallem), 1957- editor. II. Mohan, Chandrika, editor.
III. Josephrajkumar, A., editor.

LCC SB608.T8P47 2018 | DDC 632.30913   23

Published by   : **Daya Publishing House®**
                           A Division of
                           Astral International Pvt. Ltd.
                           – ISO 9001:2015 Certified Company –
                           4736/23, Ansari Road, Darya Ganj
                           New Delhi-110 002
                           Ph. 011-43549197, 23278134
                           E-mail: info@astralint.com
                           Website: www.astralint.com

Digitally Printed at   : **Replika Press Pvt. Ltd.**

Trilochan Mohapatra Ph.D
Secretary & Director General

Government of India
Department of Agricultural Research &
Education & Indian Council of Agricultural
Research Ministry of Agriculture and
Farmers' Welfare

Krishi Bhavan, New Delhi
Tel: 23382629, 23386711
Fax: 91-11-23384773
Email: dg.icar@nic.in

Foreword

Insect pests are the dominant living organism that limits production potential of plantation crops quite significantly. I am sure a keen understanding of pests is very essential to uncover the weak links so as to target such delicate sources for effective management of pests. Plantation crops survive in a very dynamic niche and foster perennial system in which mastering and cracking down key pests are very crucial in the era of doubling farmer's income. Pest management has now taken a paradigm shift towards agro-ecosystem based approaches with greater emphasis on biological suppression and volatile bio-engineering through crop-habitat diversification tools. A long-term sustainable approach is therefore the key, to combat pests in this multi-modal and multi-dimensional system.

I am quite confident that expertsin ICAR system and commodity Boards have harmoniously ventured on this book entitled "Pests of Plantations Crops" addressing all biotic constraints in industrial crops and innovative means to counter them. Cutting edge technologies on biological pest suppression, chemo-ecological approaches as well as neuropeptides were lucidly highlighted in the book for all stakeholders in accomplishing higher and sustainable production. Language is nicely edited even to cater the demand of farming community in this sector. It will be a good guide for all undergraduate, post-graduate students, researchers, plantation mangers and developmental agencies to glance for the recent updates in pest management in plantation crops.

T. Mohapatra

Preface

Pests have systematically evolved with plants in maintaining the biotic balance in the most perfect manner. Any drastic change in the biotic balance would therefore be catastrophic for the ecological well being of the biome in the system. It is always a point that one has to live with the insects and manage the population so that an economic threshold is maintained. As a perennial system, plantation crops and insects have co-existed as all the stages of crops are available round the year in the production system. A clear understanding of pests in the plantation crops is very essential to boost its production trend in the back drop of climatic change, and other exigencies. Understanding to unseat the pests in the plantation sector is a great challenge.

Several experts in the plantation sector have contributed their chapters to know the key pests in the respective field and evolve effective management strategies to knock down them to make the sector vibrant. The cutting-edge technologies narrated in the book to counter the pest attack is very lucidly scripted and pictorially demonstrated. With a clear intervention of these technologies in the plantation sector, a paradigm shift would be anticipated in the pest management scenario.

Pest management in plantation crops has now come in the form of a capsule for the stakeholders to read and digest and practice to harvest the benefits. We are sure the treatise of knowledge in the book and simple language style followed would be everlasting in the minds of the readers to know about a birds eyes view of pest management in plantation sector and the simple tips to be followed for effective management of these pests. Care was given for the long-term bio-suppression strategies to be effectively followed through area-wide and farmer-participatory approach to make IPM a reality. From a IPM concept, pest management has been translated for the agro-ecosystem based approach to a crop health concept to take a comprehensive and inclusive health care approach. We take this opportunity to thank the distinguished contributors who provided the manuscripts timely to complete the publication. It is hoped that the book will be useful to all stakeholders involved in plantation crops research and development as management in the anchor point of crop production to accomplish the glory of success in farming.

P. Chowdappa
Chandrika Mohan
A. Josephrajkumar

Contents

Nayanie S. Aratchige

3. Arecanut 79

P.S. Prathibha and Shivaji H. Thube

4. Cocoa 97

M. Sujithra and M. Alagar

5. Oil Palm   119

1. Introduction; 2. Pests of Nursery; 2.1. Insect Pests; 2.2. Vertebrate Pests; 2.3. Molluscan Pests; 3. Pests of Adult Palms; 3.1. Rhinoceros Beetle, Oryctes rhinoceros L. (Coleoptera : Scarabaeidae); 3.2. Bag Worm, Metisa plana (Lepidoptera: Psychidae); 3.3. Leaf Web Worm, Acria meyricki. (Lepidoptera: Depressaridae); 3.4. Slug Caterpillar Darna catenatus Snellen, (Lepidoptera: Limacodidae); 3.5. Coccoids (Scales and mealybugs); 3.6.. Termites (Termitidae : Isoptera); 4. Vertebrate Pests; 4.1. Avian Pests; 4.2. Mammalian Pests; References.

P. Kalidas and L. Saravanan

6. Spice Crops   143

1. Introduction; 2. Black pepper; 2.1. Major Insect Pests; 2.2. Minor Insect Pests; 3. Cardamom; 3.1. Major Insect Pests; 3.2. Minor Insect Pests; 4. Ginger and Turmeric; 4.1. Major Insect Pests; 4.2. Minor Insect Pests; References.

S. Devasahayam, T.K. Jacob and C.M. Senthil Kumar

7. Tea   163

1. Introduction; 2. Crop Loss; 3. Methods of Pest Control; 3.1. Cultural Alternatives; 3.2. Host Plant Resistance; 3.3. Plant Products; 3.4. Cultural Control; 3.6. Biological Control; 3.8. Chemical Control; 3.9. Integrated Pest Management; Acknowledgement; References.

B. Radhakrishnan

8. Cashew   177

1. Introduction; 2. Tea Mosquito Bug (TMB) - Helopeltis spp.; 2.1. Species Composition and Distribution; 2.2. Nature of Damage; 2.3. Seasonal Incidence; 3. Cashew Stem and Root Borer (CSRB); 3.2. Nature of Damage; 3.3. Biology; 3.4. Seasonal Incidence; 4. Shoot Tip Caterpillar; 5. Leaf Miner; 6. Hairy Caterpillars; 7. Leaf Thrips; 8. Leaf Beetles and Weevils; 9. Pests of Cashew Apples and Nuts; 10. Inflorescence Feeders; 10.1. Leaf and Blossom Webber; 10.2. Flower Thrips; 11. Integrated Pest Management; 11.1. Cultural Practices; 11.2. Mechanical Methods; 11.3. Host Plant Resistance; 11.4. Botanicals; 11.5. Pheromones and Kairomones; 11.6. Chemical Control; 11.7. Biological Control; 12. Conclusion; References.

P.S. Bhat, T.N. Raviprasad, K. Vanitha and K.K. Srikumar

9. Rubber   199

1. Introduction; 2. Cockchafer Grubs; 3. Bark Feeding Caterpillar; 4. Scale Insects and Mealybugs; 5. Termite (White Ant); 6. Borer Beetles; 7. Crickets; 8. Slugs and Snails; 9. Rat; 10. Wild Animals; 11. Pests of Cover Crops; 12. Pests Affecting Plantation Workers; 12.1. Mooply Beetles; 12.2. Leeches; References.

S. Thankamony

List of Contributors

Alagar, M.
Assistant Professor (Agrl. Entomology)
KVK, Sikkal (TNAU)
Nagapattinam – 611 108
Tamil Nadu
E-mail: siaamalagar@gmail.com

Bhat, P.S.
ICAR-Indian Institute of Horticultural
Research, Hessarghata P.O.
Bangalore – 560 089, Karnataka
Karnataka
E-mail: pshivarama59@gmail.com

Chandrika Mohan
Principal Scientist (Agrl.Entomology)
ICAR-Central Plantation Crops Research
Institute
Regional Station, Kayamkulam
Post Krishnapuram – 690 533
Kerala
E-mail: cmcpcri@gmail.com

Chowdappa, P.
Director
ICAR-Central Plantation Crops Research
Institute
Kasaragod – 671 124
Kerala
E-mail: pallem22@gmail.com

Devasahayam, S.
Principal Scientist (Agrl. Entomology)
ICAR-Indian Institute of Spices Research
Marikunnu P.O., Kozhikode – 673 012
Kerala
E-mail: sdsahayam@gmail.com

Jacob, T.K.
Principal Scientist (Agrl. Entomology)
ICAR-Indian Institute of Spices Research
Marikunnu P.O., Kozhikode – 673 012,
Kerala
E-mail: jacobtk@spices.res.in

Josephrajkumar A.
Principal Scientist (Agrl.Entomology)
ICAR-Central Plantation Crops Research Institute
Regional Station
Kayamkulam, Post.Krishnapuram – 690 533
Kerala
E-mail: joecpcri@gmail.com

Kalidas, P.
ICAR-Indian Institute of Oil Palm Research
Pedavegi, West Godavari (Dt.) – 534 450
Andhra Pradesh
E-mail: potinenikalidas@gmail.com

Kesavan Subaharan
Principal Scientist (Agrl.Entomology)
Division of Insect Ecology
ICAR – National Bureau of Agricultural Insect Resources
Hebbal
Bengaluru – 560 024, Karnataka
E-mail: subaharan_70@yahoo.com

Nayanie S. Aratchige
Crop Protection Division
Coconut Research Institute of Sri Lanka
Lunuwila – 61150
Sri Lanka
E-mail: nsaratchige@gmail.com

Prathibha, P.S.
Scientist (Agrl.Entomology)
ICAR-Central Plantation Crops Research Institue
Kasaragod – 671 124, Kerala
E-mail: prathibhaspillai007@gmail.com

Radhakrishnan, B.
Director
UPASI Tea Research Foundation
Tea Research Institute
Nirar Dam P.O.,
Valparai – 642 127, Coimbatore Dist.
Tamil Nadu
E-mail: dr.radhaupasi@gmail.com

Rajesh, M.K.
Principal Scientist (Biotechnology)
ICAR-Central Plantation Crops Research Institute
Kasaragod – 671 124, Kerala
E-mail: mkraju.cpcri@gmail.com

Rajkumar
Scientist (Nematology)
ICAR-Central Plantation Crops Research Institute
Kasaragod – 671 124, Kerala
E-mail: rajkumarcpcri@gmail.com

Raviprasad, T.N.
ICAR-Directorate of Cashew Research
Puttur- 574 202, Karnataka
E-mail: tnrprasaad@gmail.com

Saravanan, L.
ICAR-Indian Institute of Oil Palm Research
Pedavegi, West Godavari (Dt.) – 534 450
Andhra Pradesh
E-mail: laxmansaravanars@gmail.com

Senthil Kumar, C.M.
Senior Scientist (Agrl.Entomology)
ICAR-Indian Institute of Spices Research
Marikunnu P.O., Kozhikode – 673 012
Kerala
E-mail: senthilkumarcm@spices.res.in

Srikumar, K.K.
UPASI Tea Research Foundation
Tea Research Institute
Valparai – 642 127
Tamil Nadu
E-mail: sreeku08@gmail.com

Sujithra, M.
Scientist (Agrl. Entomology)
Division of Crop Protection
ICAR- Central Plantation Crops Research Institute
Kasaragod – 671 124, Kerala
E-mail: amitysuji@gmail.com

Thankamony, S.
Principal Scientist (Entomology)
Rubber Research Institute of India
Rubber Board PO, Kottayam – 686 009
Kerala
E-mail: thankamony1955@gmail.com

Shivaji H. Thube
Scientist (Agrl.Entomology)
ICAR-Central Plantation Crops Research Institute
Regional Station, Vittal – 574 243
Karnataka
E-mail: shivajithube@gmail.com

Vanitha K.
Scientist (Agrl. Entomology)
ICAR-Directorate of Cashew Research
Puttur – 574 202
Karnataka
E-mail: vanitha.k@icar.gov.in

2018, Pests of Plantation Crops

Editors: **P. Chowdappa, Chandrika Mohan & A. Josephrajkumar**

Published by: **ASTRAL INTERNATIONAL PVT. LTD., NEW DELHI**

Pages **1–54**

Chapter 1

Coconut

☆ *Chandrika Mohan, A. Josephrajkumar*
and P. Chowdappa

1. Introduction

Coconut (*Cocos nucifera* L.) is one of the major plantation crops in India with an annual production of 20,439.60 million nuts. Coconut plays an important role in the environment, health, food and livelihood securities of millions of people. Even though the crop is cultivated in 93 countries globally, about 78.08 per cent of the world production is contributed by the four major nations, *viz.*, Indonesia, the Philippines, India and Sri Lanka. The current average productivity of coconut in India, being 10345 nuts/ha (CDB, 2016), could be doubled through optimum plant and soil health management. Consistent flowering and fruiting pattern of coconut palm throughout the seasons is the favourable abode for a wide array of insects, mites and rodents and Kurian *et al.* (1979) enlisted 547 insects and mite species infesting coconut worldwide. Among the insects, rhinoceros beetle, red palm weevil, black headed caterpillar, eriophyid mite and white grub are the major pests of concern, which are widely distributed in all coconut growing tracts of India adversely affecting coconut industry to a larger extent.

2. Rhinoceros Beetle (*Oryctes rhinoceros* Linn.) (Coleoptera : Scarabaeidae)

The Asiatic rhinoceros beetle or black beetle was reported damaging coconut palms in 1889 by Ridley. It is native to the southern Asiatic region, but has reached Africa, Australia and Pacific Islands. A number of workers have expanded the geographical distribution of *O. rhinoceros* (Gressitt, 1953; Nirula, 1955a, b; Swan 1974; Hill, 1983). There is an exhaustive literature on rhinoceros beetles, of which comprehensive reviews have been published by Nirula *et al.* (1952), Gressitt (1953) and Bedford (1980, 2013).

Coconut palm is the primary host of rhinoceros beetle. It is a major pest in oil palm also leading to 'dead heart' symptoms. A broad list of food plants infested by *O. rhinoceros* was provided by Gressitt, (1953), which included 45 species of monocot plants, over 30 species of palms. Range of host plants invaded by black beetle are reported from India (Menon and Pandalai, 1960; Nirula *et al.*, 1952, Sivakumar and Chandrika, 2013), Indonesia (Kalshoven, 1951), the Philippines (Mackie, 1917) and Mauritius (Monty, 1978) including palmyrah, wild date, areca, date, sago palm, pandanus, pineapple, colocasia, banana, sugarcane *etc.* however, there is not much economic damage on host plants other than coconut and oilpalm.

2.1. Damage

The robust adult beetles cause damage to palms at all age groups by burrowing into the unopened spear leaf and feeds on juice from the host tissues. As the pest burrows deeper into the host (10-50 cm) it pushes out the chewed up tissues as fibres, which are seen extruding from the entry points. Once these injured leaves unfurl, they present a 'V' shaped cut pattern, reducing the functional leaf area

Figure 1.1: Rhinoceros Beetle Infestation.

a) Damage symptom on unfurled leaves, b) Spear leaf damage, c) Tusk like symptom due to beetle feeding, d) Twisted growth of juvenile palm due to beetle attack.

considerably (Figure 1.1). The damage to inflorescence is seen as round to oblong holes on the spathes which soon dry up resulting in complete loss of nuts in the affected bunch (Figure 1.2).

Attack in juvenile palms results in stunted growth and delayed flowering and the repeated attack at the growing point may even lead to palm death. In the recent past, it was found invading coconut seedlings in the nursery as well as juvenile palms through collar entry and completely damaging the seedlings. Of late, the pest was found boring into the immature tender nuts (Figure 1.2b) causing yet another route of feeding when the spear leaf is protected.

Figure 1.2
(a) Drying of spathe, (b) damage on the nut due to rhinoceros beetle.

Pest attack impedes palm growth and up to 10 per cent reduction in crop yield is reported. Black beetle infestation has to be considered serious as the damage done by this pest provides egg laying sites for another lethal pest *viz.*, red palm weevil or for entry of fungal pathogens. Attacks tend to be concentrated on the margins of palm groves and on taller, more prominent palms (Cumber, 1957; Young, 1975). One attack increases the likelihood of further attacks (Bedford, 1975; Young, 1975). Sison (1957) reported that palms with 50 per cent of all their fronds damaged had about one-fifth the number of developing nuts than on normal palms. Being a perennial crop, workers reported many difficulties in studying the effect of beetle attack on yield (Ramachandran, 1961: Ramachandran *et al.*, 1963; Young, 1975). In India damage of inflorescence is also reported in severely infested areas which cause reduction in yield up to 10 per cent (Nair, 1986) and of 5.5 to 9.1 per cent yield loss by Ramachandran *et al.* (1963). From artificially pruned leaf damage stimulation studies it was observed that damage to 50 per cent fronds corresponds to leaf area reduction of 13 per cent and decrease in nut yield by 23 per cent (Young, 1974). Pruning equivalent to1.5 attacks/month would have killed the palms if sustained long enough, even though the growing points were undamaged (Young, 1975). In south-eastern Luzon, Philippines, *O. rhinoceros* damage was significantly more common on old palms (which are predisposed to the disease), and more abundant on those infected with cadang-cadang than on neighbouring healthy palms (Zelazny and Pacumbaba, 1982). In Samoa, more attacks occur on higher than on lower palms, and that one attack is likely to be followed by others on the same palm. Attacks are apparent to an observer on the ground after about 41 days and continue to be so for up to 150 days, as the damaged fronds open (Young, 1975).

2.2. Biology

Adults of *O. rhinoceros* are large beetles 30-50 mm long and 14-21mm breadth, black or reddish black in colour, stout and possesses a characteristic cephalic horn which is relatively larger in males (Figure 1.3d). Females can be identified by the densely clothed reddish brown hairs on pygidium on the ventral surface (Figure 1.3e). The females deposit yellowish-white eggs (Figure 1.3a) in dead and decaying vegetable matter such as cattle dung, compost, heaps of sawdust, felled logs and stumps particularly of the coconut palm and oil palm. Average fecundity per female is 108 eggs. The larval and pupal stages are completed in the breeding grounds. The grubs are creamy white in colour with the body strongly arched dorsally. Grub (Figure 1.3b) period is about 130 days with three instars. The pupa is uniformly brown, slightly convex (Figure 1.3c) dorsally and the pupal period varies from 20-29 days. Adult longevity is 3 - 4 months. Adults are active during night and remain hidden during daytime in the feeding or breeding sites. The life cycle from egg to adult stage takes about 5-6 months (Nirula *et al.*, 1952; Nirula 1955a, b). The three instars of grub period last 74 to 191 days with an average of 130 days (Nirula, 1955b) on the West Coast of India.

2.3. Pest Management

Since the pest is an active flyer, integrated pest management (IPM) strategies adopted on a community basis are essential to bring an effective control of *O. rhinoceros* population. The major components of IPM package consist of sanitation, mechanical, chemical and biological methods.

2.3.1. Sanitation Method

In the IPM of *O. rhinoceros*, the best management decision to be implemented is elimination of pest multiplication sites from the coconut plantation itself that provide favourable niche for immature stages of the beetle. The dead and decaying coconut logs, heaps of fallen coconut leaves, shredded palm tissues and other organic debris in the vicinity of coconut plantations may be properly disposed off, since this act as prolific breeding grounds of the beetle (Figure 1.4a). The moisture content of the food material plays a very important role, the grub not being able to develop in absolutely dry or water soaked food materials. Farm yard manure should be properly dried and stored since low moisture content did not favour development of this pest. Composting tanks has to be iron netted to prevent beetle access.

Incorporation of the weed plant, *Clerodendron infortunatum* Linn. @ 10 per cent w/w in the farm yard manure/compost pit is suggested as a probable management strategy targeting grub stages (Chandrika and Nair, 2000) as this plant exerts insect growth regulatory properties on *O. rhinoceros*. Larval-pupal or pupal-adult intermediates, adults with malformed wings *etc.* are some of the common abnormalities elicited by the plant alkaloids on *O. rhinoceros* when the plant part is ingested by the grub along with food. These malformed adults were unable to fly and survived for only 6-8 days.

Figure 1.3a-e: (a) Egg and neonate grub, (b) Fully grown grubs, (c) Pupae, (d) Male beetle and (e) Female beetle.

2.3.2. Mechanical Control

This method involves periodic examination of the palm crown and removing the adult beetle by means of a metal hook (Figure 1.4c) during peak periods of pest abundance (June- September), but often only after damage has been done (Cherian and Anantanarayanan, 1939; Gressitt, 1953; McKenna and Shroff, 1911; Nair *et al.*, 1997). Care should be taken not to inflict any damage on the developing inner core palm tissues during this process and hole has to be filled with a mixture of neem cake and sand.

Figure 1.4

(a) Rhinoceros beetle breeding on decaying palm trunk, (b) *C. infortunatum* induced malformation in rhinoceros beetle, (c) Hooking out of beetle from infestested site.

2.3.3. Prophylactic Measures to Prevent Pest Entry

Placement of naphthalene balls @ 12g/palm (3-4 numbers) in the inner leaf axils with sand coverings to prevent quick evaporation of the compound is also found to be effective in preventing the pest incidence in young palms (Sadakathulla and Ramachandran, 1990). Application of oil cakes of neem (*Azadirachta indica* A.Juss.) or marotti (*Hydnocarpus wightiana* Bl.) in powder form @ 250g mixed with equal volume of sand, thrice a year during May, September and December to the base of three leaf axils surrounding spear leaf is an effective prophylactic method against rhinoceros beetle and red palm weevil (Chandrika *et al.*, 2001) (Figure 1.5a). Placement of two perforated sachets (Figure 1.5b) each containing 3 g chlorantraniliprole (0.4 per cent ai) or fipronil (80 per cent *ai*) or botanical cake (Figure 1.5c) (20g) developed by ICAR-CPCRI was found effective during monsoon phase. During dry period, 100 ml of water may be poured over the sachet after placement to release the molecule (Josephrajkumar *et al.*, 2015a). ICAR-CPCRI has also developed a botanical paste formulation (Figure 1.5d) for effective repulsion of the pest when swiped on the base of leaf or top most leaf axils of juvenile palms.

2.3.4. Biological Control

This method is the most important component in the IPM of *O. rhinoceros*. Insect predators are frequently observed in the breeding grounds of the beetle. They feed on the eggs and early instar larvae of the beetle. Two potential microbial agents *viz.*, *Metarhizium anisopliae* (Figure 1.7) and *Oryctes rhinoceros nudivirus* (OrNV) (Figure 1.6) cause disease to the immature and adult stages of the beetle. Use of these microbial control agents is advantageous because they are relatively host specific, does not cause environmental pollution, safe to humans and are compatible with other control methods.

Figure 1.5

(a) Leaf axil filling with oilcake sand mixture, (b) Chlorantraniliprole sachet for leaf axil filling, (c) Botanical cake and (d) Paste developed by ICAR-CPCRI for prophylactic leaf axil filling.

2.3.5. Predators

The important predators are *Santalus parallelus* Payk., *Pheropsophus occipitalis* Macleay, *P. lissoderus* Chaudior, *Chelisoches morio* (Fab.) and species of *Scarites*, *Harpalus* and *Agrypnus*; (Antony and Kurien, 1966, Kurien *et al.*, 1983, Sathiamma *et al.*, 1982). *Platymeris laevicollis* Distant (Hemiptera: Reduviidae) is an exotic predator on rhinoceros beetle. *P. laevicollis* was imported from Zanzibar to India for the control of rhinoceros beetle. As compared to the indigenous predators, *P. laevicollis* feeds on adult beetles. Egg to adult period is completed in 131-161 days. The predator is long lived (170-240 days) and fecundity high (110-170 eggs/female). The predator mass multiplied on ground roaches was field released in coconut plantations in Kerala and Karnataka @ 6 bugs/palm and could achieve significant reduction in beetle population and the damage to the palm. Leaf damage was reduced to 13.1 per cent, nil spathe damage and 1 per cent spear leaf damage as compared to 59.2 per cent, 2.5 per cent and 37.0 per cent respectively, recorded during pre release

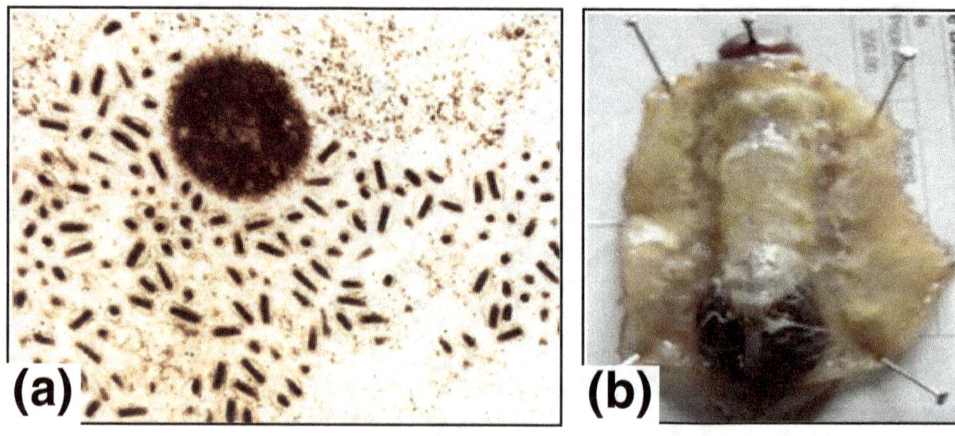

Figure 1.6

(a) OrNV particles under EM, (b) OrNV infected (fluid-filled) gut of *Oryctes* grub.

observations. But the predators failed to establish under field conditions (Antony *et al.*, 1979).

2.3.6. Pathogens

2.3.6.1. *Oryctes rhinoceros Nudi Virus*

The infection of *O.rhinoceros* by virus, Baculovirus of Oryctes (OBV) was first reported in Malaysia in 1966. Zelazny (1981) reported its presence in India and detailed work on it was done by Mohan *et al.* (1983), Mohan and Gopinathan (1989a,b) and Mohan and Gopinathan (1992). After detailed genomic characterization Mohan and Gopinathan (1992) proposed the taxonomic status of the variant for *OBV* K1, the Kerala isolate of *OBV*. After taxonomic revision of the genus Baculovirus of orcytes (OBV) was re-designated as Oryctes rhinoceros nudivirus (OrNV) through molecular characterization. *Or*NV gains entry in to the host orally through contaminated food material. It multiplies in the midgut epithelium and fat bodies of grubs and adults and also in the reproductive cells (Majumder and Jacob, 1993). Apart from *O.rhinoceros* it is pathogenic to *O.nasicornis* Linnaeus, *O.monoceros* (Olivier), *O.boas* Fabricius, *Scapanes australis grossepunctatus* Sternb., *Papuana uninodis* Prell and *Xylotrupes gideon* L. (Danger *et al.*, 1994).

All the three instars of grubs and adults of *Oryctes* are infected by the virus especially the 1st and 3rd instar grubs are more susceptible (Mohan *et al.*, 1985a), pupae are not susceptible to this disease. Infected grubs become lethargic, stop feeding and crawl to the surface of the feed. As the virus multiples, the haemolymph content increases, fat bodies disintegrate and the midgut filled with black solid food is replaced with white viscous mucoid fluid which makes the grub appear translucent, when observed against light (Figure 1.6b). Extroversion of the rectum due to increased turgor pressure is also noticed. Infected grubs die within 6 to 30 days. Diseased adults too become inactive, short-lived (by 60 per cent) and lay less number of eggs (1-2 eggs).

Diagnosis of *Or*NV: Lethargic condition, crawling to the surface of the feed, development of translucency in grubs and inactivity of adults are the key exopathological indications of this disease. Presence of this virus in the host can be detected by 3 per cent Giemsa staining of midgut fluid or midgut epithelium where in the pink colored hypertrophied nuclei with dark pink peripheral ring (Mohan *et al.*, 1983) is observed in the infected sample under microscope. Midgut slices fixed, stained and when observed under Electron Microscope shows the presence of rod-shaped viral particles (Figure 1.6a) Immunofluorescence, immuno-osmophoresis and ELSIA techniques (Mohan and Gopinathan, 1989a) can also be used to confirm the presence of this pathogen. Another sound diagnostic procedure is conducting bioassay test by inoculating healthy grubs\beetles orally with homogenized midgut of test sample and observing for typical OrNV symptoms (Mohan *et al.*, 1983). Presence of OrNV disease in the adult beetle can be diagnosed without sacrificing the host. This method involves collection of the beetles excreta in 0.01M phosphate buffer saline, centrifuging the fecal matter at 500 rpm for 10 min. and preparing 3 per cent Giemsa staining of the sediment for locating typically hypertrophied nuclei under microscope (Mohan *et al.*, 1985 b.)

Mass production, storage and Field application: Mass production and *in vivo* culturing of this bio-agent is done by rearing healthy grubs in viral contaminated food or forced- feeding using infected midgut homogenate, and maintaining them in sterilized cowdung or saw dust until the OrNV symptom develops. The infected cadavers can be stored indefinitely at –40°C. The simplest and the most economical method of dissemination of OrNV is by releasing laboratory-inoculated beetles (10-15 No./ha) preferably during dusk. The infected beetles transmit the pathogen in breeding/feeding sides by excreting viral contaminated faeces after 3rd and 9th day post OrNV inoculation (Mohan *et al.*, 1985b), where it is picked up by healthy susceptible *Oryctes*. Horizontal spread of this virus was reported to be 1 km/month (Jacob, 1996).

Dissemination of virus for pest management is effected by release of virus infected rhinoceros beetles @ 10-15 beetles/ha. The viral pathogen produces 100 per cent reduction in the egg laying capacity of female beetles and 40 per cent reduction in life span of affected population (Pillai, 1993). Introduction of virus in several islands of the South Pacific effectively suppressed the *O.rhinoceros* damage below economic threshold level. Extensive studies on the use of OrNV to suppress rhinoceros beetle population in Islands of Lakshadweep and Andaman-Nicobar had shown encouraging results during the last four decades (Mohan *et al.*, 1989; Pillai *et al.*, 1993; Jacob, 1996). The effectiveness of IPM package for management of rhinoceros beetle was well documented in mainland also (Nair *et al.*, 2010a). Thus, the success encountered by the use of this microbial pathogen has endorsed its claim as one of the landmark examples in the biological control of any insect pest.

2.3.6.2. *Metarhizium anisopliae* (Metschnikoff) Sorokin (Deuteromycotina: Hyphomycetes)

M. anisopliae commonly termed as the green muscardine fungus (GMF) is a well-known pathogen of rhinoceros beetle. The susceptibility of *O. rhinoceros* to

GMF was first reported in Western Samoa in1913 and in India by Nirula *et al.* (1955, 1956). *M. anisopliae* var. *major* (spore size 10-14 µm) is highly infective variety used widely for the control of this pest. It gains entry through membraneous joints of the cuticle of the host by mechanical and enzymatic action. The success of infection depends on an optimum temperature of 27-28°C and relative humidity of 70-90 per cent. All the stages of the host excepting the eggs are mycosed. The infected grubs become sluggish, and die within 10-15 days. Initially, the body gets hardened and a white fungal mycelial mat appears externally at all joints of the integument. After 4-7 days, green coloured spores are produced. Finally, the cadaver turns blackish green and mummified due to profuse spore growth.

Mass production: Autoclaved, filter sterile or aseptically drawn out coconut water is a good liquid substrate for the multiplication of this fungus (Danger *et al.*, 1991). It can also be mass-produced on cassava chips: rice bran mixture supplemented with urea or fishmeal extract as nitrogen source (Mohan and Pillai, 1982). Different substrates like broken rice/wheat grains, millets *etc.* were also found to be cheaper substrates for multiplication of the fungus. Area-wide farmer participatory mass production strategies using semi cooked rice media were standardized (Figure 1.7b) and validated for the biomanagement of rhinoceros beetle (Chandrika *et al.*, 2010c; Anithakumari *et al.*, 2016).

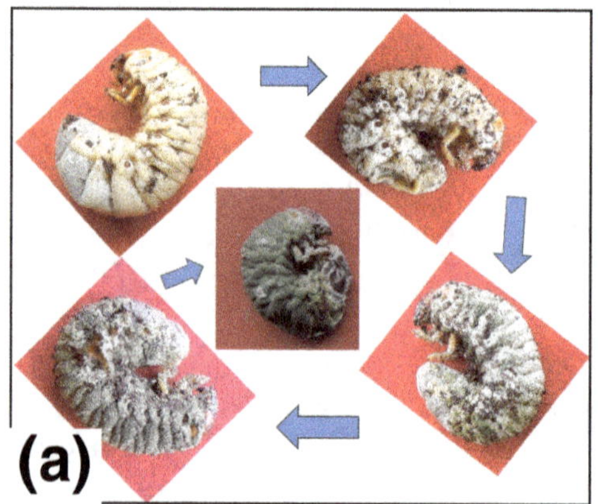
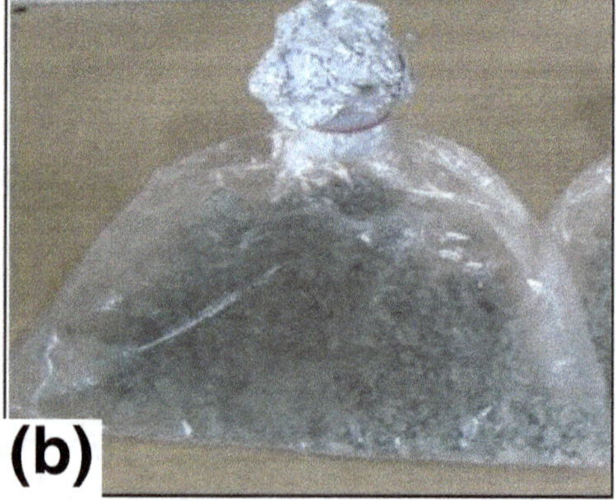

Figure 1.7

(a) Stages of GMF infection in *O.rhinoceros* grubs, (b) *M. anisopliae* culture in rice media.

For the field application of the fungus of GMF, the fungal spores are mixed with sterile water and used to drench or spread (cassava chips culture + powdered cow dung) over the breeding sites of the rhinoceros beetle @ of 5×10^{11} spores/m^3. The fungus survives in the breeding material for long periods. Use of this fungus for biocontrol of rhinoceros beetle has been popularized as a women friendly technology by ICAR-CPCRI.

2.3.7. Field Evaluation

Efficacy of these the two promising microbial pathogens *Metarhizium anisopliae* and *Oryctes rhinoceros* nudivirus in managing the pest in larger areas was demonstrated by Nair *et al.* (2010a) in an area of 2400 ha homestead coconut garden in two districts Kerala during 1999-2002 and reported 66.6 per cent to 75.1 per cent reduction of leaf damage and 79.4 per cent to 95.8 per cent reduction in spear leaf damage in a period of 3 years.

2.3.8. Semiochemicals

Aggregation pheromones have been documented for the *O. rhinoceros* and *O. monoceros* as ethyl 4-methyloctanoate (Hallett *et al.*, 1995; Gries *et al.*, 1994) which are commercially available. Specially designed PVC tube trap employing synthetic pheromone ethyl 4-methyloctonate has been found to be quite feasible for trapping black beetles in reasonable numbers. The traps are set up in the gardens @ 1 trap/ha and beetles trapped inside are collected periodically and used in virus inoculation (APCC, 2007; Nair *et al.*, 2010b).

3. Red Palm Weevil (*Rhynchophorus ferrugineus*) (Coleoptera: Curculionidae)

Red palm weevil (RPW) is a lethal pest of coconut palm widely distributed in all coconut growing regions of India. This weevil is reported to attack 17 palm species worldwide, of which date palm *Phoenix dactylifera* is the host. Although the weevil was first reported on coconut, *Cocos nucifera* from South Asia, during the last three decades it has gained foothold on date palm, in several Middle Eastern countries from where it has migrated to Africa and Europe through movement of infested planting materials.

3.1. Biology

The adult weevils measure 35 mm long and 12 mm wide and ferruginous brown in colour with a long life span extending up to 76 to 133 days. The snout is elongated in both sexes and the dorsal apical half of the rostrum in males is covered with a tuft of brown hairs (Figures 1.8a,b and Figure 1.9) The mean fecundity is

Figure 1.8

(a) Red palm weevil and (b) Colour variant from NEH region.

Figure 1.9: Size and Colour Morphs of Red Palm Weevil form different Coconut Growing Tracts of India.

Figure 1.10: Life Stages of Red Palm Weevil.

(a) Egg of red palm weevil, (b) Grubs, (c) Pupal cocoons, (d) Various stages of grubs.

about 175 eggs per female and the creamy white oval eggs are laid in small holes scooped out on soft tissues of the palm. After hatching, the grubs tunnel their way into the trunk and feed on the internal contents. The full-grown grub is stout, fleshy and apodous measuring 50 mm long and 20 mm in width. The fully-grown grubs enter the pupal stage by winding around themselves the fibrous elongate oval cocoon(Figure. 1.10). The weevils on an average take 4 months to complete its egg to adult stages depending up on weather conditions and type of food source (Nirula, 1956c, Abraham, 1994).

3.2. Damage and Symptoms

Juvenile and pre-bearing palms mostly below 20 years age are more susceptible to the pest infestation. Being an internal tissue feeder completing life stages inside the palm tissues, it is very difficult to detect the pest attack during early stages. However, on close examination of the palm, some symptoms can be detected at the early stages also. Yellowing and later wilting of the inner and middle whorl of leaves, small circular holes on the palm trunk with exudation of amber coloured viscous fluid, longitudinal splitting of leaf base, gnawing sound of grubs and presence of cocoon/chewed up fibers at palm base are the major symptoms of red palm weevil infestation. Unnoticed severe infestation results in toppling of the palm crown (Figure 11.1). In general dwarf and hybrid palms are more susceptible to red palm weevil attack than tall genotypes. The weevils orient themselves towards the palm to the fermenting odour emanated by mechanical injuries or fungal infections. Hence, the incidence of red palm weevil in coconut palm is comparatively more in areas having high incidence of rhinoceros beetle, bud rot disease and leaf rot disease. Shallow methods of planting also pave way for the pest attack (Nirula, 1956c; Abraham and Kurian, 1975; Abraham, 1994; Nair *et al.*, 1997; Rajan *et al.*, 2009, Faleiro, 2006). Pest entry through the leaf axil attachment at the trunk was observed if the bearing bunches are buckled from the point of contact due to overweight of developing nuts or succulence due to over nutrition. Access through the collar or bole region of the palm where injury is met out through tractor or tiller ploughing as well as swollen or cracked bole region through erratic nutrient uptake and improper translocation were commonly observed (Josephrajkumar *et al.*, 2014a).

3.3. Pest Management

3.3.1. Sanitation

Early detection of infestation in the field is important for any RPW-IPM programme. Being the internal tissue borer, it is very difficult to recognize pest infestation symptoms at an early stage. Hence, close palm surveillance has to be stressed for early pest detection and management. Coconut palms dead due to red palm weevil and retained in the field serve as ideal food source for a second generation or it acts as a source of inoculum for further build up of the pest in the field. Hence, field sanitation is very important to protect the palms (Abraham and Kurian, 1972; Rajan and Nair, 1997).

3.3.2. Scouting

Close and regular monitoring of the plantation, especially with previous pest history has to be practiced for early detection for saving the palm. Trained palm technicians such as Friends of Coconut Tree (FOCT) need to be effectively enriched on pest scouting and their skill upgraded through organizations such as ICAR-CPCRI, State Agricultural Universities, developmental agencies *etc.* for effective and timely monitoring, sensitization and diagnosis of this killer pest. Farmers

Figure 1.11a-d: Symptoms of Red Palm Weevil Infested Palms.

(a) Holes on palm trunk, (b) Oozing of brown viscous fluid, (c) Splitting of petiole
(d) Wilting of spear leaf, (e) Crown toppled palm.

Figure 1.11e: Symptoms of Red Palm Weevil Infested Palms (Crown Toppling).

involvement in close checkup of palms for any detectable pest entry symptoms is need of the hour to detect the attack in early curable stage. Routine close examination would reveal any of the damage symptoms so that adequate curative strategies could be resorted to.

3.3.3. Crop Geometry

Maintaining optimum palm density during planting is very important for utilizing highest benefits of light energy as well as to reduce the release of volatiles orienting the pest away from the host. Spacing for tall varieties 8 x 8 m and dwarf varieties 7 x 7 m is found ideal. Interspaces can be effectively used for raising intercrops so as to admix and diminish the volatile cues disorienting RPW away from host (Josephrajkumar *et al.,* 2014b). Gardens planted with coconut seedlings with inadequate spacing was found heavily infested during the juvenile phase especially in root (wilt) disease tracts of Kerala.

3.3.4. Prophylactic Treatment

Prophylactic leaf axil filling techniques not only prevent attack by rhinoceros beetle and bud rot infection, but also protect the palm from RPW incursion. Prophylactic leaf axil filling with oil cakes such as neem, marroti, pongamia (250 g) admixed with equal volume of river sand/naphthalene balls (12 g)/6 g chlory dust/6 g chlorantraniliprole admixed with 250g sand/palm could effectively repel rhinoceros beetle and thus reduce chance of RPW attack also. Leaf axil placement of two perforated polythene-sachet containing 3 g chlorantraniliprole safeguarded juvenile palms for about 4-6 months. In bud rot endemic zone, placement of *Trichoderma*-coir pith cake was found effective to reduce the disease and chance of RPW infestation.

The pest is attracted to kairomones emanating from fresh injuries inflicted on the palms for egg laying. Due to mechanical farm operations such as ploughing, cutting of steps for climbing the palms, tapping *etc.*, the injured palm becomes more susceptible to weevil infestation. Avoiding physical injury to palms is very critical to reduce pest incidence. While cutting fronds, leaving the basal portion at least 1 m from trunk, evading knife injury on crown region during crown clearing/tapping and careful tractor ploughing shunning away from bole and frond region to avoid injuries need to be overemphasized.

3.3.5. Curative Strategies

In cases of infestation by red palm weevil it becomes mandatory to apply chemical pesticides, either by crown application or through stem injection. After diagnosis, application of imidacloprid 18.5 SL 0.02 per cent (1 ml per litre of water) or spinosad 2.5 SC 0.013 per cent (5 ml per litre of water) or indoxacarb 14.5 EC 0.04 per cent (2.5 ml per litre of water) was found effective in the suppression of the pest (Josephrajkumar *et al.*, 2014b). Insecticide treatments are usually done after harvest of nuts and therefore a safe waiting period of 45 days is accomplished in this process before the next harvesting. In most cases young non-bearing palms are invaded by the pest. It was also found that there was no detectable residue of imidacloprid on leaves, nut and meat even after one-day after treatment up to 30 days period. Nearly one litre of the insecticidal suspension is required to treat an affected palm based on the serenity of the infestation.

3.3.6. Ecological Engineering

Judicious intercropping of palms with nutmeg, rambutan, curry leaf, banana along the interspaces disorients the pest away from the source due to crop-habitat diversification induced pest-repulsion cues. Crop heterogeneity is therefore preferred for continuous employment of farmers, income generation as well as pest regression infusing stimulo-deterrant diversionary tactics (Figure 1.12). Installation of bird perch and flowering plants like coral vines maintains pest defenders and executes ecosystem services. Damage by rhinoceros beetle as well as red palm weevil was reduced in coconut based cropping system than in mono-cropped garden.

3.3.7. Entomopathogenic Nematodes

Entomopathogenic nematode (EPN) *Heterorhabditis indica* showed high virulence (LC$_{50}$ 355.5 IJ) against *R. ferrugineus* grubs as well as greater susceptibility (82.5 per cent) of pre-pupal stage than that of grubs under laboratory trials (Figure 1.13). Synergistic interaction of *H. indica* (1500 IJ) with imidacloprid (0.002 per cent) against red palm weevil grubs was also reported. Investigations on combined application of *H. indica* and imidacloprid (0.002 per cent) for curative treatment in the field level management of red palm weevil in coconut were promising (Josephrajkumar *et al.*, 2013).

Figure 1.12: Crop Habitat Diversification for Pest Regression.

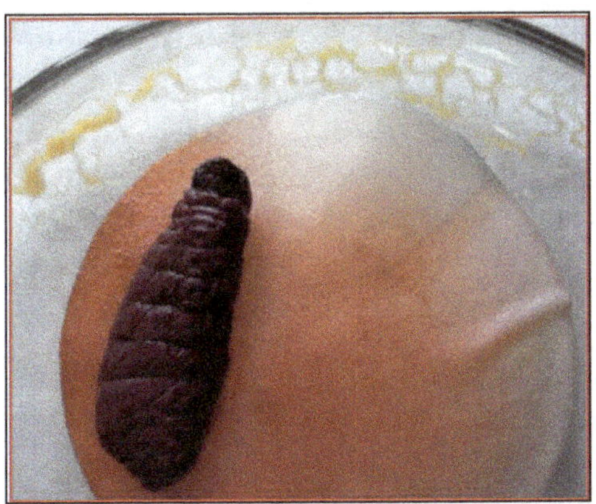

Figure 1.13: EPN Infected RPW.

3.3.8. Semiochemical

With the synthesis and availability of ferrugineol based pheromone lure (4-methyl 5-nonanone (Ferrugineone) and 4-methyl 5-nonanol (Ferrugineol) for RPW, the IPM programme was modified to incorporate pheromone traps and it was successfully utilized to combat the pest in coconut and date palm (Nair and Nair, 2002, Faleiro, 2006), provided all the precautionary steps involved in the use of pheromone traps are meticulously followed by the user. Installation of pheromone traps with ferrugineol embedded on nanoporous matrix @ 1 trap/ ha was found effective in mass trapping of weevils (Figure 1.14). Impregnation of kairamonal blends containing host-induced volatiles enhanced the weevil catches substantially. Slow and sustained release of pheromone blends for a period of six months was achieved in nanoporous matrix along with the reusable strategy of the matrix (Subaharan *et al.*, 2014). A farmer-participatory community approach would be the key factor in successful field realization. Palms around the traps would be monitored strictly to avoid slippages, if any.

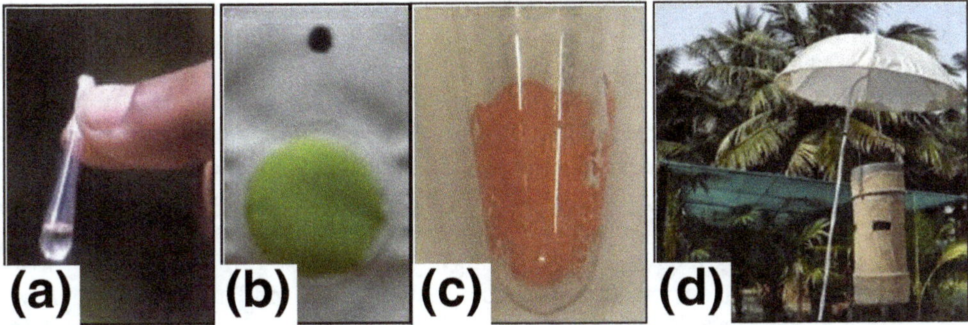

Figure 1.14: Pheromone Dispensation Strategies.

(a) Capillary vial, (b) Polymembrane matrix, (c) Nanoporus matrix and (d) Modified PVC pipe based pheromone trap.

4. Black Headed Caterpillar (*Opisina arenosella*) (Lepidoptera: Oecophoridae)

The black headed caterpillar, *Opisina arenosella* Walker is a serious defoliator pest of coconut in India and Sri Lanka (Rao *et al.*, 1948; Nirula, 1956a,b). The pest was reported to be a serious menace in other countries *viz.*, Myanmar (Ghosh, 1923), Sri Lanka (Jayaratnam, 1941) and Bangladesh (Alam, 1962).

4.1. Bioecology

Adult moth is grey coloured, 10-15 mm long with wing expansion of 20-25 mm. The male is smaller than female, with a slender abdomen ending in a short brush of scales. Female lays eggs on the abaxial leaf surface near old larval galleries which hatch in 5 days. Fecundity is about 137 eggs/female. Larval body is cylindrical, slightly compressed with three longitudinal reddish brown stripes dorsally and a black head. Average larval period is 42 days and the final instars measure about 154 mm long. There is a distinct pre-pupal stage for 2 days when the larva spins a whitish cocoon around its body and enters the pupal stage. The moth emerges out in about 12 days (Figure 1.15). The total life cycle from egg to adult takes about 8-10 weeks. The adult moths live for about 5-7 days (Nirula, 1956a; Nirula *et al.*, 1951; Chandrika and Sujatha, 2006).

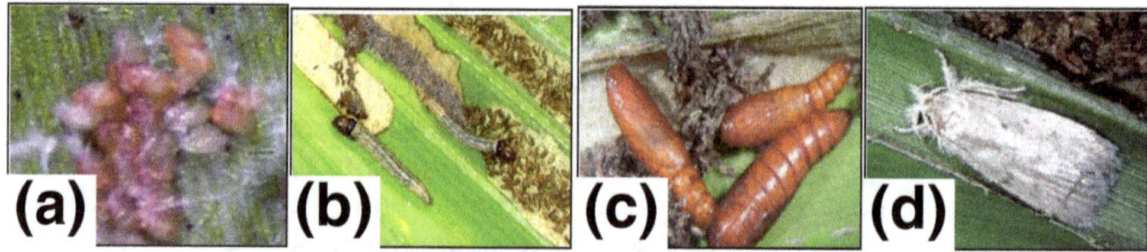

Figure 1.15: Life Stages of Coconut Black Headed Caterpillar.
a) Eggs, b) Larvae, c) Pupae and d) Adult moth.

4.2. Damage

The caterpillars of *O. arenosella* make silken webs supported with excreta and leaf bits and hiding in these galleries feed on the chlorophyll containing abaxial leaf tissues. The damage symptom is quite prominent with conspicuous grey patches on the upper frond surface. Infestation starts from the outer whorl of fronds and proceeds inwards. Damage results in drying of leaves, reduction in rate of production of spikes, increased premature nut fall and retarded growth (Lever, 1969). In extreme cases drying of 80-90 per cent of leaves on the palm crown leaves a burnt appearance to the plantation (Figure 1.16). In coconut frond, drying occurs due to many reasons like lightning, drought and diseases *etc.* and hence close examination of leaves for presence of pest stages is essential to identify pest problem. As the pest attack directly affects chlorophyll content of leaves, it get reflected in yield and a crop loss of up to 45 per cent in terms of nut yield was recorded from infested palms in the succeeding year of severe pest incidence apart from rendering the leaves unsuitable for thatching and other purposes (Chandrika *et al.*, 2010a).

Figure 1.16: Coconut Palms Infested by *O. arenosella*.

Sporadic pest outbreaks occur usually under favourable conditions. The pest infestation on coconut palms is usually severe during dry spells of summer which gets worsened in drought and water stress conditions especially in certain localities of Karnataka and Andhra Pradesh. Generally in the West Coast of India, pest infestation reaches peak in hot summer months of February to May. On the East Coast maximum population is reported from April to June. After the onset of monsoon, there will be decline in the population of the pest, but there is a chance of pest build up in the endemic areas from November – December. Hence, monitoring the endemic areas regularly at monthly intervals from November onwards helps in locating pest attack at a very early stage itself.

4.3. Pest Management

The pest can be effectively managed by the biological control methods. However, an Integrated Pest Management (IPM) strategy is recommended in severe out break conditions.

4.3.1. Mechanical Method

Cutting and burning the infested leaves/leaflets at the early infestation stage can be practiced to prevent pest buildup. In case of very severe infestation also, removal and burning of fully dried 2-3 outer whorl of leaves is recommended to kill the pupae and other pest stages. Careless discarding of the pest affected leaves in the vicinity of healthy palms, can lead to newer infestations. This aspect has to be well taken care of while transporting pest infested leaves/leaflets to pest free areas as such or using pest-infested leaves for wrapping other commodities for transporting to newer areas.

4.3.2. Chemical Method

Since a very rich natural enemy fauna is associated with the pest in the field, chemicals are generally not being encouraged for pest management of *O. arenosella*. In case of very severe outbreaks, one spray of chlorantraniliprole @ 0.1ml/L of water (0.002 per cent) is recommended. Spray solution has to reach the underside of fronds to drench the larval galleries of the pest. Due to the difficulty experienced in palm climbing and spraying tall palms, chemical spraying recommendation is at a low profile in IPM.

4.3.3. Biological Method

Parasitoids and predators play an important role in the natural biological suppression of *O. arenosella* (Pillai and Nair, 1993). A checklist of the parasitoids, predators and pathogens of *O. arenosella* occurring in India and SriLanka was prepared by Dharmaraju (1962). Information on natural enemies of the pest was documented by various workers (Mohamed *et al.*, 1982, Narendran, 1985; Sathiamma, 1993; Pillai and Nair,1993). As the perennial nature of the crop permits a continuous interaction between the natural enemy and the pest without ecological upheavals, bio intensive-IPM has been recommended as the ideal and sustainable solution for the management of *O. arenosella* (Cock and Perera, 1987; Sathiamma, 1993). Field performance on biological suppression of coconut leaf eating caterpillar through release of stage specific parasitoids was established as early as 1920s (Nirula, 1956a; Rao *et al.*, 1948) and successful field biosuppression of this pest by release of parasitoids is well documented (Sathiamma *et al.*, 1996; Cock and Perera, 1987; Sathiamma, 1993; Mohanty *et al.*, 2000; Chandrika and Sujatha, 2006; Venkatesan *et al.*, 2006; Sujatha and Chalam, 2009).

Among coconut pests, the black headed caterpillar in the natural environment is attacked by the highest number of parasitoids and predators. The major parasitoids include the larval parasitoids *Apanteles taragamae* Wilkinson, *Bracon hebetor* Say, *Goniozus (Perisierola) nephantidis* Mues., the pre pupal parasitoid *Elasmus nephantidis* Rohw., and the pupal parasitoids *Brachymeria nosatoi* Habu, *B.nephantidis* Gahan, *B.atteviae* Joseph, *B.lasus* Walker, *Antrocephalus hakonensis* Ashmead, *Trichospilus pupivorus* Ferr., *Xanthopimpla punctata* F.,and *X.nana nana* Schulz.

Among the 40 species of parasitoids recorded from India (Pillai and Nair, 1993), the larval parasitoids *G. nephantidis* (Bethylidae), *B. brevicornis* (Braconidae), the prepupal parasitoid, *E. nephantidis* (Elasmidae), and the pupal parasitoid *B. nosatoi* (Chalcididae) are the most promising ones (Figure 1.17). The major desirable attributes of these parasitoids are their greater host searching ability, production of higher proportion of females, occurrence throughout the year and their distribution in all pest infested areas. Techniques have been developed for mass production of the promising parasitoids. The pest-infested area should be monitored regularly and parasitoid releases should be initiated at the post-monsoon period during November

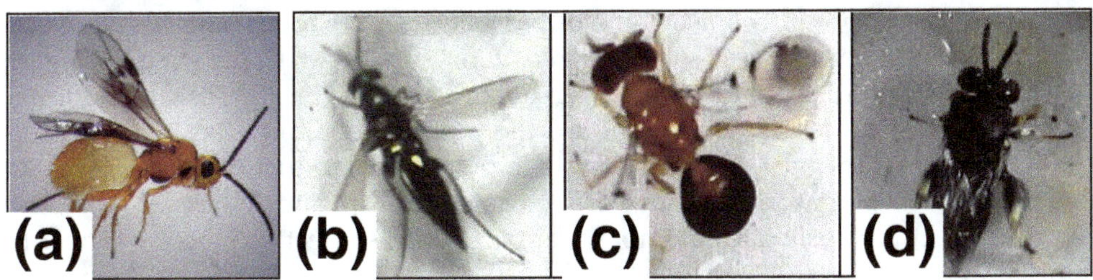

Figure 1.17: Parasitoids of *O.arenosella*
(a) *B. brevicornis*, (b) *E. nephantidis*, (c) *T. pupivora* and (d) *B. nephantidis*.

– December at the very beginning of pest incidence. Releases of parasitoids are to be synchronized with the stage of pest in the field at fixed dosages at fortnightly intervals till the pest population is suppressed. Presence of larvae in the leaflets should be confirmed before release of parasitoids to avoid unnecessary wastage of parasitoids. The parasitoid *G. nephantidis* is released if the pest is at 3rd instar larval stage or above at the rate of 20 parasitoid/palm and *B. brevicornis* at the rate of 30 parasitoid/palm. The pre-pupal parasitoid, *E. nephantidis* and pupal parasitoid, *B. nosatoi* are also very effective in managing the pest. They are released at the rates of 49 and 32 per cent respectively for every 100 pre-pupa and pupae estimated to be present on the palm (Sathiamma *et al.*, 1996; Chandrika *et al.*, 2010b). *G. nephantidis* adults could be released at the trunk (at 1.2 metre height from the ground level) of the coconut palm for the management of *O. arenosella* instead of releasing at the crown region of the palm or arbitrarily on unit area basis (Venkatesan *et al.*, 2003). Feeding the parasitoids with honey and exposing the newly emerged parasitoids to the host odours (smell of the volatiles of the injured *O. arenosella* larvae and host feeding gallery volatiles) was found to improve the host searching efficiency of *G. nephantidis* (Subaharan *et al.*, 2005). *G. nephantidis* and *B. brevicornis* could easily be mass multiplied on larvae of the rice moth *Corcyra cephalonica*. The pre-pupal parasitoid, *E. nephantidis* is a highly host and stage specific parasitoid and always requires a steady supply of pre-pupa of *O. arenosella* for mass multiplication. Techniques were evolved for mass multiplication of the promising parasitoids (Sathiamma *et al.*, 1999). Venkatesan *et al.* (2006) demonstrated that four releases of *G. nephantidis* @ 10 adults/palm at fifteen days interval could suppress the pest population and the cost of release of the parasitoid/ha is Rs. 2100.

The cost of production is calculated on the basis of the production of the host insect as well as the parasitoid. Cost-Benefit Ratio (CBR) of the production of *G. nephantidis* was 1:1.45. This technology is highly feasible and can be adopted for the large scale production of the parasitoid. Localized production of *G. nephantidis* could be taken up at village or district level especially by farmers, unemployed graduates, private and public sector units and NGOs. This module can be scaled up to any level (Venkatesan *et al.*, 2007).

Insect and spider predators are abundant in the coconut ecosystem. The dominant insect predators are the carabid beetles *Parena nigrolineata*, *Calleida splendidula*; anthocoreid *Cardiastethus exiguus*, Chrysopids Ankylopteryx sp. *Chrysopa* sp. *etc.* A total of 26 species of spiders were recorded along the pest of which *Rhena*, *Sparassus* sp. and *Cheiracanthium* sp. are the major predators (Sathiamma *et al.*, 1987). Predatory ants also play major role in population reduction of *O. arenosella* in the field. Although some pathogens such as *Serratia* sp. and *Aspergillus* sp. were reported on *O. arenosella* they are not so far exploited as effective biocontrol agents (Sathiamma *et al.*, 2000).

4.4. Area-wide Farmer Participatory Demonstration

Field release of the three stage-specific parasitoids *viz.* *G. nephantidis*, *E. nephantidis* and *B. nosatoi* at fixed norms and intervals in a heavily infested coconut garden (2.8 ha) for a period of five years resulted in highly significant reduction (94 per cent) in *O. arsenosella* population (Sathiamma *et al.*, 1996). The control plot where no release was made, the population was highly fluctuating. In the parasitoid released plot the percentage of parasitism also remained high 3.7-47.6 (*G. nephantidis*), 0-55.6 (*E. nephantidis*) and 0-71.4 per cent (*B.nosatoi*) and the parasitoids continue to exert check on the buildup of the population (Sathiamma *et al.*, 1996). Follow up observation revealed that even after three years no build up of the pest was noted in the released site. Chandrika and Nair (2002) had reported 52.6 and 94.7 per cent reduction in pest population after one and two years respectively, of parasitoid release in a *O. arenosella* infested tract in Kerala.

Chandrika *et al.* (2008) has reported 93-100 per cent reduction in *O. arenosella* population in coastal Kerala and Karnataka in a period of two years with regular monitoring and release of stage specific parasitoids *viz.*, *G. nephantidis*, *B. brevicornis*, *E. nephantidis* and *B. nosatoi*. Mohanty *et al.* (2000) had reported biological pest suppression of *O. arenosella* in coastal districts of Orissa by the release of parasitoids. Sujatha and Chalam (2009) had reported biological suppression of *O.aerenosella* in Andhra Pradesh by release of parasitoids. ICAR-CPCRI has undertaken biosuppression of the pest in Arsikere, Karnataka resulting in complete recovery of palms from black headed caterpillar attack (Figures 1.18 and 1.19).

5. Nut Infesting Eriophyid Mite (*Aceria guerreronis* Keifer) (Eriophyidae : Acarina))

Among the various acarine fauna affecting coconut palm, the eriophyid mite, *Aceria guerreronis* is the most destructive pest. The history of the occurrence of *A. guerreronis* on coconut starts with the first report from the Guerrero State, Mexico

**Figure 1.18: Black Headed Caterpillar Infested Area–
Jajur Village, Arsikere during 2013.**

**Figure 1.19: Palms Completely Recovered from
Black Headed Caterpillar Infestation.**

by Keifer in 1965. *A. guerreronis* is reported globally from 30 countries of Tropical America, Africa and Asia (Mariau, 1969; Mariau, 1977, 1986; Hall and Bacerril, 1981; Medina and Abreu, 1986; Griffith, 1984; Howard *et al.*, 1990; Flechtmann, 1989; Moore and Alexander, 1985; Fernando *et al.*, 2000; Seguni, 2002). In India, the first report of this mite was from Amballur Panchayat in Ernakulam district of Kerala by Sathiamma *et al.* (1998). Within a short period of time the mite had spread rapidly and at present its occurrence is seen in entire coconut growing tracts of India covering West and East Coast, North-East regions and Lakshadweep Islands (Nair, 2000; Ramaraju *et al.*, 2000; Mallik *et al.*, 2003; Khan *et al.*, 2003, Mullakoya, 2003). An exhaustive bibliography of coconut eriophyid mite was published by Rethinam and Muhartoyo (2003).

5.1. Bioecology

Coconut mite is a microscopic creamy white, vermiform organism measuring 200-250 microns in length and 36-52 microns in breadth. The body is elongated, cylindrical, finely ringed and bears two pairs of legs at the anterior end (Figure 1.22). Fecundity is 100-150 eggs. The eggs hatch into protonymphs which moult to deutronymphs and finally to adults and the total life cycle is completed in 7-10 days. Under favourable conditions, the high reproductive potential and shorter life cycle of the mite results in the enormous multiplication of the colonies. In India, the pest activity has been observed throughout the year with the population peak during the summer months (Rajan *et al.,* 2010). Observations on the population of the mite within various age groups of the nuts showed that third and fourth bunches harbour maximum mite population.

Figure 1.20: Eriophyid Mite Damage Symptoms on Nuts of Varying Maturity.

5.2. Nature of Damage and Crop Loss

In coconut, mites infest the developing young buttons after pollination and are seen in the floral bracts (tepals) and the soft meristematic portions beneath the perianth. Entry of the mite into the developing nuts takes place during the early phase of the development immediately after fertilization. The dispersal of the pest takes place mainly through wind. Honeybees and other insects visiting inflorescence of coconut also act as agents for dispersal. Appearance of elongated white streaks below the perianth of the developing nut is the first external symptom of mite infestation on young buttons (Figure 1.21). Further these streaks form triangular

Figure 1.21: Progression of Eriophyid Mite Infestation.

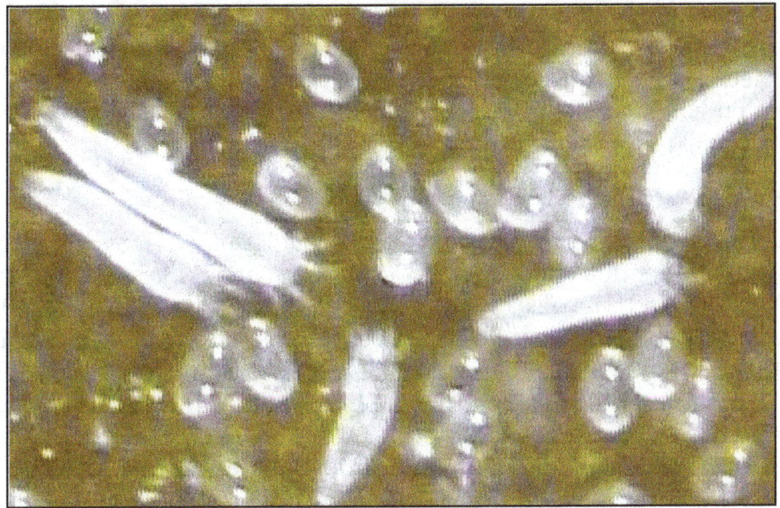

Figure 1.22: Eriophyid Mite Colony under Microscope.

yellow patches. Draining of sap by the feeding activity of the mite colony results in drying of infested tissues and subsequently browning of affected portion. As the nut grows, warts and longitudinal fissures appear on the nut surface (Figure 1.20). In severe infestation, the husk develops cracks, cuts and gummosis. Shedding of buttons and young nuts and malformation of nuts as a result of retarded growth are the other indications associated with severe attack of the pest (Nair, 2000).

In India, during 1998 when the pest outbreak was reported almost 70 per cent of nuts were affected showing malformation and reduction of nut size (Nair, 2000). Surveys carried out in Alappuzha district, Kerala during 2000 has shown significant reduction in crop loss due to mite incidence indicating an average loss of 30.94 per cent in terms of copra and 41.74 per cent in husk production (Muralidharan *et al.*, 2001). An average loss of copra yield to the tune of 27.5 per cent was reported from Tamil Nadu (Ramaraju *et al.*, 2000) and 18-42 per cent in Karnataka (Mallik *et al.*, 2003). Naseema Beevi *et al.* (2003) reported 26-53 per cent reduction in fibre length due to mite infestation. But observations recorded during subsequent years revealed overall reduction in incidence and intensity of pest in areas of its initial occurrence (Nair *et al.*, 2003). Rajan *et al.* (2007) reported that the loss in terms of copra in Southern districts of Kerala ranged from 8-12 per cent compared to an average loss of 25 per cent in initial years. With acute summer and drastic increase in temperature pest incidence was found to be higher especially in dwarf genotypes and segregants leading to excessive miniature nuts.

5.3. Varietal Preference

The nut traits *viz.*, colour, shape and size influence the severity of mite infestation. A coconut variety exhibiting complete resistance to eriophyid mite is not reported from any country. However, varieties like Malayan Yellow Dwarf (MYD), Malayan Red Dwarf, Rennal Tall, Cameroon Red Dwarf, Equatorial Green Dwarf and Hybrid [MYD x West African Tall (WAT)] were reported to show different degrees of tolerance to mite attack in different countries of the world (Mariau, 1986). Field observations indicated lesser mite incidence in Chowghat Orange Dwarf (COD) variety. Kalpaharitha (a selection of Kulasekaram Tall) recorded lowest mite incidence in the field at ICAR-CPCRI and could be a preferred choice in endemic zones (Josephrajkumar *et al.*, 2016).

5.4. Pest Management

Over five dozen systemic and contact insecticides have been evaluated world over and recommended from time to time for management of coconut mite (Nair *et al.*, 2005). In India also, a wide spectrum of pesticides have been tried by various research agencies including both Central Institutes and State Agricultural Universities (Nair *et al.*, 2005; Ramaraju *et al.*, 2000; Saradamma *et al.*, 2000; Mallik *et al.*, 2003). Owing to the concern over environment contamination by repeated chemical pesticides application, currently botanical pesticides *viz.*, neem based biopesticides are recommended for management of the pest in the field.

5.4.1. Cultural

Removal of dried spathes, inflorescence parts, fallen nuts *etc.* and burying them in the soil or burning them reduces the pest inoculum and consequent infestation.

5.4.2. Botanicals

Spraying 2 per cent neem oil –garlic-soap mixture or azadirachitn 10,000 ppm @ 0.004 per cent or root feeding with neem formulations containing azadirachtin 50,000 ppm at 7.5 ml or azadirachtin 10,000 ppm at 10 ml with equal volume of water three times during March-April, October-November and December-January is recommended for the management of the pest (Mallik *et al.*, 2003; Nair *et al.*, 2003; Rajan *et al.*, 2009). Three sprayings of palm oil (200 ml) and sulphur (5g) emulsion on the terminal five pollinated coconut bunches during January-February, April-May and October-November evinced significant reduction (67.4 to 69.8 per cent) of mite incidence (Josephrajkumar *et al.*, 2016).

Preparation of Spray Solution

To prepare one litre of 2 per cent neem oil-garlic soap emulsion, 20 ml pure neem oil, 20 g cleared garlic pearls and 5g washing soap are required. Dissolve the soap in 500 ml of water and add neem oil to this solution and mix it well. Grind garlic pearls well, mix it well in 500 ml of water and add this to the soap-neem oil mixture by sieving through a cloth to remove debris of garlic pearls. The mixture is stirred well and can be used for spraying. The pesticide mixture shall be used on the day of preparation. Pesticide should be applied as fine droplets on 2 to 6 months old bunches so as to provide its penetration into the perianth lobes.

Root Feeding Method

In root feeding method, an active semi hard, pencil thick and brownish coloured root was selected and a slanting cut of 45° at the tip portion was made with a sharp knife. Botanical formulation containing 7.5 ml of Azadirachtin or 10 ml Azadirachtin was mixed with equal volume of water and dispensed in a polythene pouch. The cut end of the root was immersed in the botanical formulation up to the bottom of the pouch and tied with a twine. Care should be taken to avoid any injury or spillage of the pesticide solution and cover the root gently with leaf mulch or loose soil.

5.4.3. Biological Methods

Presently, emphasis is given for development of biocontrol strategies as they are safe and ecofriendly and vital in sustainable management of the pest. The fungal pathogen, *Hirsutella thompsonii* (Figure 1.23) has received considerable attention throughout the world as the most effective natural enemy of eriophyid mite of coconut (Kumar, 2002; Beevi *et al.*, 1999; Kumar *et al.*, 2001). Application of talc-based preparation of *H. thompsonii* @ 20 g/l/palm containing 1.6×10^8 cfu with a frequency of three sprayings per year reduced mite population significantly (Chandrika *et al.*, 2014). The predatory mite *Neoseiulus baraki* is effectively utilized for biomanagement of the pest in Sri Lanka. Release of 5000 *N. baraki* at 3-4 month intervals on to quarter of the coconut plantation for 2 years has been recommended to control the coconut mite (Aratchige *et al.*, 2012).

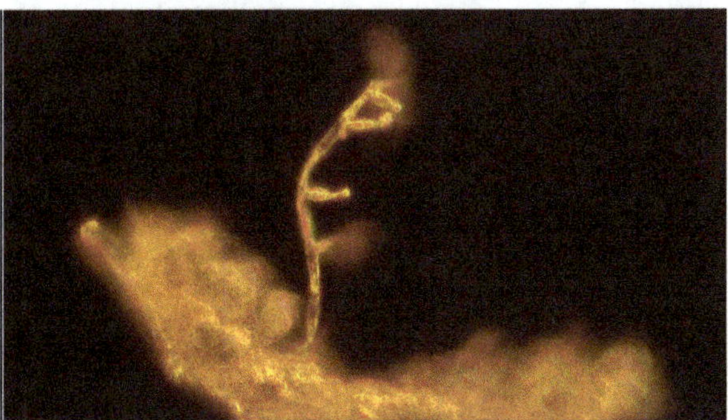

Figure 1.23: Mycelia of *H. thompsonii* Emerging from Infected Mite.

5.4.4. Palm Health Management

The nutritional status of the palm plays a significant role in the management of the pest. The nutrient management package consists of balanced application of NPK fertilizers at recommended doses in two splits (Urea 1.0 kg, super phosphate 2.0 kg, muriate of potash 2.5 kg); application of neem cake @ 5 kg per palm per year; *in situ* growing of green manure crops like cow pea or Sunn hemp (seed rate of 100g/ palm basin) in the garden and its incorporation in coconut basin and conservation of soil moisture by appropriate mulching methods (Rajan *et al.,* 2009). IPM package was demonstrated in farmer's fields at Krishnapuram village, Kerala covering 25 ha area of coconut gardens in 208 farmer holdings. Here the integrated nutrient management technology was implemented along with recommended practice of azadirachtin spraying thrice a year and the mite incidence could be brought down to 15.3 per cent from 68 per cent over a period of three years (Rajagopal *et al.,* 2003).

6. White Grub (*Leucopholis coneophora* Burm.) (Coleoptera: Scarabaeidae)

6.1. Damage and Symptoms

The major white grub species infesting coconut is *Leucopholis coneophora* and occurs mainly in sandy loam soil and attains pest status in discontinuous patches along the Western coastal tracts especially of Kerala and Karnataka. Two closely related species *viz., Leucopholis burmeisteri* Brenske and *Leucopholis lepidophora* Blanchard were also reported to infest coconut especially in Karnataka. The subterranean grubs are highly polyphagous and feed on tuber crops, banana, arecanut, rhizomes and vegetables *etc.* which are grown as intercrops in coconut gardens. In nursery seedlings the grubs (Figure 1.24) feed on tender roots and also tunnel into the bole and collar regions resulting in the drying of the spindle leaves followed by gradual death of the seedlings. Continuous feeding by the grubs on mature palms results in yellowing of leaves, premature nut fall, tapering of stem, delayed flowering, retardation of growth and reduction in yield. The pest has an annual life cycle. Adult beetles are chestnut brown coloured (Figure 1.24) and they emerge out of soil after pre-monsoon showers in May–June. Peak grub population is seen in the coconut basin during August–October (Abraham and Mohandas, 1988; Chandrika and Vidyasagar, 1993).

6.2. Pest Management

Deep ploughing during pre and post-monsoon periods exposes the grubs to predators. Mechanical collection and destruction of beetles during peak emergence period is deemed as an effective management practice and was found to be the highly

Figure 1.24: Grubs and Adult of *L.coneophora*.

significant than light trap collection. Drenching the root zone with chlorpyrifos 20EC @ 2.5ml/L or imidacloprid 17.8 SL @ 675 ml/ha or bifenthrin 10 EC @ 20 litre/ha during May-June and September- October are recommended for management of root grubs of coconut (ICAR-CPCRI, 2015). The insecticide has to be applied evenly in the active root zone of the palm. A scoliid wasp, *Campsomeriella collaris* (Fabricius), eugregarine protozoan pathogen *Pseudomonocystis* sp., and *Codyceps* sp. were found to parasitize the grubs in the field under natural condition. Predatory birds *viz.*, king fisher, brahminy kite and crow were effective predators of cockchafers during ploughing and adult emergence period. Drenching aqua suspension of entomopathogenic nematode, *Steinernema carpocapsae* in the interspaces at 5-10 cm depth with 1.5 billion IJ/ha and need based repeated application is effective for reducing grub population.

7. Coreid Bug (*Paradasynus rostratus* Dist.) (Hemiptera: Coreidae)

The coreid bug, *Paradasynus rostratus* Distant is a potential emerging pest of coconut palm. Eggs are laid on the spadix and leaf sheath and the emerged first and second instar nymphs congregate on spadix and buttons. Nymphs and adults puncture the meristematic regions of tender buttons (1-3 months old) injecting toxin around the feeding site causing necrosis and these spindle-shaped depressions could be visible when the perianth of shed button is removed (Figure 1.25). Infested female flowers get dried and stay attached to the inflorescence. Most of the infested buttons and tender nuts shed down. Retained nuts on the bunch develop furrows and crinkles on their husks and are malformed. In many cases gummosis can be seen on such damaged nuts. Crop loss inflicted by coreid bug on coconut include production of barren inflorescence (8-9 per cent), shedding of button and immature nuts (18-66 per cent), qualitative and quantitative loss in fruit components [reduction in nut water (52 per cent) and husk weight (38.2 per cent)], incomplete kernel formation in severely infested nuts and puny nuts (12 per cent) in severely infested plantations.

7.1. Pest Management

Regular crown cleaning has to be undertaken to destroy eggs and immature stages of the pest. Spraying of azadirachtin 300 ppm @ 0.0004 per cent (13 ml/1) or chlorantraniliprole (0.018 per cent) @ 0.3ml/1 on young pollinated coconut bunches during May-June and September-October was found effective for satisfactory control of the pest in the field (Chandrika *et al.*, 2016b). Among the natural enemies, the weaver ant, *Oecophylla smaragdina* is found to be the most efficient predator of coreid bug in the field. *Chrysochalcisea indica. Chrysochalcissa oviceps* and *Gryon homeoceri* were identified as potential egg parasitoids.

8. Scale Insects

In coconut, four species of armoured scales *viz.*, *Aonidiella orientalis*, *Aspidiotus destructor*, *Lepidosaphes megregori*, *Chionaspis* sp. and three species of soft scales *viz.*, *Ceroplastes floridensis*, *Coccus hesperidum*, *Vinsonia stellifera* were recorded from Kerala, Tamil Nadu and Minicoy (Lever, 1969; Rajan *et al.*, 2009; Howard *et al.*,

Figure 1.25: Life Stages and Damage Symptoms of Coreid Bug.
(a) Eggs, (b, c) Nymphs, (d) Adult, (e) Drying of inflorescence, (f, h) Infested buttons,
(g) Infested mature nuts, (i) Barren nuts.

2001) (Figure 1.26). Scale insects are distributed throughout tropical and subtropical regions of the world, particularly on Islands and is present in all coconut growing tracts globally. Young coconut palms aged 10-15 years are more vulnerable to

Figure 1.26: Scale Insects

(a) *Aspidiotus destructor*; (b) *Vinsonia stellifera*, (c) *Lepidosaphes megregori*, (d) *Ceroplastes floridensis*.

coconut scale damage. The scale infests mainly on the undersurface of leaves, but occasionally they attack frond stalks, flower clusters and young fruits.The bright yellow colour of affected palms is clearly visible from a great distance. In extreme cases, the leaves dry out and entire fronds drop off. The coconut scale *Aspidiotus destructor* Sign. is one of the emerging problem especially in tune with climate change. The scale insect forms encrustations on the lower surface of the leaves. They suck sap from the leaves and the encrustations block the stomata. Affected leaves became yellow in color. *Aspidiotus destructor* completes the egg to adult period in 30-35 days. They also infest the inflorescence and nuts, which under severe infestation results in bottom shedding and decrease in nut production. Periodic outbreaks have been reported from several places during the hot months of the year. *Lepidosaphes megregori* Banks is the pink colored hard scale infesting the upper leaf surface and also on nuts. *Aonidiella orientalis* Newstead is another species infesting the upper leaf surface and also on nuts. Other scale insects recorded on coconut are *Hemiberlesia lataniae* Sign., *Coccus hesperidim* L. infesting nuts and spikes, *L. taplevi* Williams infesting leaves, spike and nut and *Pseudaulacaspis cockerelli* Cooley infesting leaves and nut (ICAR-CPCRI, 1987; Jalaluddeen *et al.*, 1991).

Strict surveillance on the transport of infested plant parts across borders should be accomplished. Heavily infested twigs and branches are to be pruned off to eliminate scales when infestations are on limited parts of the plant. Three sprays of 2.5 per cent fish oil rosin soap or neem oil 0.5 per cent were found to be effective in reducing the population of *A. destructor*. Coccinellid beetles, *Chilocorus nigritus, Cryptognatha nodiceps, Pseudoscymnus anomalus, Pseudoscymnus dwapikalpa, Scymnus luteus, Rhyzobius* spp. and *Telsimia nitida* suppresses the pest population by predation. A moderate level outbreak of *Aspidiotus destructor* in Kerala was effectively suppressed by the lady beetle and immature grubs of *Sasajiscymnus dwapikalpa* (Figure 1.27).

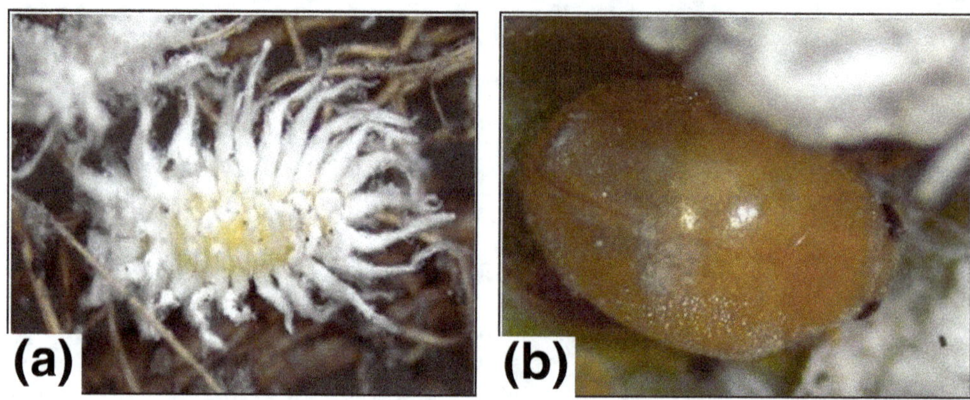

**Figure 1.27: Predators of Scale Insect–*Sesajiscymnus dwapikalpa*
(a) Grub and (b) Beetle.**

9. Mealybugs

Mealybugs are soft-bodied, sap feeding insects with white, powdery to granular waxy filaments protruding from margins and tail end. The insects appear as covered with cotton which are hydrophobic sappy secretions from numerous fine pores of diverse structure over the insect's body (Howard *et al.*, 2001). Presence of triocular pores with three elongated loculi is typical of mealybugs. Antennae are present and legs are well developed. Mature females measure 3-5 mm in length and are elliptical from the dorsal view and convex from side view. Nymphs and adult mealybugs suck the sap by inserting their long and thin stylets into the epidermis. Growing points of palms like spindle leaves, floral parts, leaflets and roots are primarily infested by mealybugs. They are mostly aggregated, feed continuously at the same site for a longer period of time and remain sedentary in most cases to disturbances due to slow retraction of stylets. Feeding symptoms include chlorosis, stunting of seedlings, leaf deformation, early fruit drop, heavy build up of honey dew and severe infestation leads to stunted growth of seedlings. Ants are consistently attracted by honey dew which also assist in phoretic dispersal of crawlers.

Palmivorous mealybug species are distributed widely but 50 per cent are in the genera *Dysmicoccus, Planococcus, Pseudococcus* and *Rhizoecus*. The most common reported mealybugs of palms are known primarily as pests of other crops on account of their wide adaptability, dissemination, establishment and long-term survival in diverse localities (Howard *et al.*, 2001). Six species of mealybugs are associated with coconut in India. They are *Palmicultor palmarum, Pseudococcus longispinus, Pseudococcus cocotis, Dysmicoccus* sp., *Nipaecoccus nipae* and *Rhizoecus* sp. (Figure 1.28) (Josephrajkumar *et al.*, 2012; Chandrika *et al.*, 2016). Destruction of highly infested plant parts at the initial stages of infestation and removal of alternate weed hosts in the immediate vicinity is practiced for pest management. As the pest is naturally suppressed by predators especially coccinellid beetles, conservation of coccinellid lady beetles in the ecosystem is recommended. In case of pest outbreak, regular monitoring and spot application with 0.5 per cent neem oil emulsion two-times in fortnightly intervals during summer was recommended to avoid further spread of mealybugs from infested fields.

Figure 1.28: Mealybugs.

(a-b) *Palmicultor palmarum*, **(c)** *Dysmicoccus finitimus*, **(d)** *Pseudococcus cryptus*, **(e)** *Nipaecoccus nipae*.

10. Whiteflies

Two types of whiteflies *viz.*, areca whitefly, *Aleurocanthus arecae* and spiralling whitefly *Aleurodicus* dispersus (Figure 1.29) have been recorded from coconut in India (David and Manjunatha, 2003; Chandrika *et al.*, 2007; Josephrajkumar *et al.*, 2010). Recently outbreak of a new invasive rugose spiralling whitefly *Aleurodicus rugioperculatus* Martin in parts of Kerala, Tamil Nadu, Andhra Pradesh and

Figure 1.29. Whiteflies.

(a, d) *Aleurocanthus arecae*, (b, e) *Aleurodicus dispersus* and (c, f) *Aleurodicus* sp. *rugioperculatus.*

Karnataka in the year of deficit monsoon and weather changes is documented. Though observed in large-area, it was effectively suppressed by the parasitoid, *Encarsia guadeloupae* (Chandrika *et al.*, 2016, Shanas *et al.*, 2016). Adults of *A. arecae* are small (1-3 mm), fly-like, fragile and often smoky-greyish in colour. They are found feeding on mature coconut leaflets. Eggs are laid in circular to spiral rings on the abaxial surface of leaves. Immature stages secrete waxy substances in the exuviae. Nymphs and adults insert the stylets on plant tissues, feed on the phloem sap and secretes honeydew. Puparia are black coloured. These sugar-rich excreta support sooty mould fungus interfering with photosynthesis (David and Manjunatha, 2003). The typical spiralling fashion of egg laying and feeding damage by *A. dispersus* was also recorded on various cultivars of coconut ranging from 1-6 colonies/leaflet. Dwarf coconut varieties *viz.*, Laccadive Green Dwarf, Laccadive Orange Dwarf and Laccadive Yellow Dwarf evinced more number of colonies than the tall cultivars *viz.* Benaulim, Laccadive Tall and Laccadive Micro. The susceptibility of dwarf cultivars was mainly attributed due to low canopy level that is in close proximity with other host plants in the immediate vicinity, whereas, in tall accessions the canopy is well isolated from other host plants. Young palms of tall varieties were also attacked by the pest (Josephrajkumar *et al.*, 2010). Whiteflies are naturally under ckeck attributing to the presence of effective bio-suppression agents

Lady beetles *viz.*, *Seragium parcesetosum*, *Jauravia pallidula* and a hump-backed nitidulid predator, *Cybocephalus* sp. were found predaceous on adults and nymphs of whiteflies. Eggs of *A. arecae* were also fed by an anthocorid bug in Kerala. Natural biological suppression is found to be very successful and no intervention with insecticides is recommended at this point of time (Chandrika *et al.*, 2007). *Encarsia* sp. nr. *haitiensis* and *Encarsia guadeloupae* were identified as potential parasitoids against *A. dispersus* (Ramani, 2000). Two different species of lady beetles *Chilocorus subindicus* and *Scymnomorphus* sp. were found predatory on spiralling whitefly as well as on coconut scale insects in the Minicoy island. Conservation of these lady beetles is therefore required for the natural suppression of the spiralling whitefly (Josephrajkumar *et al.*, 2010).

11. Slug Caterpillars, Darna (*Macroplectra*) *nararia*, *Contheyla rotunda*, *Latoia lepida* (Limacodidae : Lepidoptera)

Early-instar caterpillar feeds on undersurface of coconut leaflets by scrapping the surface tissues giving a glistening appearance on the feeding area. Leaf spot-like black halo marking develops on the feeding areas which later coalesce and form bigger lesions (Figure 1.30). Late instar caterpillars feed voraciously the leaf tissues leaving only the midribs and the feeding injury is often intensified by grey leaf blight fungus, *Pestalotiopsis palmarum*. Scorched/burnt appearance of leaves is the characteristic symptom observed in the field on severe infestation. In severely infested palms, premature drooping of leaves and shedding of nuts were also observed bringing drastic reduction in nut yield (Rajan *et al.*, 2011). High temperature (>39°C) and relative humidity (>85 per cent) flares up the pest as noticed in East Godavari during April-May. Establishment of light traps in endemic tracts could help in monitoring of the pest as well as reduce the population of moths (Sujatha *et al.*, 2011). Larvae of *M. nararia* are parasitized by *Eurytoma tatipakensis* Kur., *Euplectromorpha natadae* Kur. and *Secodes narariae* Kur under natural condition. Good nutrition as well as irrigation is required to recoup the infested palms which take about 20-24 months.

12. Rodents, *Rattus rattus wroughtoni*

Rats damage tender coconuts by scooping a small hole about 5 cm diameter near the stalk region (Figure 1.31) and these damaged nuts are retained in the bunch

Contheyla rotunda Grub and Adult Moth   **Latoia lepida**

Infested coconut garden   Slug caterpillar   Silken pupa

Glistening feeding damage by early instar larvae   Aggregation of feeding caterpillars

Figure 1.30: *Darna nararia* Infestation on Coconut.

for 2-6 days and later shed off. Fallen nuts with typical hole are seen around the basin of the palm. Three to six months old tender nuts are mostly preferred by this mammalian pest. Rats also damage leaf stalks, unopened spathe, female flowers and mature nuts in the field as well as the stored nuts. The intensity of damage is more during summer and early monsoon (April-June) and less during post-monsoon (August-October). Further, the damage increases when certain intercrops such as cocoa and cassava are cultivated along with coconut.

In coconut plantations, the black rats generally live on the crowns of the coconut palm by constructing nests. Hence, removal of dried leaves, spathes and matrix regularly expose the nesting placing of these rats to predators. A habitat alteration discourages rats from population build up on the crown. Application of 10 g Bromadiolone (0.005 per cent) blocks two times at an interval of 12 days on the crown of one tree out of every five trees is recommended for effective control of black rat. This method is highly cost-effective. If the damage is restricted to certain

Figure 1.31: Tender Coconuts Damaged by Rats.

palms, only such palms require baiting (Bhat *et al.*, 1993). Planting coconut seedlings in correct spacing as well as destruction of fallen fronds and other palm residues at regular intervals ward off the rat activity from coconut gardens. This method is highly cost-effective. If the damage is restricted to certain palms, only such palms require baiting. Wrapping the trunk of coconut trees using polythene sheets was found to reduce the damage by rats in Minicoy. Hanging a used fertilizer bag on the top of the crown could avoid nut damage effectively. Rat snakes and Barn owls are the common predators that control the rat population.

13. Menace of Invasive Pests

Accidental introduction of plant and animal pests, diseases and invasive alien genotypes poses threat to biosecurity. In this context we have to be alert and prepared on the major invasive pests of coconut in our bordering countries *viz.*, coconut leaf beetle (CLB), *Brontispa longissima* Gestro (Chrysomelidae: Coleoptera) and armoured scale insect, *Aspidiotus rigidus* Reyne (Diaspididae: Hemiptera) which are posing economic loss to coconut industry (Josephrajkumar *et al.*, 2015b,c).

13.1. *Brontispa longissima* Gestro (Chrysomelidae : Coleoptera)

The outbreak of the *B. longissima* in Myanmar and Maldives in recent years poses a great threat and concern to the nearby countries such as India, Sri Lanka and Bangladesh. Coconut leaf beetle (CLB) was originally described in 1885 from Aru Islands in Indonesia and from Papua New Guinea. Over a period of 130 years, it has widely spread in over 25 countries in Asia, Australia and Pacific Ocean Islands attacking a number of of cultivated and wild ornamental palm species in addition to coconut palms. Adult beetles measure 7.5-10.0 mm long and 1.5-2.0 mm wide, with a conspicuous orange to reddish pronotum (Figure 1.32). The anterior part of elytra is also orange to reddish in colour. Grubs and adult beetles inhabit the developing unopened still folded heart leaves of coconut palm and feed on leaf tissues (Figure 1.32).

Figure 1.32: *Brontispa longissima* and Infested Palm.

Shipments of ornamental palms from countries having the pest infestation have been the main source of spread within the Asia Pacific region. Pest management is mainly effected by release of biocontrol agents. Two parasitoids of coconut leaf beetle *viz., Tetrastichus brontispae* Ferriere (Hymenoptera: Eulophidae), a pupal parasitoid and *Asecodes hispinarum* Boucek (Hymenoptera : Eulophidae), a larval parasitoid have been successfully used in several countries to control the beetle.

13.2. *Wallacea* sp.

The chrysomelid beetle *Wallacea* sp. feeding on the spear leaf region of coconut seedlings (Figure 1.33) was recently recorded from South Andaman and little Andaman Islands (Prathapan and Shameem, 2015). Though 80-90 per cent of seedlings were infested by the pest damaging 40 per cent leaf area, seedling mortality was not observed. The feeding niche of *Wallecae* sp. confining on coconut spear leaf is a matter of concern, however, the pest was not observed from any adult palm during the snap survey conducted in October 2014 by ICAR-CPCRI (ICAR-CPCRI, 2015). Invasive nature of *Wallacea* sp. is under scrutiny, as a close relative *Wallaceana* sp. reported from Indonesia. Adult beetles are brownish with six rows of constrictions on each elytron and measured 4.72 mm long and 0.9 mm wide. Grubs and adults remain within the folds of the spindle leaves and feed from within. Typical feeding damage was seen within the leaf folds before unfurling along with faecal matters. In severe cases, the feeding streaks coalesce forming broader lesion with brown margin (Figure 1.33). Though a few feeding adult beetles were observed in between the leaf folds of emerged leaves, the grubs were mostly confined within the spindle region only. Domestic quarantine need to be strengthened to avoid entry of *Wallacea* sp. in the mainland.

13.3. *Aspidiotus rigidus*

Hard scale, *A. rigidus*, is a close relative of *Aspidiotus destructor*. Though *A. destructor* is under check by natural enemies, *A. rigidus* is ravaging Philippines incurring huge loss to coconut growers in that country. Bioinvasion of *A. rigidus* in Philippines has infested approximately 7,80,000 trees affecting 50-70 per cent of the coconut farms in Batangas and the nearby provinces (Watson *et al.*, 2014). So far

Figure 1.33: (a) *Wallacea* sp. infested coconut seedling, (b) Feeding damage lesions, (c) Grub, (d) Pupa and (e) Adult.

there is no report of *A. rigidus* infesting coconut palm from India. Strict quarantine regulations have to be imposed as these pests can be passively carried through any inert packaging materials, nuts *etc.*

Incursion management of invasive pests involves strengthening quarantine, surveillance and monitoring as well as sensitization campaign. Creation of an incursion management team comprising of experts from all disciplines as well as an emergency preparedness module would be the need of the hour to tackle accidental introduction of invasive pests in to the country.

14. Conclusion

Coconut, the perennial palm crop is subject to infestation by coleopteran, lepidopteron and hemipetran insects during its long life span including fatal pest like red palm weevil. IPM in harmony with complete palm health care has to be resorted to manage the immediate pest problem as well as to bring the palm to the sustainable potential productive stage. Very minimal chemical intervention is recommended in coconut IPM, that too to life saving spot application for red palm weevil infestation. Proven cases of biomanagement of pests *viz.*, rhinoceros beetle and black headed caterpillar have confirmed a meaningful way of pest management utilizing the indigenous fauna for the biological suppression of the pests of coconut. The current scenario warrants an augmentative release of these promising bioagents in areas wherever pest infestation is found. Natural enemy fauna build up was witnessed in many cases of upsurge of minor pests *viz.*, scale insects, mealybugs and whiteflies.

Spider fauna plays an important role in the natural suppression of pests in the field. Conservation of these promising biocontrol agents of the pests has become quite imperative. Palm health management by proper adoption of nutritional care, soil and water conservation modules is imperative for crop protection. Biological control of pests assume special significance as it can bring safe, economic and useful results in preventing the loss due to the important pests and there by strengthen the coconut production in the country. Even though awareness on green and clean cultivation is building up among the farming community, pest management in synergy with environmental health aspect has to be adequately stressed among the coconut growers and the farming community. Coconut being a long standing palm, allows light penetration for growing sufficient intercrops if proper spacing is envisaged at the planting time itself. Volatile cues from intercrops help in reducing pest orientation towards the palm in addition to optimizing income per unit area. Availability of biocontrol agents to stakeholders has to be assured by field level capacity building programmes on mass production technologies. A planned and holistic programme through awareness creation, capacity building on incursion management and strict quarantine are essentially warranted to combat accidental invasion of new species.

References

Abraham, C.C. (1994). Pests of coconut and arecanut. pp. 715-716. In: *Advances in Horticulture* Vol. 10, K. L. Chadha and P. Rethinam (Eds.). Malhotra Publishing House, New Delhi, India.

Abraham, V. A. and Kurian, C. (1972). *Chelisoches morio* F. (Forficulidae: Dermaptera) a predator on eggs and early instar grubs of the red palm weevil *Rhynchophorus ferrugineus* F. (Curculionidae: Coleoptera) *Journal of Plantation Crops* 1 (Supplement): 147-152.

Abraham, V.A and Kurian, C. (1975). An integrated approach to the control of *Rhynchophorus ferrugineus* F., red weevil of coconut palm. pp. 1-5. In: *Proceedings of the 4th session of the FAO technical working party on coconut production, protection and processing,* 14-25 September, Kingston, Jamaica

Abraham, V.A. and Mohandas, N. (1988). Chemical control of white grub *Leucopholis coneophora* Burm., a pest of coconut palm. *Tropical Agriculture* 65(4): 355-357.

Alam, M. (1962). A list of insects and mites of Eastern Pakistan. *Report of Agriculture Department.* 107p.

Anithakumari, P., Muralidharan, K. and Chandrika M. (2016). Impact of area-wide extension approach for bio-management of rhinoceros beetle with *Metarhizium anisopliae. Journal of Plantation Crops* 44(1). 2016: DOI: http://dx.doi.org/10.19071/jpc.2016.v44.i1.3010

Antony, J. and Kurien, C. (1966). Biology of the histerid beetle *Hister (Santalus) orientalis* Payk., a predator of *Oryctes.* In: *Proceedings 53rd Indian Science Congress* Part III, 362 pp.

Antony, J., Daniel, Mariamma, Kurien, Chandy and Pillai, G.B. (1979). Attempts on introduction and colonisation of the exotic reduviid predator, *Platymeris laevicollis* Dist. for the biological suppression of the coconut rhinoceros beetle *Oryctes rhinoceros* L., pp. 445-454. In: *Proc. PLACROSYM II,* C.S. Venkataram and others (Eds.).

APCC (2007). Final Technical Report 2004-2007 CFC/DFID/APCC/FAO project on Coconut Integrated Pest Management. Asian and Pacific Coconut Community, Jakarta, Indonesia, p 506.

Aratchige, N.S., Fernando, L.C.P., Kumara, A.D.N.T., Jayalath, K.V.N.N., Perera, K.F.G.,Suwandarathne, N.I. (2012). Biological control of coconut mite: Effect of inundative release of the predatory mite, *Neoseiulusbaraki* (Acari: Phytoseiidae). *In:* Hettiarachchi,L.S.K.,Abeysinghe,I.S.B.(Eds). Proc. 4thSymp.on Plantation Crop Research – Technological Innovations for Sustainable Plantation Economy. Tea Research Institute of Sri Lanka, St. Coombs, Talawakelle, 22100, Sri Lanka. pp. 125-133.

Bedford, G. O. (1975). Observations on the biology of *Xylotropes gideon* (Coleoptera: Scarabaeidae; Dynastidae) in Melanesia. *Ausralian J. Entomol. Soc.* 14(3): 213-216.

Bedford, G. O. (1980). Biology, ecology and control of palm rhinoceros beetles. *Ann. Rev. Entomol.* 25: 309-339.

Bedford, G.O. (2013). Biology and Management of Palm Dynastid Beetles: Recent Advances. *Ann. Rev. Entomol.* Vol. 58: 353-372.

Beevi, S.P., Beena, S.; Lyla, K.R.; Varma, A.S.; Mathew, M.P. and Nadarajan, L. (1999). *Hirsutella thompsonii var. synnematosa* samson. McCoy and M'Donnell on coconut mite *Aceria (Eriophyes) guerreronis* (Keifer) – a new report from India. *Journal of Tropical Agriculture* 27: 91-93.

Bhat, S.K., Vidyasagar, P.S.P.V. and Sujatha, A (1993). Biology and control of *Rattus rattus wroughtonii* Hinton, a pest of coconut in India. In : Advances in Coconut Research and Development 1993 (Eds.) MK Nair, HH Khan; P Gopalasundaram; EVV Bhaskara Rao, Oxford and IBH Publishing Co. Pvt. New Delhi, 759 p (International symposium on Coconut Research and Development (ISOCRAD II), 26-29 November 1991, CPCRI, Kasaragod) p.535-541.

CDB, (2016). Coconut Development Board, Kochi, Kerala http://coconutboard.nic. in/accessed on 21/11/2016

Chandrika M. and Nair, C.P.R. (2000). Effect of *Clerodendron infortunatum* on grubs of coconut rhinoceros beetle, *Oryctes rhinoceros* L. In: Proc. *PLACROSYM*-13 *Recent Advances in Plantation Crops Research* (Eds.) N. Muralidharan and R. Rajkumar. pp 297-299.

Chandrika M. and Vidyasagar P.S.P.V. (1993). Bioecology of coconut white grub *Leucopholis coneophora* Burm. in Kerala. *Journal of Plantation Crops* (Suppl.) 21: 167-172.

Chandrika M., Nair, C.P.R. and Rajan, P. (2001). Scope of botanical pesticides in the management of *Oryctes rhinoceros* L. and *Rhynchophorus ferrugineus* Oliv. affecting coconut palm. *Entomon* 26 (Spl. Issue): 47-51.

Chandrika M., Nair, C.P.R., Nampoothiri, C.K. and Rajan, P. (2010a) Leaf eating caterpillar (*Opisina arenosella*)-induced yield loss in coconut palm. *International Journal of Tropical Insect Science* 30(3): 132-137.

Chandrika M., Nair, C.P.R. and Rajan, P. (2010b) Large scale field validation of biocontrol strategy for management of coconut leaf eating caterpillar *Opisina arenosella* Walker utilizing indigenous parasitoids. 337-341 In: *Organic horticulture - Principles, Practices and Technologies* (H.P. Singh and George V. Thomas Eds) Westville Publishing House, New Delhi. 440pp

Chandrika M., Rajan, P. and Anithakumari, P. (2010c) Farm level production of the green muscardine fungus for management of rhinoceros beetle. Technical booklet, CPCRI, Regional Station, Kayamkulam, 8 p.

Chandrika M., Rajan, P. and Nair, C.P.R. (2007). Incidence of whitefly (*Aleurocanthus arecae* David) on coconut foliage. *Journal of Plantation Crops* 35(2): 119-121.

Chandrika M. and Sujatha, A. (2006). The Coconut leaf caterpillar, *Opisina arenosella* Walker *CORD*: 22 (Special Issue): 25-78.

Chandrika M., Thomas, G.V., Josephrajkumar, A. (2014). Mite management of coconut in India. *In*:Akter, N., Azad, A.K., Alam, M.N. (eds). *Proc. Mite management of coconut in SAARC member countries*. SAARC Agriculture Centre (SAC), Dhaka, Bangladesh. 10-11 August, 2014. pp. 43-58.

Chandrika M., Rajan, P. and Josephrajkumar, A. (2016a) Plantation Crops pp 543-556, In: *Mealybugs and their management in Agricultural and Horticultural Crops* (Eds: M.Mani and C.Shivaraju), Springer (India) Pvt Ltd.

Chandrika M., Renjith, P.B., Josephrajkumar, A., Sunny Thomas and Shanavas, M. (2016b) Evaluation of botanical formulations and chlorantraniliprole for management of coreid bug infestation in coconut. P 91. *In: Third International Symposium on Coconut Research and Development (ISOCRAD 3).* Eds. P. Chowdappa, K. Muralidharan, K. Samsudeen and M.K. Rajesh, December 10-12, 2016, ICAR-CPCRI, Kasaragod.

Cheriyan, M.C. and Ananthanarayanan, K.P. (1939). Studies on the coconut palm beetle (*Oryctes rhinoceros* L.) in South India. *Indian Jounal of Agricultural Sciences* 9(3): 541-559.

Cock, M.J.W. and Perera, P.A.C.R. (1987). Biological control of *Opisina arenosella* Walker (Lepidoptera: Oecophoridae). *Biocontrol News and Information.* 8(4): 283-310

Cumber, R. A. (1957). Ecological studies of the rhinoceros beetle *Oryctes rhinoceros* (L.) in Western Samoa. *Tech. Pap. South Pac. Comm.* No. 107: 32 pp.

Danger, T.K., Geetha, L., Jayapal, S.P. and Pillai, G.B. (1991). Mass production of the entomopathogen *Metarhizium anisopliae* in coconut water wasted from copra making industry. *Journal of Plantation Crops* 19: 54-69.

Danger, T.K., Solomon, J.J. and Pillai, G.B. (1994). Infection of the coconut palm beetle *Xylotrupes gideon* (Coleoptera: Scarabaeidae) by a non-occluded *Baculovirus. Z. Pflankranh. Pflschutz.* 101: 561-566.

David, B.V. and Manjunatha, M. (2003). Description of a new species *Aleurocanthus* (Quaintance and Baker) (Aleyrodidae : Homoptera) from *Areca catechu* in India and comments on *Aleurocanthus nubilans* (Buckton). *Zootaxa,* 173: 1-4

Dharmaraju, E. (1962). A checklist of parasites, the hyperparasites, predators and pathogens of the coconut leaf eating caterpillar *Nephantis serinopa* Meyrick recorded in Ceylon and in India and their distribution in these countries. *Ceylon Coconut Quarterly* 13(3-4): 10 pp.

Faleiro, J.R. (2006). A review of the issues and management of red palm weevil *Rhyncophorus ferrugineus* (Coleoptera: Rhynchophoridae) in coconut and date palm during the last one hundred years. *International Journal of Tropical Insect Science,* 26: 135-154.

Fernando L.C.P., Wickramananda I.R. and Aratchige, N.S. (2000). Status of coconut mite, *Aceria guerreronis* in Sri Lanka. In: *Proc. of International Workshop on Coconut mite (Aceria guerreronis).* Eds. L.C.P.Fernando, G.J. de Moraes and Wickranananda,I.R., CRI, Sri Lanka. pp. 1-8.

Flechtmann, C. H. W. (1989). *Cocos weddelliana* H. Wendl. (Palmae: Arecaceae), a new host plant for *Eriophyes guerreronis* (Keifer, 1965). (Acari: Eriophyidae) in Brazil. *Int. J. Acarol.* 15(4): 241.

Ghosh, C.C. (1923). *Oryctes rhinoceros* and other important palm pests in Burma. pp. 99-103. In: *Report of Proceedings of the V Entomological Meeting,* Pusa, 1923.

Gressitt, J. L. (1953). The coconut rhinoceros beetle (*Oryctes rhinoceros*) with particular reference to the Palau Islands. *Bull. Bernice P. Bishop Mus.* No. 212: 157 pp.

Gries, G., Gries, R., Perez, A, L., Oehlschlager, A. C., Gonzales, L. M., Pierce, H. D. Jr., Zebeyou, M. and Kouame, B. (1994). Aggregation pheromone of the African rhinoceros beetle, *Oryctes monoceros* (Olivier) (Coleoptera: Scarabaeidae). *Zeitschrift fur Naturforsch. Section C, Biosciences* 49(5 and 6): 363-366.

Griffith, R. (1984). The problem of the coconut mite, *Eriophyes guerreronis* (Keifer), in the coconut groves of Trinidad and Tobago, pp. 128-132. In: *Proc. 20ᵗʰ Annual Meeting of the Caribbean Food Crops Society*, R. Webb, W. Knausenberger and L. Yntema (Eds.). East Caribbean Center, College of the Virgin Islands and Caribbean Food Crops Society, St. Croeix, Virgin Islands, USA.

Hall, R. A. and Becerril, A. E. (1981). The coconut mite, *Eriophyes guerreronis*, with special refer-ence to the problem in Mexico, pp. 113-120. In: *Proc. 1981 British Crop Protection Conf. Pests and Diseases.* British Crop Protection Council. Farnham, UK.

Hallett, R. H., Perez, A. L., Gries, G., Gries, R. Pierce, H. D. Jr., Oehlschlager, A.C., Gonzalez, L.M., Borden, J.H. and Yue, J.M. (1995). Aggregation pheromone on coconut rhinoceros beetle, *Oryctes rhinoceros* (L.) (Coleoptera: Scarabaeidae). *J. Chem. Ecol.* 21(10): 1549-1570.

Hill, D.S. (1983). *Agricultural Insect Pests of the Tropics and Their Control (Second Edition).* Cambridge University Press, Cambridge, 746 pp.

Howard, F.W, Abreu Rodriguez, E. and Denmark, H.A. (1990). Geographical and seasonal distribution of the coconut mite, *Aceria guerreronis* (Acari: Eriophyidae), in Puerto Rico and Florida, USA. *J. Agric. Univ.* (Puerto Rico) 74(3): 237-251.

Howard, F.W., Moore, D., Giblin-Davis, R.M. and Abad, R.G. (2001). *Insects on palms,* CAB International, Wallingford, Oxon, UK pp 42-70.

ICAR-CPCRI (1987). *Annual Report* 1987. Central Plantation Crops Research Institute, Kasaragod, 174 pp.

ICAR-CPCRI (2015). Annual Report 2014-15, ICAR-Central Plantation Crops Research Institute, Kasaragod – 671124, Kerala, India, 126 p.

Jacob, T.K. (1996). Introduction and establishment of Baculovirus for the control of rhinoceros beetle *Oryctes rhinoceros* (Coleoptera: Scarabaeidae) in the Andaman Islands (India). *Bulletin of Entomological Research* 86: 257-262.

Jalaluddeen, S.M., Thirumoorthy, S., Mohanasundaram, M., Chinnaiah, C. and Chinnaswamy (1991). Coccid complex of coconut in Tamil Nadu. *Indian Coconut Journal* 22 (7): 17.

Jayaratnam (1941). A study of the control of coconut caterpillar Nephantis serinopa Meyr.in Ceylon with special reference to its eulophid parasite *Trichospilus pupivora* Ferr. *Trop.Agricult.* (Ceylon) 96: 3-21.

Josephrajkumar, A., Rajan, P., Chandrika M. and Thomas, R.J. (2012). New distributional record of buff coconut mealybug (*Nipaecoccus nipae*) in Kerala, India. *Phytoparasitica* 40: 533–535 (DOI: 10.1007/s12600-012-0260-2).

Josephrajkumar, A., Rajan, P., Chandrika, M. and Jacob, P.M. (2010). Re-invasion and bio-suppression of spiralling whitefly, *Aleurodicus dispersus* Russell on coconut in Minicoy Island. *Indian Coconut Journal* 53(2): 20-23.

Josephrajkumar, A., Sunny Thomas, Shanavas, M. and Chandrika M. (2014a) Managing the hidden villain in coconut garden. *Kerala karshakan e-journal* (September issue) 2(4): 21-27.

Josephrajkumar, A., Chandrika M., Sunny Thomas, Namboothiri, C.G.N. and Shanavas, M. (2014b) Subduing red palm weevil attack on coconut through fine-tuned management approaches. *In: Book of Abstracts National Conference on Sustainability of coconut, arecanut and cocoa farming Technological Advances and Way forward.(Eds.)* Muralidharan, K., Rajesh M.K., Muralikrishna K.S., Jesmi Vijayan, Jeyasekhar, S., CPCRI, Kasaragod, August 22-23: 86.

Josephrajkumar A., Chandrika M. and Rajan, P. (2013). Evaluation of entomopathogenic nematodes against red palm weevil, *Rhynchophorus ferrugineus* (Olivier) and synergistic interaction with the neonicotinoid, imidacloprid. *In Souvenir and Abstracts: New Horizons in Insect Science* (Eds.) A.K. Chakravarthy, C.T Ashok Kumar, Abraham Verghese, N.E. Thiagaraj, International Congress on Insect Science, February 14-17, 2013, Bengaluru, 48-49p.

Josephrajkumar, A., Chandrika M. and Krishnakumar, V. (2015a) Management of rhinoceros beetle. *Indian Coconut Journal* 58(7): 21-23.

Josephrajkumar, A., Chandrika M. , Jerald, B.A. and Krishnakumar, V. (2015b) Surveillance note on invasive pests of coconut and strategies on incursion management. Paper presented at *International conference - Innovative Insect Management Approaches for Sustainable Agro Eco System* (IIMASAE) held Agriculture College and Research Institute, Madurai 27-30 January 2015 Srinivasan *et al.* (Eds) Book of Abstracts p 587-589.

Josephrajkumar, A., Chandrika M., Merinbabu, Jerard, B.A., Krishnakumar, V., Hegde, V and Chowdappa, P. (2016). *Invasive Pests of Coconut.* Technical Bulletin No. 93, Centenary Publication 12: 28p.

Kalshoven, L.G.E. (1951). Pests of Crops in Indonesia (De plagen van de cultuurgewassen in Indonesie). The Hague- van Hoeve 2: 513-1065.

Keifer, H.H. (1965). Eriophyid studies B-14. *Calif. Dept. of Agric., Bureau of Entomology,* 20 pp.

Khan, H.H., Sujatha, S., Sivapuja, P.R. (2003). Regionally different strategies in managing eriophyid mite. In: Coconut Eriophyid Mite- Issues and strategies *Proceedings of the international Workshop on coconut mite* held at Bangalore, (Eds.) H.P. Singh and P. Rethinam, Coconut Development Board, p. 89-96.

Kumar, P.S. (2002). Development of a biopesticide for the coconut mite in India pp 335-340. The BCPC conference : Pests and Diseases. Volume 1 and 2 proceedings of an International Conference. 18-21 November 2002, Brighton, UK.

Kumar, P.S.; Singh S.P. and Gopal, T.S. (2001). Natural incidence of *Hirsutella thompsonii* Fisher on the coconut eriophyid mite *Aceria guerreronis* Keifer in certain districts of Karnataka and Tamil Nadu. *Journal of Biological control.* 15(2): 151-156.

Kurien, C., Sathiamma, B. and Pillai, G.B. (1979). World distribution of pests of coconut. *FAO Technical document* 119: 53 p.

Kurien, C., Pillai, G.B., Antony, J., Abraham, V.A. and Natarajan, P. (1983). Biological control of insect pest of coconut, pp. 361-375. In: *Coconut Research and Development. Proceedings of the International Symposium I*, N.M. Nayar (Ed.). December 27-31, 1976. Central Plantation Crops Research Institute, Kasaragod. Wiley Eastern, New Delhi.

Lever, R.J. A.W. (1969). Pests of the coconut palm. Rome, Italy, FAO Agricultural Series No. 77: 190 pp.

Mackie, D.B. (1917). *Oryctes rhinoceros* in the Philippines. *Philipp. Agric. Rev.* 10: 315-334.

Majumdar, N.D. and Jacob, T.K. (1993). Virus sensitive sites in the chromosomes of rhinoceros beetle (*Oryctes rhinoceros* L.) *Nucleus* (Calcutta) 36 (1-2): 66-68.

Mallik, B., Chinnamadegowda, C., Jayappa, J., Guruprasad, H. and Onkarappa, S. (2003). pp. 27-34 Coconut eriophyid mite in India –issues and strategies In: *Coconut eriophyid mite –issues and strategies* (Proceedings of the International Workshop on coconut mite, Bangalore) Eds. H.P. Singh and P.Rethinam, Coconut Development Board, Kochi.

Mariau, D. (1969). *Aceria guerreronis* Keifer: a recent pest of coconut in Dahomey [*Aceria guerreronis* (Keifer) Recente ravageur de 1a cocoteraies africaines et americaines]. *Oleagineux* 24(5): 269-272.

Mariau, D. (1977). *Aceria (Eriophyes) guerreronis*, an important pest of African and American coconut groves [*Aceria (Eriophyes) guerreronis:* un important ravageur des cocoteraies africaines et americaines]. *Oleagineux* 32(3): 101-111.

Mariau, D. (1986). Behaviour of *Eriophyes guerreronis* Keifer with respect to different varieties of coconut [Comportement de *Eriophyes guerreronis* Keifer a legard de differentes varietes de cocotiers]. *Oleagineux* 41(11): 499-505.

McKenna, J. and Shroff, K.D. (1911). The rhinoceros beetle (*Oryctes rhinoceros* Linn.) and its ravages in Burma. *Burma Dept. Agric. Bull.* No. 4, 6 pp.

Medina, G.S. and Abreu, E. (1986). The coconut palm eriophyid mite *Eriophyes guerreronis* (Keifer) in Puerto Rico. *J. Agric. Univ. Puerto.* 70(3): 223-224.

Menon, K. P. V. and Pandalai, K. M. (1960). *The Coconut Palms – A Monograph.* Indian Central Coconut Committee, Ernakulam, India, 384 pp

Mohamed, U.V.K., Abdurahiman, U.C. and Remadevi, O.K. (1982). Coconut caterpillar and its natural enemies. Dept. of Zoology, University of Calicut, Calicut. Zoological Monograph No. 2, 162 pp.

Mohan, K.S. and Gopinathan, K.P. (1989a). On quantitation of serological cross reactivity between two geographical isolates of *Oryctes* baculovirus by a modified ELISA. *Journal of Virological Methods* 24: 203-214.

Mohan, K.S. and Gopinathan, K.P. (1989b). Characterisation of viral proteins of *Oryctes* baculovirus and comparison between two geographical isolates. *Archives of Virology* 109: 207-222.

Mohan, K.S. and Gopinathan, K.P. (1992). Characterisation of the genome of *Oryctes* baculovirus, a viral biocide of the pest *Oryctes rhinoceros*. *Journal of Biosciences* 17(4): 421-430.

Mohan, K.S. and Pillai, G.B. (1982). A selective medium for isolation of *Metarhizium anisopliae* from cattle dung. *Transactions of the British Mycological Society* 78(1): 181-182.

Mohan, K.S., Jayapal, S.P. and Pillai, G.B. (1983). Baculovirus disease in *Oryctes rhinoceros* population in Kerala. *Journal of Plantation Crops* 11(2): 154-161.

Mohan, K.S., Jayapal, S.P. and Pillai, G.B. (1985a). Response of *Oryctes rhinoceros* larvae to infection by *Oryctes* baculovirus. *Journal of Plantation Crops* 13(2): 116-124.

Mohan, K.S., Jayapal, S.P. and Pillai, G.B. (1985b). Diagnosis of baculovirus infection in coconut rhinoceros beetles by examination of excreta. *Z. Pflkranh. Pflschutz.* 93(4): 379-383.

Mohan, K.S., Jayapal, S.P. and Pillai, G.B. (1989). Biological suppression of coconut rhinoceros beetle *Oryctes rhinoceros* (L.) in Minicoy, Lakshadweep by *Oryctes* baculovirus - impact on pest population and damage. *Journal of Plantation Crops* 16 (Suppl.): 163-170.

Mohanty, J.N., Prakash, A., Rao, J., Pawar, A.D., Patnaik, N.C. and Gupta, S.P. (2000). Parasitization of the coconut black-headed caterpillar, *Opisina arenosella* Walk. in Puri district of Orissa by field release of hymenopterous parasitoids. *Journal of Applied Zoological Researches* 11(1): 17-19.

Monty, J. (1978). The coconut palm rhi-noceros beetle *Oryctes rhinoceros* (L.) in Mauritius, and its control. *Rev. Agr. Sucr. Ile Maurice* 57(2): 60-76.

Moore, D. and Alexander, L. (1985). Dominica: Coconut mite on coconut. *FAO Plant Protection Bulletin* 33(3): 119.

Mullakoya, P. (2003). Status report on the incidence of coconut mite in Union territory of Lakshadweep. In: Coconut Eriophyid Mite- Issues and strategies *Proceedings of the international Workshop on coconut mite* held at Bangalore, (Eds.) H.P. Singh and P. Rethinam, Coconut Development Board, p. 97-98.

Muralidharan, K., Mathew, C.J., Thampan, C., Amarnath, C.H., Anithakumari, P., Mohan, C., Vijayakumar, K., Sairam, C.V., Nair, C.P.R., Arulraj, S. (2001). Pilot sample survey on the incidence and yield loss due to eriophyid mite on coconut in Alappuzha District. *Indian Coconut Journal* 21: 28–32.

Nair, C.P.R. (2000). Status of coconut eriophyid mite, *Aceria guerreronis* K. in India. In: *Proceedings International Workshop on Coconut Eriophyid Mite,* CRI, Sri Lanka, p. 9-12.

Nair, C. P. R., and Rajan, P. Chandrika M. and Murali Gopal (2010a). Management of Rhinoceros beetle *Oryctes rhinoceros* L. in coconut gardens by biocontrol methods 342-347. In: Organic horticulture - Principles, Practices and Technologies (H.P. Singh and George V. Thomas Eds) Westville Publishing House, New Delhi. 440 pp.

Nair, C.P.R. and Nair, S.S. (2002). An IPM module for red weevil. *Paper presented at National Conference on Coastal Agricultural Research, Goa 6-7,* April, 2002

Nair, C.P.R., Chandrika M., Rajan, P., Sandhya, S.R. and Anand Gopinath (2003). Technological advances in the management of coconut eriophyid mite, *Aceria guerreronis* Keifer. In: Coconut Eriophyid Mite- Issues and strategies *Proceedings of the international Workshop on coconut mite* held at Bangalore, (Eds.) H.P. Singh and P. Rethinam, Coconut Development Board, p. 99-103.

Nair, C.P.R., Rajan, P, Chandrika M. and Josephrajkumar, A. (2010b). Rhinoceros beetle and Red Palm Weevil-two major pests of coconut palm in India.178-184. In*: In a nutshell -Essays on Coconut* (Ed. C.V. Ananada Bose). Coconut Development Board, Kochi, 243 pp.

Nair, C.P.R., Rajan, P. and Chandrika M.. (2005). Coconut eriophyid mite, *Aceria guerreronis* Keifer – An overview. *Indian Journal of Plant Protection* 33(1): 1-10.

Nair, C.P.R.,Daniel, M. and Ponnamma K.N. (1997). Integrated Pest management in Palms Coconut Development Board, Cochin, 30 pp.

Nair, M.R.G.K. (1986). *Insects and Mites of Crops of India.* Indian Council of Agricultural Research (ICAR). New Delhi, India, 83 pp.

Narendran, T.C. (1985). A taxonomic revision of the Chalcid parasites (Hymenoptera: Chalcididae) associated with *Opisina arenosella* Walker (Lepidoptera: Xylorictidae). *Entomon.* 10: 83-86

Naseema Beevi., Mathew, T.B., Hebsy Bai and Saradamma, K. (2003). Status of eriophyid mite in Kerala –resume of work done. pp. 64-75. In: *Coconut eriophyid mite –issues and strategies (Proceedings of the International Workshop on coconut mite,* Bangalore) Eds. H.P. Singh and P. Rethinam, Coconut Development Board, Kochi.

Nirula, K.K. (1955a). Investigations on the pests of the coconut palm. Part I. *Oryctes rhinoceros* L. *Indian Coconut Journal* 8(4): 161-180

Nirula, K.K. (1955b). Investigation on the pests of the coconut palm. Part II. *Oryctes rhinoceros* L. *Indian Coconut Journal* 9(1): 30-79.

Nirula, K.K. (1956a). Investigations on the pests of coconut palm. Part III. *Nephantis serinopa* Meyrick. *Indian Coconut Journal* 9: 101-131.

Nirula, K.K. (1956b). Investigations on the pests of coconut palm. Part III. *Nephantis serinopa* Meyrick. *Indian Coconut Journal* 9: 174-199.

Nirula, K.K. (1956c). Investigation on the pests of coconut palm, Part-IV. *Rhynchophorus ferrugineus*. Indian Coconut Journal 9: 229-247

Nirula, K.K., Antony, J. and Menon, K.P.V. (1951). Investigations on the pests of coconut palm. 2. The coconut caterpillar *Nephantis serinopa* Meyr. *Indian Coconut Journal* 4: 217-234.

Nirula, K.K., Antony, J. and Menon, K.P.V. (1952). The rhinoceros beetle (*Oryctes rhinoceros* L.) life history and habits. *Indian Coconut Journal* V (2).

Nirula, K.K., Radha, K. and Menon, K.P.V. (1955). The green muscardine disease of *Oryctes rhinoceros* L. I. Symptomatology, epizootology and economic importance. *Indian Coconut Journal* 9: 3-10.

Nirula, K.K., Radha, K. and Menon, K.P.V. (1956). The green muscardine disease of *Oryctes rhinoceros* L. II. The causal organism. *Indian Coconut Journal* 9(2): 83-89.

Pillai, G. B., Sathiamma, B. and Dangar, T. K. (1993). Integrated control of rhinoceros beetle, pp. 455464. In: *Advances in Coconut Research and Development*, M. K. Nair, H. H. Khan, P. Gopalasundaram and E. V. V. Bhaskara Rao (Eds.). Oxford and IBH Publishing Co. Pvt. Ltd, New Delhi.

Pillai, G.B. (1993). Biological control of insect pests of plantation crops. In: *Organics in soil health and crop production* (Ed.) P.K. Thampan, Peekay Tree Crops Development Foundation, Cochin pp. 235-252

Pillai, G.B. and Nair, K.R. (1993). A checklist of parasitoids and predators of *Opisina arenosella* Wlk. on coconut. *Indian Coconut Journal* 23(9): 2-9.

Prathapan, K.D. and Shameen, K.M. (2015). *Wallacea* sp. (Coleoptera: Chrysomelidae) – A new spindle infesting leaf beetle on coconut palm in the Andaman and Nicobar Islands. *J. Plantn. Crops*, 43(2): 162-164.

Rajagopal, V., Anithakumari, P., Rohini Iyer and Nair, C.P.R. (2003). Strategic approaches of management in coconut. In: Coconut Eriophyid Mite- Issues and strategies Proceedings of the international Workshop on *Coconut Mite* held at Bangalore, (Eds.) H.P. Singh and P. Rethinam, Coconut Development Board, p. 22-26.

Rajan, P, Chandrika M. and Josephrajkumar, A. (2010). Coconut eriophyid mite, *Aceria guerreronis* keifer. 191-198. In: *In a nutshell -Essays on Coconut* (Ed. C.V. Ananda Bose)*. Coconut Development Board, Kochi, 243 pp.

Rajan, P. and Nair, C.P.R. (1997). Red palm weevil, the tissue borer of coconut palm. *Indian Coconut Journal* 27(12): 2-4.

Rajan, P., Chandrika M., Nair, C.P.R. (2007). *Technical folder on Coconut mite, Aceria guerreronis* Keifer, CPCRI, Regional Station, Kayangulam, Kerala.

Rajan, P., Chandrika M., Nair, C.P.R. and Josephrajkumar, A. (2009). *Integrated pest management in coconut. Technical bulletin No. 55*, CPCRI, Regional Station, Kayangulam, Kerala, India, p. 20.

Rajan, P., Josephrajkumar, A. and Sujatha, A. (2011). Gradient Outbreak of Coconut Slug Caterpillar, *Macroplectra nararia* Moore in East Coast of India. *Cord.* 27(1): 61-69.

Ramachandran, C. P. (1961). Assessment of damage to coconut due to *Oryctes rhinoceros* L. - a preliminary approach to the problems. *Indian Coconut Journal* 14(4): 152-162.

Ramachandran, C. P., Kurien, C. and Mathew, J. (1963). Assessment of damage to coconut due to *Oryctes rhinoceros* L. Nature of damage caused by the beetle and factors involved in the estimation of loss. *Indian Coconut Journal* 17(1): 312.

Ramani, S. (2000). Fortuitous introduction of an aphelinid parasitoid of the spiralling whitefly *Aleurodicus dispersus* Russell (Aleyrodidae : Homoptera) into the Lakshadweep Islands, with notes on host plants and other natural enemies. *Journal of Biological Control* 14: 55-60.

Ramaraju, K., Natarajan, K., Sundara Babu, P.C., Palanisamy, S. and Rabindra, R.J. (2000). Studies on coconut eriophyid mite *Aceria guerreronis* Keifer in Tamil Nadu, India. In: *Proceedings International Workshop on Coconut Eriophyid Mite*, CRI, Sri Lanka, p. 13-31.

Rao, Y.R., Cheriyan, M.C. and Ananthanarayanan, K.P. (1948). Investigations of *Nephantis serinopa* Meyr. in South India and their control by biological methods. *Indian Journal of Entomology* 10(2): 205-247.

Rethinam, P. and Muhartoyo, (2003). Coconut mite, (*Aceria guerreronis* Keifer) – Bibliography *Cord* 19(1): 69-84.

Sadakathulla, S. and Ramachandran, T.K. (1990). Efficiency of naphthalene balls in the control of rhinoceros beetle attacks in coconut. *Cocos* 8: 23-25.

Saradamma, K., Beevi, S.N., Mathew, P.B., Sreekumar, V., Jacob, A. and Vishalakshy, A. (2000). Management of coconut eriophyid mite in Kerala. *Paper presented at Entomocongress 2000*, Perspective plan for the new millennium, 5-8, November, Thiruvananthapuram, India.

Sathiamma, B. (1993). *Opisina arenosella* Wlk., the leaf eating caterpillar of coconut palm. Central Plantation Crops Research Institute, Kasaragod, *Technical Bulletin* 10: 12.

Sathiamma, B., Abraham, V.A. and Kurien, C. (1982). Integrated pest management of the major pests of coconut. *Indian Coconut Journal* 12(6-9): 27-29.

Sathiamma, B., Jayapal, S.P. and Pillai, G.B. (1987). Observations on spiders (Order Araneae) predacious on the leaf eating caterpillar, *Opisina arenosella* Wlk. (*Nephantis serinopa* Meyrick) in Kerala. *Entomon* 12(1): 45-47.

Sathiamma, B., Chandrika, M. and Gopal, Murali, (2000). Biological potential and its exploitation in coconut pest management. pp. 261-283. In: *Biocontrol Potential and its Exploitation in sustainable Agriculture* Vol 2. (Eds. R. K. Upadhay, K.G. Mukerji and B.P. Chamola). Kluwer Academic/Plenum publishers, New York.

Sathiamma, B., Chandrika, M., Gopal, Murali, Abraham, V.A. and Radhakrishnan Nair, C.P. (1999). Biological suppression of coconut pests. Central Plantation Crops Research Institute, Kasaragod. Technical Bulletin No. 37, 13 pp.

Sathiamma, B., Nair, C.P.R. and Koshy, P.K. (1998). Out break of a nut infesting eriophyid mite *Eriophyis guerreronis* (K.) in coconut plantations in India. *Indian Coconut Journal* 29(2): 1-3.

Sathiamma, B., Sabu, A.S. and Pillai, G.B. (1996). Field evaluation of the promising species of indigenous parasitoids in the biological suppression of *Opisina arenosella* Wlk. the coconut leaf eating caterpillar. *Journal of Plantation Crops* 24: 9-15.

Seguni, Z. (2002). Incidence, distribution and economic importance of the coconut eriophyid mite, *Aceria guerreronis* Keifer in Tanzania, pp. 54-57. In: *Proceedings International Workshop on Coconut Mite (Aceria guerreronis)*, L.C. P. Fernando, G. J. de Moraes and I. R. Wickramananda (Eds.). CRI, Sri Lanka, 6-8 January 2000, 117 pp.

Shanas, S., Joseph Job, Tom Joseph and Anju Krishnan, G. (2016). First report of the invasive rugose spiraling whitefly, *Aleurodicus rugioperculatus* Martin (Hemiptera: Aleyrodidae) from the Old World. *Entomon* 41(4): 365-368.

Sison, P. (1957). Further studies on the biology, ecology and control of the black beetle of coconut, *Oryctes rhinoceros* Linn. *Proc. 8th Pac. Sci. Congr.* (1953). 3A: 1299-1318.

Sivakumar, T. and Chandrika M. (2013). Occurrence of rhinoceros beetle, *Oryctes rhinoceros* L., on banana cultivars in Kerala *Pest Management in Horticultural Ecosystems* 19(1): 99-101.

Subaharan K., Ponnamma, K. N., Sujatha, A., Basheer, B. M. and Raveendran, P. (2005). Olfactory conditioning in *Goniozus nephantidis* (Muesebeck) a parasitoid of coconut black headed caterpillar, *Opisina arenosella* Walker. *Entomon.* 30 (2): 165-170.

Subaharan, K., Eswarmoorthy, M., Pavan Kumar, B.V.V.S.,Vibina Venugopal, Rajamanickam, K., Ganesan,S., Srinivasan,T., Chalapathi Rao N.B.V., and Sanjeev Gurav. (2014). Nanomatrix for delivery of semiochemicals of coconut red palm weevil, *Rhychophorus ferrugineus. In*: National Symposium on Emerging Trends in Pest Management, TNAU Coimbatore, 22-24 January, 2014. pp 42.

Sujatha, A. and Chalam, M. S. V. (2009). Status of coconut blackheaded caterpillar, *Opisina arenosella* Walker and evaluation of bio-agents. *Annals of Plant Protection Sciences.* 17(1): 65-68.

Sujatha, A., Emmanuel, N. and Arul Raj, S. (2011). Light trap - induced suppression of coconut slug caterpillar, *Macroplectra nararia* Moore menace in East coast of India, *Journal of plantation Crops* 39(3): 390–395.

Swan, D. I. (1974). A review of the work on predators, parasites and pathogens for the control of *Oryctes rhinoceros* (L.) (Coleoptera: Scarabaeidae)on coconut in the Pacific area. *Commonw. Inst. Biol. Control Misc. Publ.* 7: 63.

Venkatesan, T., Jalali, S.K., Murthy, K.S. and Bhaskaran, T.V. (2007). Economics of production of *Goniozus nephantidis* (Muesebeck), an important parasitoid of coconut black-headed caterpillar, *Opisina arenosella* (Walker) for bio-factories. *Journal of Biological Control* 21(1): 53-58.

Venkatesan, T., Jalali, S.K., Srinivasa Murthy., Rabindra, R.J. and Rao, N.S. (2006). Field evaluation of different doses of *Goniozus nephantidis* (Muesebeck) for the suppression of *Opisina arenosella* Walker on coconut. *International Journal on Coconut R and D (CORD)* 22 (special issue): 78-84.

Venkatesan, T., Jalali, S.K., Srinivasa Murthy., Rabindra, R.J. and Rao, N.S. (2003). A novel method of field release of *Goniozus nephantidis* Muesbeck, an important primary parasitoid of *Opisina arenosella* Walker on coconut. *Journal of Biological Control* 17(1): 79-80.

Watson, G.W., Adalla, C.B., Shepherd, B.M. and Carner, G.R. (2014). *Aspidiotus rigidus* Reyne: a devastating pest of coconut in Philippines. *Agriculture and Forest Entomology.* doi: 10.1111/afe.12074.

Young, E. C. (1974). The epizootiology of two pathogens of the coconut palm rhinoceros beetle *J. Invertebr. Pathol.* 24(1): 82-92.

Young, E. C. (1975). A study of rhinoc-eros beetle damage in coconut palms. *South Pac. Comm. Tech. Pap.* No. 170: 63.

Zelazny, B. (1981). India-presence of baculovirus of *Oryctes rhinoceros. FAO Plant Protection Bulletin* 29 (3/4): 77-78.

Zelazny, B. and Pacumbaba, E. (1982). Phytophagous insects associated with cadangcadang infected and healthy coconut palms in Southeastern Luzon, Philippines. *Ecol. Entomol.* 7(1): 113-120.

2018, Pests of Plantation Crops *Pages* **55–77**
Editors: **P. Chowdappa, Chandrika Mohan & A. Josephrajkumar**
Published by: **ASTRAL INTERNATIONAL PVT. LTD., NEW DELHI**

Chapter 2

Integrated Pest Management of Coconut in Sri Lanka

☆ *Nayanie S. Aratchige*

1. Introduction

Agriculture contributes about 10.1 per cent of the Gross Domestic Production (GDP) of Sri Lanka and tea, rubber and coconut sectors account for 0.9, 0.2 and 0.8 per cent of the GDP respectively at current prices (Anon, 2014a). Coconut is grown in 394, 840 ha (Anon, 2014b) with an annual national production of 3027 million coconuts (Ranasinghe *et al.*, 2016). The area under coconut is second only to the staple food, paddy, which is cultivated in more than 700,000 ha in Sri Lanka (Anon, 2014a). Coconut is predominantly grown in North-Western and Western provinces (coconut triangle) and Southern province (mini coconut triangle) of Sri Lanka. According to the world ranking in coconut production, Sri Lanka holds the fifth position (Anon, 2016a).

Coconut is a major livelihood crop of nearly 4 million people in Sri Lanka, where 70 per cent of the total produced is used for domestic consumption(Anon, 2014b), mainly as oil and milk, while the rest is exported as fresh coconut, desiccated coconut, coconut cream, milk powder, traditional oil and virgin coconut oil, coconut water, charcoal, fibre and fibre products. Coconut accounts for US$ 205 million of export earnings and 379 million coconuts have been exported as copra, oil and desiccated coconut in 2013 (Anon, 2014b).

The Sri Lankan coconut industry faces a number of challenges, both biotic and abiotic, and pests are among the major limiting factors of coconut production. During its long lifespan, different parts of the coconut palm are damaged by an array of pests. They are invertebrates such as insects and mites and vertebrates, mainly mammalian pests. However, not all pests cause damages of economic significance. The damage done by some pests cause severe economic losses as their populations are usually high in plantations, they are present all over the country and usually damage the crop throughout the year. They are not successfully controlled by natural enemies present in the environment and therefore, in most of the cases, intervention by the growers is necessary to control them. These pests are regarded as major pests of coconut. On the other hand, minor pests usually do not damage the coconut palm to an extent that will cause an economic crop loss to the grower. Their populations are usually below the economic threshold level and very well controlled by their natural enemies. Therefore, intervention of the grower is not necessary to control the minor pests. If the balance between the pest and the natural enemy populations is disturbed, pest population may rise above the economic threshold level at which point the growers' intervention is needed. Except for few, minor pests are usually restricted to certain areas and infestations are reported during certain times only. Therefore, the research on pests of coconut in Sri Lanka mainly targets on developing the management strategies of major pests only. Red palm weevil (*Rhynchophorus ferrugineus* Olivier), rhinoceros beetle (*Oryctes rhinoceros* L.), coconut black-headed caterpillar (*Opisina arenosella* Walker), plesispa beetle (*Plesispa reichei* Chapuis) and coconut mite (*Aceria guerreronis* Keifer) are the most destructive, major pests of coconut in Sri Lanka. Management of these pests has always been a formidable challenge to the coconut growers.

2. Red Palm Weevil, *Rhynchophorus ferrugineus* Olivier

Red palm weevil (RPW), a widely distributed pest in all coconut growing areas in Sri Lanka, is a serious pest that can cause fatal damage to coconut. Damage is done by the larva (grubs) of this pest. Coconut palms of 3-15 years are usually more prone to damage by RPW (Menon and Pandalai, 1960; CABI, 2015a). Adult RPW, which is characterized by dark red colour body with black spots on the thorax and a snout in the head, is about 2-3 cm long and 1 cm wide. Adult females of RPW are attracted to the fermenting odour that emanates from the fresh wounds on the trunk, pith, roots or rarely on unopened inflorescences of coconut and preferentially lay eggs on them (Menon and Pandalai, 1960; CABI, 2015a). Larvae of RPW burrow into the stem feeding on the soft tissue inside the trunk of the coconut palm. Pupation takes place inside a fibrous cocoon and adult RPW emerges from it. Several generations of RPW can damage the coconut palm simultaneously. Yellowing and drooping of leaves, slanted bud, small holes of about 0.5-1.5 cm, reddish/dark brown viscous fluid oozing out of those holes ('stem bleeding') and the crunching noise of the RPW larvae feeding on soft tissues inside the trunk are the symptoms of the infestation.

Because of the concealed nature of the damaging instar of this pest, management of RPW is enormously difficult. Current methods to control the pest have been focused on integrated pest management (IPM) with a wide range of preventive and curative measures in Sri Lanka. The IPM package for RPW in Sri Lanka includes surveillance, cultural methods, chemical methods and pheromone trapping (Anon, 2012a). However, the palm loss due to RPW damage is inordinate mainly because the coconut growers find it difficult to identify the damage at an early stage which is of paramount importance in managing the pest. Detection of the infested palms in the early stage by the conventional methods (checking for external symptoms, crunching sounds *etc.*) is time consuming, labour intensive and costly (Siriwardena *et al.*, 2010). Coconut Research Institute of Sri Lanka (CRISL), in collaboration with the University of Moratuwa, Sri Lanka has developed a portable acoustic device (Figure 2.1) for early detection of RPW damage in coconut palms (Siriwardena *et al.*, 2010). The device consists of a sensor which is made of piezoelectric transducer squashed in together with two circular stainless steel discs to acquire the sound signals from an infested palm, a signal processing unit that captures and amplifies the sound received by the sensor while filtering out the other noise signals in the environment and an output device that helps the user to determine whether the palm is infested or not (Siriwardena *et al.*, 2010). As an additional feature, a pre-recorded play-back sound clip of the RPW larval feeding sound is included in the device for the user to familiarize with the correct sound. This device is a compact, light-weight, pocket-held and robust device which is becoming popular among the coconut growers in Sri Lanka.

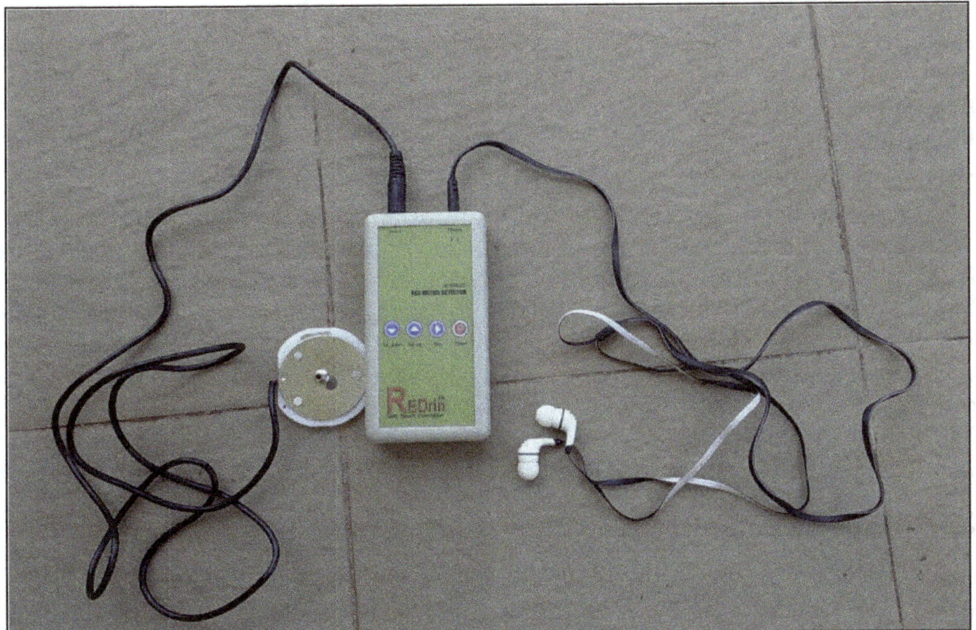

Figure 2.1: Portable Acoustic Device ReDrin™ for Early Detection of Red Palm Weevil Damage in Coconut Palms.

Cultural methods (maintenance of clean plantations by burning dead palms due to RPW, cleaning of stems of young coconut palms by removing unnecessary/ wounded fronds and other debris for early identification of the damage and maintenance of young coconut palms with minimum wounds and if fresh wounds are detected, application of coal tar or burnt engine oil) are recommended to prevent the RPW damage in coconut palms (Anon, 2012a). Once the larvae are detected inside the palm trunk, the only possible way of controlling the pest is by the application of systemic insecticides. Trunk injection and root feeding of Monocrotophos have been recommended to control RPW larvae inside the palms (Anon, 2012a).

Monocrotophos is a highly toxic insecticide and is classified in the WHO Class 1b (Anon, 2009). Due to its high toxicity, use of monocrotophos is not permitted in Sri Lanka except for few pests in coconut including RPW for which more than 95 per cent of the imported monocrotophos in to the country is used. To replace monocrotophos, alternative insecticides are being tested by the CRISL and recently two insecticides, namely phenthoate (Anon, 2012a) and a mixture of thiamethoxam 20 per cent and chlorantraniliprole 20 per cent have been recommended to control RPW in coconut. Though not as effective as monocrotophos, both insecticides, if properly administered, can kill RPW larvae inside the palm trunk.

Males and females of RPW are attracted to an aggregation pheromone, ferrugeneol (4-methyl-5-nonanol) (Gunnawardena and Badarage, 1995a; Rajapakse *et al.*, 1998) and its activity can be enhanced by N-pentanol (Gunnawardena and Badarage, 1995b). The pheromone that is artificially synthesized in the laboratory is used in a trap that has been designed using the pheromone and a 5 L plastic bucket with a raised lid and a food bait of either sugar cane or toddy or yeast (2g) + sugar (150g) + water (1L) mixture (Anon, 2012a). Insecticide solution or a soap solution is added to the bucket to kill trapped weevils. The pheromone trap is installed in the plantation, preferably at the periphery of the plantation or slightly away from the young plantation or in adult plantation to catch the weevils. The recommended trap density in Sri Lanka is 5-6 traps per ha of coconut plantation.

Long term control of economically important pests is more achievable by area-wide pest management programs using coordinated pest control tactics in large areas such as production regions than aiming at controlling the target pest temporarily in a small scale basis (*e.g.* field by field) (Elliot *et al.*, 2008; Faust, 2008; Hoddle *et al.*, 2013). Piloted by the Coconut Cultivation Board, the main extension body of the coconut industry in Sri Lanka, area-wide (>250 ha) national extension campaigns and trapping programs, advocating the integrated management of RPW has resulted in 40-50 per cent reduction in damage incidence, reflected by the drop in rate of pesticide application (M Silva, Coconut Cultivation Board, pers. comm).

3. Rhinoceros Beetle, *Oryctes rhinoceros*

Rhinoceros beetle is widely distributed in many coconut growing countries in the world (CABI, 2015b). It can cause economic damage to the seedlings and young palms, resulting in retardation of growth and sometimes even death of the young seedlings. On adult palms, an average of 10 per cent of fronds damaged by rhinoceros beetle damage-simulated cuts has resulted in 3 per cent reduction of the leaf area and a 4-5 per cent loss of nut production (Zelazny, 1979). Damage is caused by the adult through feeding on the growing frond of the coconut palm.The pest has a wide host range including *Agave sisalana* (sisal hemp), *Ananas comosus* (pineapple), many palm species, *Carica papaya* (papaya), *Colocasia esculenta* (taro), *Musa paradisiaca* (plantain), Pandanus (screw-pine) and *Saccharum officinarum* (sugarcane) (CABI, 2015b).

Adult rhinoceros beetle is easily identified by its black colour and the horn on its head ('cephalic horn') which is longer in males than in females (CABI, 2015b).

Adult female lays about 90-100 cream/white eggs during its life period of 2-4 months in decaying organic matter such as cow dung pits, coir pits and organic manure (Goonewardene, 1958; Hurpin and Fresneau, 1967). Larvae feed on the organic matter where the pupation also takes place (CABI, 2015b). Adult rhinoceros beetles make short, straight, horizontal burrows towards the centre of the coconut frond and then vertical tunnels towards the bud region (Bedford, 1976; Gressitt, 1953; Catley, 1969). Fibrous materials that have come out from the burrow or sometimes the adult rhinoceros beetle could be seen at the opening of this burrow. The damage of adult rhinoceros beetle is seen as geometric 'V' shaped cuts and holes on the frond once it is opened (Wood 1968; Sadakathulla and Ramachandran, 1990). When the flag leaf is damaged by the rhinoceros beetle it will be hanging from the top. Crooked bud leaf with retarded growth of the seedling occurs when the young seedling of less than a year is damaged and if the growing point of the young seedling is seriously injured, it will be fatal to the seedling.

Management of rhinoceros beetle is hard because they are robust, long-lived, fecund, cryptic (both adults and immatures) and nocturnal. Availability of breeding sites in abundance, difficulty in applying control measures in tall palms, inefficiency in transmitting entomopathogenic agents (while mating and at breeding sites only) among the populations and inability to mass produce rhinoceros beetles for laboratory assays also contribute to the difficulty in controlling the pest. In Sri Lanka, IPM is recommended for effective control of the rhinoceros beetle. Current IPM based strategies for rhinoceros beetle in Sri Lanka include removal of breeding sites, extracting beetles using metal hooks, application of coal tar or burnt engine oil on innermost leaf bases to repel the adult beetles, placing naphthalene balls at the leaf bases, insecticides and biological control.

Metarhizium anisopliae (Metsch.) Sorokin one of the biological control agents that have been recommended to control both adults and larvae of rhinoceros beetle in Sri Lanka. Known commonly as the green muscardine fungus (GMF), it has been recommended to apply into the breeding grounds or the infested beetles be released into the field (Suwandharathne and Kumara, 2007; Fernando and Aratchige, 2014). No significant difference in the mycelial growth and spore production of Sri Lankan and a Philippine isolate of GMF was observed, however, the mycelial growth of the local isolate significantly differed among the temperatures of 25, 27.5, 30 and 32.5°C with the highest growth at 27.5°C and lowest at 32.5°C (Subhathma *et al.,* 2012). Sri Lankan isolate produced more spores compared to the Philippine isolate and it was higher at 25°C and 27.5°C in the Sri Lankan isolate and at 25°C in the Philippine isolate (Subhathma *et al.,* 2012). These authors also confirmed that the cumulative mortality of the third instar larvae was higher with the local isolate (67 per cent) than with the Philippine isolate (42 per cent) and LT_{50} values were 14 and 18 days respectively for the two isolates. It is evident that local isolate of GMF is not suitable under all environmental conditions in Sri Lanka and it is more suitable for wet- and intermediate-zones than the dry-zone where the ambient temperature sometimes exceeds 30°C (Subhathma *et al.,* 2012). GMF is easily mass produced on-farm using maize grit (Suwandharathne and Kumara, 2007) and is applied to breeding grounds or artificial impregnation boxes of 1m x 1m x 0.5m which is filled with breeding medium (a mixture of coir/saw dust and cow dung) to transmit the disease to other individuals (Singh and Arancon Jr., 2007).

Another biological control agent, the *Oryctes* virus (OrV), is also recommended to control rhinoceros beetle in Sri Lanka. Adult rhinoceros beetles are force-fed or allowed to wade through a suspension OrV and released into the field to infect the field population of rhinoceros beetle. Both GMF and OrV are not readily transmitted by themselves and this is considered a drawback of using them. Another point to be added is the low survival of GMF particularly in dry areas. Performance of GMF on rhinoceros beetles depends on the interaction between temperature and humidity (Anon, 2015). The LTe_{50} of the fungal conidia negatively correlates with the relative humidity and the humid environments coupled with high temperatures are detrimental to the spore viability (Anon, 2015). However, high humidity with average temperatures and over casted weather with intermittent rains are favourable for disease development (Nirula, 1957). Apart from that, the limited mobility of GMF between the breeding grounds also hurdles its use. Maintenance of mother cultures of OrV has been difficult as it needs refrigeration facilities and therefore the involvement of either government or private sector institutions to assist the growers is crucial in practicing biological control of the rhinoceros beetle.

Mass trapping using the aggregation pheromone, ethyl-4-methyl octanoate which attracts both male and female beetles, is an effective way to reduce the rhinoceros beetle population (CABI, 2015b; Bedford 2013; Norman and Basri; 2004; Oehlschlager, 2007; Bedford, 2014). Use of this pheromone at the trap density of 1 per ha is recommended with two types of traps (Anon, 2012b). One type of trap, 'PVC tube trap' is made of 1.5 m long and 10-15 cm diameter PVC pipe. Two windows, 10×10 cm^2 each are made on either side at a distance of 0.5m and 0.75 m from one end. Close to the other end, few holes are made to drain excess rain water. This end is closed with an end-cap or any other closely fitting lid and fix on the ground vertically by the aid of a pole or may be attached to a trunk of an adult palm. Pheromone sachet is fixed on one window using a wire (Singh and Arancon, 2006). The beetles attracted to the pheromone fall into the pipe and are killed manually or using an insecticide/soap solution. The other type of trap (cross vane trap) is made of a plastic bucket of 10 liters and two metal sheets of the diameter of the bucket and a height of 0.5 m, arranged at an angle of 90° by making a slit on each sheet halfway (Anon, 2016b). Cross vanes are pushed into the opening of the bucket and made to stand erect by removing a triangular portion of the metal. Few holes on the lower side of the bucket are made to drain excess water. A soap-water solution is filled up to the level of the holes. The pheromone sachet/vial is hung on the middle of the vanes. The beetles attracting to the pheromone, hit the metal vanes and drop into the soap-water solution and die.

Use of pheromone traps has proven to be effective in trapping rhinoceros beetles and thereby decreasing the damage levels with different success rates in adult plantations (Suwandharathne, unpublished data). However, practically the growers are encouraged to use the traps in large plantations or if growers of small plantations desire to use these traps, to install them in few adjoining plantations simultaneously and practice area-wide trapping. It is also advised to install the traps preferably along the periphery of the plantation or in non-coconut areas. However, placing pheromone traps at the border of the plantation has found to effective only in blocks without infested breeding sites (Desmier *et al.*, 2001; Loring, 2007).

4. Coconut Black-headed Caterpillar, *Opisina arenosella*

Coconut black-headed caterpillar, a foliage pest is an outbreak pest in many coconut growing countries (Perera *et al.*, 1988). Adult coconut caterpillar is a brown moth of about 1.2 cm. Adult females lay cream colour eggs on lower surface of the coconut leaflet (Perera, 1987; Perera *et al.*, 1988) and they hatch into larvae, which is the damaging instar. In Sri Lanka, the pest is not widely prevalent in all coconut growing areas but restricted to certain areas mainly in the dry and intermediate-zones. However, recent information has shown pest outbreaks in new areas.

Coconut plantations damaged by the black-headed caterpillar can be recognized from the distance by the burnt appearance of the palms. When the damaged fronds are closely observed, green (at the initiation of the damage) or brown patches on the upper surface of the leaflets and brown galleries on the lower surface of the leaflets are seen. In an outbreak, sometimes the larvae can damage the outer skin of the fruit too. Cultural (cutting and burning of damaged fronds or portion of fronds if 5 fronds in less than 30 palms are damaged), biological and chemical control are recommended to control the coconut black-headed caterpillar in Sri Lanka (Anon, 2012c). As biological control agents, larval (*Eriborus trochanteratus* Morley, *Bracon hebetor* Say and *Goniozus nephantidis* Muesebeck) and pupal (*Brachymeria nephantidis* Gahan) parasitoids are used to control the pest. Spraying of Carbosulfan 20SC is recommended for seedlings (Anon, 2012c).

5. Coconut Mite, *Aceria guerreronis*

Coconut mite was first described from the specimens collected in Guerrero, Brazil by Keifer (1965) and later it has been reported in many coconut growing countries in America and Africa (Navia *et al.*, 2013). In late 1990's the pest expanded its range eastward towards Asia, where the first occurrence was reported from Sri Lanka and India, almost simultaneously (Fernando *et al.*, 2002; Sathiamma *et al.*, 1998). In Sri Lanka, the coconut mite infestation was first reported from the Kalpitiya Peninsula (North Western Province, dry-zone) (Fernando *et al.*, 2002). Within two years, it has spread to almost all coconut growing areas in the dry- and intermediate-zones of the country and few coconut growing areas in the wet-zone (Fernando *et al.*, 2002; Fernando and Aratchige, 2010; Aratchige, 2014). At present, the coconut mite has invaded all districts except Nuwara Eliya which is mainly a hilly area where coconut is not as extensively grown as in other districts. The incidence of coconut mite varies from district to district with higher incidences in dry- and intermediate-zones than in the wet-zone (Fernando and Aratchige, 2010; Aratchige, 2014).

Coconut mite is microscopic (205-255 µm in length and 36-52 µm in width), worm-like and white (Keifer, 1965) and feeds on the meristematic zone of the coconut fruit causing physical damage. Symptoms of the damage are first seen as a triangular white patch on the fruit surface next to the margin of the perianth and later as the infested fruit grows, the surface becomes necrotic and suberized with deep, longitudinal fissures and gummy exudates. In severe damage, fruit is distorted, stunted and unevenly grown. It is the only coconut fruit infesting mite in Sri Lanka that causes severe crop losses to the growers. Its damage can lead to small and deformed fruits (Alam and Islam, 2014) and premature fruit drop (Doreste, 1968; Nair, 2002; Wickramananda *et al.*, 2007). It can also cause reduction in copra yield (Hernández, 1977; Howard *et al.*, 2001; Moore and Howard, 1996; Ramaraju *et al.*, 2005; Muralidharan *et al.*, 2001; Alam and Islam, 2014), coconut fiber length and tensile strength (Naseema Beevi *et al.*, 2003) and yield of brown and white fibre from fruits (Kumar and Ramaraju, 2010).

Fernando and Aratchige (2010) reported that the palms infested by the coconut mite in Sri Lankan plantations varies between 2-100 per cent and as high as 86 per cent of the total fruits harvested (Wickramananda *et al.*, 2007). The percentage of small-sized fruits and deformed fruits are considerably higher in infested palms (0.72-25.5 per cent and 0.33-6.9 per cent respectively) compared to uninfested palms (<1 per cent) (Waidyarathne, unpublished data). An estimated loss of 15.8 per cent of total crop loss has been observed when the losses due to button and immature fruit fall, size reduction in the harvested fruits and fruit deformation were taken collectively (Wickramananda *et al.*, 2007). The rainfall of the same month and the previous month did not significantly affect the coconut mite densities; however, the drought length (*i.e.* the number of days without rainfall of >5 mm) affected the densities of coconut mite (Aratchige *et al.*, 2012a). Longer the dry period, higher were the coconut mite densities. In general, the peak densities were observed during February-March and June-September *i.e.* during the period of either decreasing or low rainfall and the populations of coconut mite remained low during rainy seasons (Aratchige *et al.*, 2012a).

Due to its concealed habitat, insufficient exposure of coconut mite to harsh environmental conditions, pesticides and natural enemies, management of this pest is difficult. Moreover, it is capable of breeding in extremely high populations establishing permanent infestations in new areas, mostly in population outbursts in a short time. Its inefficiency in host finding is compensated by its high reproductive rates and ability to find refuge in secluded habitat, *i.e.* under the perianth of the coconut fruit. Feasible application of control measures for this pest is also a constraint due to the tall stature of many commercially grown coconut cultivars. However, both chemical and non-chemical methods have been tested in many countries to manage the coconut mite.

Monocrotophos (Fernando *et al.*, 2002), NeemAzal (Wickramananda *et al.*, 2003), neem oil and garlic mixture (Anon, 2001) and a mixture of 30 per cent used engine oil in water, soap powder and wheat flour (Chandrasiri and Fernando, 2004) have been recommended as chemical control measures to manage the coconut mite in Sri Lanka. Though these chemicals were effective in controlling the pest, growers' acceptance was low due to practical difficulty in application of the chemicals and varying degree of control. An emulsion of 20 per cent vegetable oil and 0.5 per cent sulphur WP was found to be effective in controlling the coconut mite (Fernando and Chandrasiri, 2010). Spraying of this emulsion on the immature fruit surface resulted in mean mortality of 87 per cent and 98 per cent reduction in the coconut mite population, compared to the control palms within first 20 days after spraying. The spraying of this emulsion at 6-month intervals significantly increased the undamaged fruits and decreased the damaged, small-sized fruits in the harvest. Results of a cost-benefit analysis has shown that benefit: cost ratio varies between 0.56-4.15 suggesting that the spraying of the emulsion is profitable to the growers. Application of this emulsion was less effective on *Neoseiulus baraki* Athias-Henriot, the predatory mite of the coconut mite (Fernando and Chandrasiri, 2010). Based on these results, spraying of 20 per cent palm oil and 0.5 per cent sulphur in an aqueous emulsion has been recommended to control coconut mite. This is the most effective chemical control recommendation against the coconut mite.

Biological control of the coconut mite had drawn relatively little attention in the world until late 1990's (Moore and Howard 1996; Moraes and Zacarias, 2002). Since its detection in Sri Lanka and India in late 1990s, this area of research has been intensified in those countries. Later countries such as Brazil, Benin, Tanzania and Oman have also focused their research in this direction. In Sri Lanka, local and exotic predatory mites and an entomopathogenic fungus have been evaluated as biological control agents.

At the outset of the outbreak of coconut mite in Sri Lanka, two exotic predatory mites, *N. cucumeris* Oudemans and *N. californicus* McGregor were tested but they were found to be ineffective as they were unable to establish their populations in the field (Anon,1999).Later, another predatory mite, *Proctolaelaps bickleyi* Bram was also imported and tested which ended up in a failure due to its inability to creep under the perianth and predation on the local predatory mites (Anon, 2005).Therefore, research on biological control using predatory mites in Sri Lanka mainly focused on the local predatory mite, *N. baraki*, which is the most common predatory mite found in association with the coconut mite especially in the dry- and intermediate-zones of the country (Fernando and Aratchige, 2010). The key factors for selecting it as a potential predatory mite were its flat and elongated idiosoma with short distal setae and short legs (Moraes and Zacarias, 2002; Moraes *et al.*, 2004) enabling it to creep under the perianth, its close contact with the coconut mite (Fernando *et al.*, 2003; Aratchige, 2007; Aratchige *et al.*, 2012b) and its ability to feed and develop on coconut mites (Anon, 2003). Series of experiments were conducted in Sri Lanka to evaluate its bio-ecological aspects and potential to control the coconut mite (Fernando *et al.*, 2003, 2010; Aratchige *et al.*, 2008, 2010, 2012a, 2012b; Kumara *et al.*, 2014).

Population dynamics studies conducted in Sri Lanka shows that the fluctuation pattern of *N. baraki* differs spatially and temporally. The population densities of *N. baraki* are not significantly regulated by the amount but by the frequency of rainfall of the same month, drought length and the density of the coconut mite, one month prior to that of *N. baraki* (Aratchige *et al.*, 2012a). The authors have suggested that *N. baraki* populations are more influenced than the coconut mite populations by external abiotic factors probably due to more active movement of the former than the latter in search of food and shelter.

For testing of *N. baraki* for its potential as a predatory mite of the coconut mite, essentially a mass production protocol was sought. Except coconut mite, out

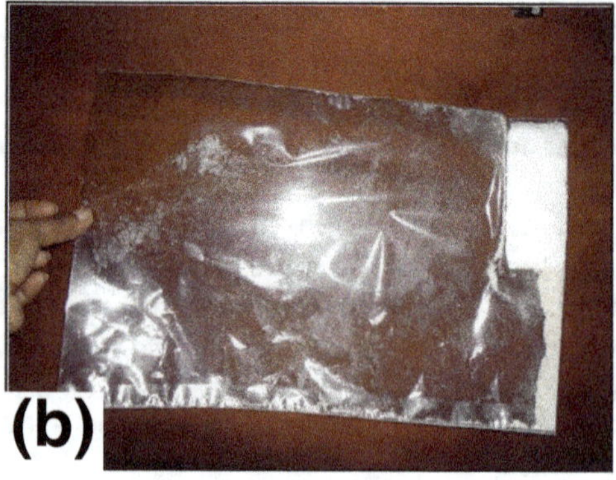

**Figure 2.2: (a) Tray-type arena for the mass production of *N. baraki*,
(b) Sachet-type rearing method of *N.baraki*.**

of the several food sources that were tested (Fernando *et al.*, 2004; De Silva and Fernando, 2008; Anon, 2001, 2002, 2003) *Tyrophagus putrescentiae* Shrank (Acari: Ascidae) was more suitable (Fernando *et al.*, 2004). Therefore, *T. putrescentiae* was subsequently used in studies to develop mass rearing methods (Aratchige *et al.*, 2010; Kumara *et al.*, 2014) and in field releases (Fernando *et al.*, 2010; Aratchige *et al.*, 2012b). *N. baraki* can be mass produced on *T. putrescentiae* reared on a black polyvinyl sheet in tray-type rearing arenas (Aratchige *et al.*, 2010). Insect glue layer applied along the periphery of the polyvinyl sheet avoids the escape of the mites from the arena and protects the cultures from external contaminants. A glass sheet on a small piece of synthetic net and a piece of foam wrapped in a wet tissue on the glass sheet are also kept inside the arena to supply shade, oviposition sites and drinking water respectively. Rice bran and flour mixture (1:1) is sprinkled on the polyvinyl sheet and *T. putrescentiae* and *N. baraki* are mass produced on the same arena (Figure 2.2). A 240-fold increase of *N. baraki* is obtained from this method in 5 weeks (Aratchige *et al.*, 2010). Laborious nature of removal of the insect glue layer from the polyvinyl sheet is one of the drawbacks of this method which was excluded in the sachet-type rearing unit (Kumara *et al.*, 2014). In this method, *N. baraki* is mass produced on *T. putrescentiae* inside a two-ply polypropylene sachet, which contains rice bran and flour mixture (1:1) and a piece of wet tissue paper in a partially separated compartment. A 260-fold increase of the original population could be obtained from this method in 6 weeks (Kumara *et al.*, 2014). Using this method, approximately 2,000 sachets could be accommodated in an insectary room of 4 x 3.5 m² (air-conditioned, 25°C) which has 6 racks with 3 shelves of 2 m x 0.5 m x 1.5 m each (Figure 2.3).

The first attempt to evaluate the single release of *N. baraki* in controlling the coconut mite was reported by Fernando *et al.* (2010) in which a single inundative release of 10,000 *N. baraki* significantly increased its population and decreased the coconut mite population. During the post release period of 6 months, mean numbers of 8.99 *N. baraki* and 1264.77 coconut mites per fruit were observed in the released palms compared to the unreleased palms (6.19 *N. baraki* and 1815.0 coconut mites per fruit). However, the marked reduction in the coconut mite populations was not

Figure 2.3: Mass Rearing Insectary Room of *N. baraki*.

mirrored in the reduction of the damage due to coconut mites. Authors suggested that multiple releases of *N. baraki* would be a better alternative due to the occurrence of continuous infestation of coconut mites in its several generations over a long period of 2-6 months on fruits.

Release of 5000 *N. baraki*/palm at 2-4 month intervals for 2 years on to quarter of a plantation of 5ac resulted in a higher percentage of normal-sized fruits at harvest (85.6 and 88.4 per cent in two released blocks compared to 79.1 and 80.1 per cent in unreleased blocks) and a lower percentage of small-sized fruits (13.3 and 10.1 per cent in two released blocks compared to 20.0 and 17.2 per cent in unreleased blocks) (Aratchige *et al.*, 2012b). Release of *N. baraki* in this manner has shown to be economically effective as it has resulted in benefit: cost ratio of 1.76-3.11 (Aratchige *et al.*, 2012b).

In Sri Lanka, release of 5000 *N. baraki* at 3-4 month intervals (depending on the rains) on to quarter of the coconut plantation at least for 2 years has been recommended to control the coconut mite (Aratchige *et al.*, 2012b) and this is the first recommendation of using predatory mites for the control of coconut mite in the world. The Coconut Research Institute of Sri Lanka conducts a national level project in collaboration with the Coconut Cultivation Board (CCB) and 2 state-owned private companies, Chilaw Plantation PLC. and Kurunegala Plantation PLC. and under this project,13 mass producing laboratories of *N. baraki* are being operated to supply *N. baraki* sufficiently to the growers. After commencing release of *N. baraki* from one laboratory managed by the CCB, a 35-43 per cent reduction in the rejected fruits in the harvest has been observed in 2013, 2014 and 2015 in one block of 10ac compared to 2010-2012 pre-release period (Wickramasinghe I, unpublished data).

The natural incidence of the *Hirsutella thompsonii* Fisher, an entomopathogenic fungus of coconut mite was low (Edgington *et al.*, 2008) and out of the isolates collected from different geographical regions of Sri Lanka, only four isolates, namely IMI 390486, IMI 391722, IMI 391942 and IMI 391723, performed better when the growth characteristics and sporulation in culture were compared. Isolate IMI

391722 outperformed the others showing the highest vegetative growth between 20-35°C and the second highest spore production at 15-35°C (Edgington *et al.*, 2008). The highest efficacy in reducing the pest population and the highest percentage of fruits with mycosis were achieved when two applications of spore suspensions of the above isolate were made at two week intervals on the five youngest coconut bunches (Fernando *et al.*, 2007). However, this effect was relatively short and the results were not consistent (Fernando *et al.*, 2007) and the use of *H. thompsonii* for the control of coconut mite was discontinued in Sri Lanka.

Variation in the varieties in responses to the coconut mite damage is one of the aspects that have been less studied in Sri Lanka and in other countries. Certain varieties have shown less damage by the coconut mite (Julia and Mariau, 1979; Schliesske, 1988; Suarez, 1990; Muthiah and Natarajan, 2004; Nair, 2002; Ramaraju *et al.*, 2002; Thirumalai Thevan *et al.*, 2004; Varadarajan and David, 2003; Moore and Alexander, 1990; Perera *et al.*, 2013; Mohan *et al.*, 2014). Preliminary study to investigate the differences between levels of tolerance in the varieties in terms of symptoms initiation, subsequent expression of symptoms, coconut mite population levels, distance between fruit surface and the perianth of the fruit and the crop loss due to mite damage revealed that the above parameters were least in Yellow Dwarf x Sri Lanka Tall hybrid (DYT) suggesting DYT as a putative tolerant coconut variety for coconut mite damage in Sri Lanka (Perera *et al.*, 2013). In a recent study, variety Yellow Dwarf showed the least damage with moderate coconut mite densities while the varieties Gonthambili and Ran thambili were moderately damaged by the coconut mite probably due to round shaped fruits that restrict access to coconut mites. Variety Ran thambili showed the least coconut mite densities while the variety Gonthambili was infested with a moderate coconut mite density. Sri Lanka Green Dwarf, followed by the Sri Lanka Tall variety were the most susceptible variety to the coconut mite damage (Aratchige and Perera, unpublished data).

6. Plesispa Beetle, *Plesispa reichei*

Plesispa beetle is a recent invasion into Sri Lanka and is believed to be an accidental introduction into the country (Anon, 1999). It was first reported in Sri Lanka in 1999 from a coconut seedlings nursery in Gampaha district in the Western Province (Anon, 1999). The infestation has expanded rapidly into many other coconut growing areas through transportation of infested seedlings (Suwandharathna *et al.*, 2011). It is one of major pests of coconut cultivation in Philippines, Indonesia, Thailand, Singapore and Malaysia (Sivapragasam and Hong, 2007). Both adults and larvae feed on the leaflets of unopened fronds in seedlings in nurseries and young plantations retarding the growth. The damage is observed as brown patches or streaks on leaflets when they are opened.

In order to understand the biology of plesispa beetle, research has been intensified in Sri Lanka to determine its development and life cycle (Kumari *et al.*, 2009; Suwandharathne *et al.*, 2011; Bentharage *et al.*, 2012). The incubation period of the eggs varied between 7 - 9 days at 25 -31°C; higher the temperature, lower was the incubation period (Suwandharathne *et al.*, 2011; Bentharage *et al.*, 2012). The hatchability of plesispa beetle eggs was 81 per cent (Kumari *et al.*, 2009) and it was higher when the eggs were laid on the spear leaf than on the other leaves (Bentharage *et al.*, 2009). Although the incubation period of eggs was not influenced by the maturity of leaves on which they were laid, egg production of females was increased when they were allowed to feed on mature leaves, *i.e.* first and second open leaves (Bentharage *et al.*, 2009). Authors have suggested that the nutritional differences in mature leaves might have contributed to the increased egg production on them.

Five larval instars of plesispa beetle have been reported (Bentharage *et al.*, 2012). Temperature has a significant effect on development of 2nd and 4th instar larvae, pupal period and the total larval period while 1st and 3rd instar larval development was not affected by the varying temperature between 25 -31°C (Kumari *et al.*, 2009; Suwandharathne *et al.*, 2011). The duration of the life cycle (egg to adult) was 39.0, 30.2 and 29.7 days at 25, 28 and 31°C respectively suggesting that the warmer temperatures favour the rapid development (Kumari *et al.*, 2009; Suwandharathne *et al.*, 2011). Irrespective of the instar, larval development was slower on spear leaves compared to the first and second open leaves (Bentharage *et al.*, 2012). Leaf area consumption by the adults is also affected by the leaf growth stage; adults collected from the field feed more leaf area when they were allowed to feed on second open leaves (Bentharage *et al.*, 2012).

Chemicals have been recommended to control plesispa beetles in coconut in Sri Lanka but economically and environmentally acceptable alternatives are being assessed to aid in the management. Classical biological control is being evaluated with exotic parasitoid (*Tetrastichus brontispae*) (Suwandharathne, 2013).

7. Future Challenges and Research Needs

Coconut is an inimitable part of the Sri Lankan society as it is a basic sustenance of oil and a source of milk, drink and fuel and a part of many cultural events in Sri Lanka. It also adds a scenic beauty to the golden beaches around the country which is attracted by both local and foreign tourists. It is also used as raw material for many industries such as fibre, charcoal, desiccated coconut, coir products, timber and handicrafts and an important foreign income generator. Pests are a major limiting factor of the production of coconut in Sri Lanka and their damages continue to persist with severe crop losses, regardless of the developments in the knowledge and applications of the control of pests of coconut. Many factors contribute to this, among which slow diffusion of technology contributes to insufficiency in timely application of control measures against the pests. In this regard, government intervention is necessary to strengthen the extension system of the coconut industry to disseminate the knowledge to the stakeholders on control of pests.

Pest management is always a high priority in coconut plantations. Integrated pest management of coconut needs decision making guidelines for the implementation of the most appropriate control strategies of the pest. This is in particular needed because coconut is a crop with a tall stature and therefore control of pests is always difficult with high costs. Compared to other crops grown in Sri Lanka, growers' direct involvement in coconut plantation management is lower and the plantations are mainly managed by caretakers. Majority of the coconut growers in Sri Lanka maintain small-scale plantations where they demand for more convenient and practical control measures for which new avenues should be looked into and existing recommendations are needed to be broadened and fine-tuned.

Rapid worldwide demand for organic coconut production has prompted Sri Lankan coconut growers more motivated to produce coconuts organically which has claimed for research-based information to address issues, including pest management, to develop the industry as a viable and sustainable entity. Though research-based information on pest management are lacking within the context of organic coconut cultivation per se, integrated pest management practices that have been developed for conventional coconut farming have provided the framework for pest management tools in organic coconut farming. However, more research is needed to fill the knowledge gap in certain areas in the pest management in organic coconut cultivation. Soil fertility management has always been claimed to be beneficial in reducing the pest damages and though it has been a widely studied area of research in coconut, it has not been tested within the context of pest management even in conventional coconut cultivation. Information on mulching materials that will deter the development of pest insects (plant materials with insecticidal properties, insect repellents) will also be useful in prevention of development and survival of pests on organic mulches, but this concept has not been tested in the coconut system. One major challenge faced by the organic coconut farmers and the researches is the unavailability of sufficient and effective biological control agents for some major pests such as red palm weevil, plesispa beetle and termites. Urgent attention is needed to identify and develop mass rearing and inoculation methods of biological control agents of such pests. Unavailability of sufficient information on botanicals for the control of coconut pest is also another issue in controlling the pests in organic coconut farming. Though research-based information are available on a broad range of botanicals with insecticidal and repellent properties, such products have not been tested for pests of coconut in organic farming. Extracts of neem (*Azadirachta indica*) kernels, milk weed (*Calotropis procera*), periwinkle (*Catharanthes roseus*), tobacco (*Nicotiana tabacum*) and spinosads (insecticidal agent obtained from *Saccharopolyspora spinosa*) may have great potential as botanicals for pest management in organic coconut farming but research are needed in this direction.

To improve the potency of pheromone traps, it is suggested to combine them with other different techniques and also to use them in area-wide programs in trapping red palm weevil and rhinoceros beetles which may require institutional/ governmental involvement in technology dissemination and resource supply and growers participation in executing control measures. However, the rationale for such programs should be based on sound scientific evidence on the effect of area-wide trapping pests on the damage reduction and thereby the crop loss, because the integration of different techniques of pest control and area-wide management program are often costly. This is one of the areas where sound research-based information are lacking. Research on pheromones of RPW, rhinoceros beetle and black-headed caterpillar have also been widened to identify suitable, cost effective dispensers and formulations of pheromones with emphasis on nanotechnology.

In rhinoceros beetle control using OrV and GMF, basic information on the density dependence of the biological control agents on effectiveness in controlling the pest needs to be meticulously studied because, it could be expected that in low populations of beetles, spread of the OrV and GMF in the beetle population to be low due to low contacts of infected beetles with the healthy ones. The dispersal capacity of the beetle is another consideration that is to be taken in to account when using biological control using the OrV and GMF because its behavior is expected to be different when host plants are readily available in compared to the situation where host plants are relatively less densely present. As evidenced by Zelazny *et al.* (1989), OrV isolates could differ in its virulence when using against the adults and larvae of rhinoceros beetle and this aspect needs to be seriously taken in to consideration if the virus is included in the integrated management program of the pest. For rhinoceros beetle, effect of mass trapping of adults via pheromones could be enhanced when combined with biological control agents. This is possible with auto disseminated traps where beetles are lured to traps with pheromone mixed with spore solutions of GMF (Anon, 2015).

Rhinoceros beetle damage could also lead to RPW damage as both pests coexist in many coconut plantations in Sri Lanka and that the adult female RPW lays eggs on wounds made by the rhinoceros beetle. Therefore, use of pheromone traps could target both pests at a time using their respective pheromones (Hallett *et al.*, 1999). However, this type of recommendation should be based on thorough investigation of the population dynamics and the flight activity of the two insects.

Combination of traps such as barrel traps filled with a breeding substrate and coupled with solar-powered UV LEDs (Moore, 2013 a, b; 2014), pan traps (Moore, 2014) and fishnet traps (Moore, 2014) are other options for trapping rhinoceros beetles and with some modifications could also be used for red palm weevil as well. The inadequacy of the effectiveness of natural enemies of many of the pests of coconut highlights the needs for expanding the scouting of natural enemies probably in the area of origin of the pest. Given the exotic nature of the plesispa beetle in Sri Lanka, classical biological control appears to be promising as natural enemies of the pest are implausible to occur in Sri Lanka.

Use of resistant host plants is not currently recommended against any coconut pests in Sri Lanka. Improved cultivars and their parent materials are being evaluated for resistant to coconut mite damage. However, it is intended to expand this area of research in to the other coconut pests as well. It is also envisioned to determine the mode of resistance with possible molecular background and to include such traits if available, in the coconut breeding program.

Growing coconut with lot of uncertainties due to changed pest scenario has generated the need for reliable and timely forecast of pests. Nevertheless, this need has not been adequately addressed and there is practically no organized system of forecasting pests of coconut or other crops in Sri Lanka. In this regard, comprehensive scientific information on variations in pest severity with fluctuations of weather parameters, extent of pest damages and their threshold levels and monitoring systems of pest advancements are not available for most of the pests in

Sri Lanka. This statistics would also provide baseline information for understanding the outbreaks of coconut pests in new areas.

Coconut-based agro forestry systems have great potential for supplying multiple benefits to the grower. One important consideration within the context of coconut-based agro forestry systems is the management of pests in the system (Aratchige, 2011). Pest management in an agro forestry system may be different from that in a mono-crop system as the interactions among the components in a former system are more complicated than in the latter system. This is one of the areas which has not drawn attention in Sri Lanka and therefore, needs to be studied in future.

Acknowledgement

Author wishes to acknowledge Prof. H P M Gunasena for his comments on this manuscript.

References

Alam, N.S. and Islam, M.N. (2014). Insect pest management of coconut in Bangladesh with special emphasis on Eriophyid mite *Aceria guerreronis*. In: Akter, N., Azad, A.K.,Alam, M.N. (eds). *Proc. Mite management of coconut in SAARC member countries.* SAARC Agriculture Centre (SAC), Dhaka, Bangladesh. 10-11 August, 2014. pp 26-38.

Anon. (1997). *Annual Report,* Coconut Research Institute of Sri Lanka.

Anon. (1999). *Annual Report,* Coconut Research Institute of Sri Lanka.

Anon. (2001). *Annual Report,* Coconut Research Institute of Sri Lanka.

Anon. (2002). *Annual Report,* Coconut Research Institute of Sri Lanka.

Anon. (2003). *Annual Report,* Coconut Research Institute of Sri Lanka.

Anon. (2005). *Annual Report,* Coconut Research Institute of Sri Lanka.

Anon.(2009). The WHO recommended classification of pesticides by hazard and guidelines to classification. World Health Organization.

Anon. (2012a). Advisory leaflet on Red weevil and its control (B3). Coconut Research Institute of Sri Lanka.

Anon. (2012b). Advisory leaflet on black beetle and its control (B4). Coconut Research Institute of Sri Lanka.

Anon. (2012c). Advisory leaflet on coconut caterpillar and its control (B2). Coconut Research Institute of Sri Lanka.

Anon. (2014a). Annual Report of the Central Bank of Sri Lanka. National Output and expenditure.http://www.cbsl.gov.lk/pics_n_docs/10_pub/_docs/efr/annual_report/AR2014/English/6_Chapter_02.pdf.Accessed on 20-12-2015.

Anon. (2014b). SriLanka Socio Economic Data. Central Bank of Sri Lanka. Vol. xxxvii. http://www.cbsl.gov.lk/pics_n_docs/10_pub/_docs/statistics/other/Socio_Econ_Data_2014_e.pdf. Accessed on 23-12-2015.

Anon. (2015). USDA new pest response guidelines. *Oryctes rhinoceros* (L.) Coleoptera: Scarabaeidae. http://www.aphis.usda.gov/import_export/plants/manuals/emergency/downloads/nprg-o_rhinoceros.pdf. Accessed on 9-02-2016.

Anon. (2016a). FAOSTAT.http://faostat3.fao.org/browse/Q/QC/E. Accessed on 16-02-2016.

Anon. (2016b). Advisory leaflet on black beetle and its control (B4). Coconut Research Institute of Sri Lanka.

Aratchige, N.S. (2007). Predators and the accessibility of herbivore refuges in plants. PhD Thesis, University of Amsterdam.

Aratchige, N.S. (2011). Integrated pest management in coconut based agroforestry systems. *In:* Pushpakumara, D.K.N.G., Gunasena, H.P.M., Gunethilake, H.A.J. and Singh, V.P. (Eds). Increasing coconut land productivity through agroforestry interventions. *Proc. Symp. held at the Coconut Research Institute of Sri Lanka* on 15th March 2011. pp. 57-66.

Aratchige, N.S.,Sabelis, M.W. and Lesna, I. (2008). Effect of coconut mite-induced changes in coconuts on the searching behaviour of predatory mites, *Neoseiulus baraki*. In: Nainanayake, N.P.A.D. and Everard, J.M.D.T. (Eds). *Proc. 2nd Symp. on Plantation Crop Research – Export competitiveness through quality improvements.* Coconut Research Institute, Lunuwila, Sri Lanka. pp. 187-198.

Aratchige, N.S., Fernando, L.C.P., de Silva, P.H.P.R., Perera, K.F.G., Hettiarachchi, C.S., Waidyarathne, K.P. and Jayawaredena S.M.V. (2010). A new tray-type arena to mass rear *Neoseiulus baraki*, a predatory mite of coconut mite, *Aceria guerreronis* in the laboratory. *Crop Prot.* 29: 556-560.

Aratchige, N.S., Fernando, L.C.P., Waidyarathne, K.P. and Chandrasiri, K.A.S. (2012a). Population dynamics of *Aceriaguerreronis* (Acari: Eriophyidae) and its predatory mite, *Neoseiulusbaraki* (Acari: Phytoseiidae) in two coconut growing areas in Sri Lanka. *Exp. Appl. Acarol.* 56: 319-325.

Aratchige, N.S., Fernando, L.C.P., Kumara, A.D.N.T., Jayalath, K.V.N.N., Perera, K.F.G. and Suwandarathne, N.I. (2012b). Biological control of coconut mite: Effect of inundative release of the predatory mite, *Neoseiulus baraki* (Acari: Phytoseiidae). In: Hettiarachchi, L.S.K., Abeysinghe, I.S.B. (Eds). *Proc. 4th Symp. on Plantation Crop Research – Technological Innovations for Sustainable Plantation Economy.* Tea Research Institute of Sri Lanka, St. Coombs, Talawakelle, 22100, Sri Lanka. pp. 125-133.

Bedford, G.O. (1976). Observations on the biology and ecology of *Oryctes rhinoceros* and *Scapanes australis* (Coleoptera: Scarabaeidae, Dynastinae): Pests of coconut palms in Melanesia. *J. Aust. Ent. Soc.* 15: 241-251.

Bedford, G.O. (2013). Biology and management of palm dynastid beetles: recent advances. *Ann. Rev. Entomol.* 58: 353-372.

Bedford, G.O. (2014). Advances in the control of rhinoceros beetle, *Oryctes rhinoceros* in oil palm. *J. Oil Palm Res.* 26(3): 183-194.

Bentharage, B.D.C.D., Aratchige, N.S., Suwandarathne, N.I. and Wijayaratne, L.K.W. (2012). Development and leaf area consumption rate of two-colour coconut leaf beetle, *Plesispa reichei* on selected coconut leaf stages *In: Proc. Res. Symp. Uva Wellassa University*. November 22-23, pp. 106-109.

CABI (2015a). Invasive Species Compendium. CABI. http://www.cabi.org/isc/ datasheet/47472#20127201272. Accessed on 31-12-2015.

CABI (2015b). Invasive Species Compendium. CABI. http://www.cabi.org/isc/ datasheet/37974. Accessed on 05-01-2016.

Catley, A. (1969).The coconut rhinoceros beetle *Oryctes rhinoceros* (L.) (Coleoptera: Scarabaeidae: Dynastinae). *PANS*. 15: 18-30.

Desmier, R., Asmady, H. and Sudharto, P.(2001). New improvement of pheromone traps for the management of the rhinoceros beetle in oil palm plantations. *In*: Cutting-edge technologies for sustained competitiveness: *Proc. PIPOC International Palm Oil Congress*, Agriculture Conference, Kuala Lumpur, Malaysia, 20-22 August 2001. pp. 624-632.

De Silva, P.H.P.R. and Fernando, L.C.P. (2008). Rearing of coconut mite *Aceria guerreronis* and the predatory mite *Neoseiulus baraki* in the laboratory. *Exp Appl Acarol*. 44: 37-42.

Doreste, S.E. (1968). El ácaro de la flor del cocotero (*Aceria guerreronis* Keifer) en Venezuela. *Agr. on Trop*. 18: 370–386.

Edgington, S., Fernando, L.C.P. and Jones K. (2008). Natural incidence and environmental profiling of the mite-pathogenic fungus *Hirsutella thompsonii* Fisher for control of coconut mite in Sri Lanka. *Int. J. Pest Manag*. 54: 123–127.

Elliot, N.C., Onstad, D.W. and Brewer, M.J. (2008). History and ecological basis for area wide pest management In: Koul, O., Cuperus, G., Elliot, N. (Eds.) Area wide Pest Management: Theory and Implementation CABI, Oxfordshire, UK pp. 15-33.

Faust, R.M. (2008). General introduction to area wide pest management In: Koul, O., Cuperus, G. Elliot, N. (Eds.) Area wide Pest Management: Theory and Implementation CABI, Oxfordshire, UK pp. 1-14.

Fernando, L.C.P., Wickramanada, I.R. and Aratchige, N.S. (2002). Status of coconut mite, *Aceria guerreronis* in Sri Lanka. In: Fernando, L.C.P., Moraes, G.J. de, Wickramanada, I.R. (eds). Proc. Int. Workshop on Coconut Mite (*Aceria guerreronis*). Coconut Research Institute, Lunuwila, Sri Lanka, pp. 1–8.

Fernando, L.C.P., Aratchige, N.S., and Peiris, T.S.G. (2003). Distribution pattern of coconut mite *Aceria guerreronis* and its predator *Neoseiulus* aff. *Paspalivorus* on coconut palms. *Exp. Appl. Acarol*. 31: 71–78.

Fernando, L.C.P., Aratchige, N.S., Kumari, S.L.M.L., Appuhamy, P.A.L.D. and Hapuarachchi, D.C.L. (2004). Development of a method for mass rearing of *Neoseiulus baraki*, a predatory mite on the coconut mite, *Aceria guerreronis*. *COCOS*. 16: 22-36.

Fernando, L.C.P., Manoj, P., Hapuarachchi, D.C.L. and Edgington, S. (2007). Evaluation of four isolates of *Hirsutella thompsonii* against coconut mite (*Aceria guerreronis*) in Sri Lanka. *Crop Prot.* 26: 1062–1066.

Fernando, L.C.P. and Aratchige, N.S. (2010). Status of coconut mite *Aceria guerreronis* and biological control research in Sri Lanka. *In*: Sabelis, M.W., Bruin, J. (eds). Trends in Acarology. Springer, Amsterdam, pp. 379–384.

Fernando, L.C.P., Waidyarathne, K.P., Perera, K.F.G. and de Silva, P.H.P.R. (2010). Evidence for suppressing coconut mite *Aceria guerreronis* by inundative release of the predatory mite, *Neoseiulus baraki*. *Biol Control*, 53: 108–111.

Fernando, L.C.P. and Aratchige, N.S. (2014). Pest management in organic coconut production. In: Gunasena, H.P.M., Gunethillake, H.A.J., Everard J.M.D.T. (Eds). Organic coconut production in Sri Lanka, pp. 85-118.

Goonewarden, H.E.F. (1958). The rhinoceros beetle (*Oryctes rhinoceros* L.) in Ceylon. Part 1: Introduction, distribution and life history. *Trop. Agric.* 114: 39-60.

Gunnawardena, N.E. and Badarage U.K. (1995a). 4-methyl-5-nonanol (ferruginol) as an aggregation pheromone of the coconut pest *Rhynchophorus ferrugineus* (Coleoptera: Curculionidae) synthesis and use a preliminary field assay. *J. Nat. Sc. Council of Sri Lanka.* 23: 71-79.

Gunnawardena, N.E. and Badarage U.K. (1995b). Enhancement of the activity of ferruginol by N-pentanol in an attractant baited trap for the coconut pest *Rhynchophorus ferrugineus* (Coleoptera: Curculionidae) synthesis and use a preliminary field assay. *J. Nat. Sc. Council of Sri Lanka.* 23: 81-86.

Gressitt, L. (1953).The coconut rhinoceros beetle (*Oryctes rhinoceros*) with particular reference to the Palau Islands. *Bull. Bernice P. Bishop Mus.* No. 212, p157.

Hallett, R.H. Oehlschlager, C. and Borden, J.H. (1999). Pheromone trapping protocols for the Asian palm weevil, *Rhynchophorus ferrugineus* (Coleoptera: Curculionidae). *Int J. Pest Manag.* 45(3): 231-237.

Hernández Roque, F. (1977). Combate químico del eriófido del cocotero *Aceria (Eriophyes) guerreronis* (K.) en la Costa de Guerrero. *Agric Tec México* 4: 23–38.

Hoddle, M.S., Al-Abbad, A.H., El-Shae, H.A.F., Faleiro, J.R., Sallam, A.A. and Hoddle, C.D. (2013). Assessing the impact of area wide pheromone trapping, pesticide applications and eradication of infested date palms for *Rhynchophorus ferrugineus* (Coleoptera: Curculionidae) management in Al Ghowaybah, Saudi Arabia. *Crop Prot.* 53: 152-160.

Howard, F.W., Moore, D., Giblin-Davis, R.M. and Abad, R.G. (2001). Insects on palms. CABI Publishing, Wallingford.

Hurpin, B. and Fresneau, M. (1967). Contribution a la luttecontre les Oryctesnuisibles aux palmiers. Elevageen laboratoire de *O. rhinoceros*. *Oléagineaux* 22: 1-6.

Julia, J.F. and Mariau, D. (1979). Nouvelles recherché en Côte d'Ivoire sur *Eriophyes guerreronis* K., acarien ravageur des noix du cocotier. *Oléagineux* 34: 181–189.

Keifer, H.H. (1965). Eriophyid Studies B-14. California Department of Agriculture, Bureau of Entomology, Sacramento (Special publication).

Kumar, P.P. and Ramaraju, K. (2010). Impact of nut-infesting eriophyid mite, *Aceria guerreronis* Keifer, on the out-turn and quality of coconut fiber. In: de Moraes, G.J., Castillo, R.C., Flechtmann, C.H.(eds). *Proc. XIII Int. Congress of Acarol.* Abstracts. Recife, p. 214.

Kumara, A.D.N.T., Fernando, L.C.P., Suwandharathne, N.I. and Aratchige, N.S. (2014). Use of polypropylene bags for mass rearing *Neoseiulus baraki*. (Acari: Phytoseiidae), a predatory mite of *Aceria guerreronis* Keifer (Acari: Eriophyidae). *Biocon Sc. Tech.* DOI:10.1080/0958307.2014.918245.

Kumari, A.M.M.T., Suwandharathne, N.I. and Fernandopulle, M.N.D. (2009). Determination of the pupal development of the two colour leaf beetle, *Plesispa reichei* (Chapuis) under three different temperatures. In: Proc. 9th Agricultural Res. Symp., Faculty of Agriculture and Plantation Management, Wayamba University of Sri Lanka. 11-12 August, 2009, pp. 177-180.

Menon, K.P.V. and Pandalai, K.M. (1960). Pests. In: The Coconut palm. A Monograph. Erankulam, South India: Indian Central Committee, pp. 261-265.

Mohan, C., Thomas, G.V. and Josephrajkumar, A. (2014). Mite management of coconut in India. In: Akter, N., Azad, A.K., Alam, M.N. (eds). *Proc. Mite management of coconut in SAARC member countries.* SAARC Agriculture Centre (SAC), Dhaka, Bangladesh. 10-11 August, 2014, pp. 43-58.

Moore, A. (2013a). Development of barrel traps. Research in support of the Guam coconut rhinoceros beetle eradication project. Cooperative extension service, University of Guam.

Moore, A. (2013b). Solar powered ultraviolet light emitting diode for CRB pheromone traps. CRB technical report series No. 29. Research in support of the Guam coconut rhinoceros beetle eradication project. Cooperative extension service, University of Guam.

Moore, A. (2014). Final report for US Forest Service Grant 11-DG-11052012-101. Support for the Guam coconut rhinoceros beetle eradication project. University of Guam.

Moore, D. and Alexander, L. (1990). Resistance of coconuts in St. Lucia to attack by the coconut mite, *Eriophyes guerreronis* Keifer. *Trop Agric.* 67: 33–36.

Moore, D. and Howard, F.W. (1996).Coconuts. *In*: Lindquist, E.E., Sabelis, M.W., Bruin, J. (eds). *Eriophyoid mites: Tbiology natural enemies and control.* Elsevier, Amsterdam, pp. 561–570.

Muralidharan, K., Mathew, C.J., Thampan, C., Amamath, C.H., Anithakumari, P., Mohan, C., Vijayakumar, K., Sairam, C.V., Nair, C.P.R. and Arulraj, S. (2001). Pilot sample survey on the incidence and yield loss due to eriophyid mite on coconut in Alappuzha District. *Indian Coconut Journal* 21: 28–32.

Moraes, G.J. and de, Zacarias, M.S. (2002). Use of predatory mites for the control of eriophyid mites. In: Fernando, L.C.P., Moraes, G.J. de, Wickramanada, I.R.

(eds). *Proc. Int. Workshop on Coconut Mite* (*Aceria guerreronis*). Coconut Research Institute, Lunuwila, Sri Lanka, pp. 78–88.

Moraes, G.J. de, Lopes, P.C. and Fernando, L.C.P. (2004a). Phytoseiid mites (Acari: Phytoseiidae) of coconut growing areas in Sri Lanka, with descriptions of three new species. *J. Acarol. Soc. Jpn.* 13: 141–160.

Muthiah, C. and Natarajan, C. (2004). Varietal reaction and nutrient management of coconut eriophyid mite. *The Planter* 80: 159–169.

Nair, C.P.R. (2002). Status of coconut eriophyid mite *Aceria guerreronis* Keifer in India. In: Fernando, L.C.P., Moraes, G.J. de, Wickramanada, I.R. (eds). Proc. Int. Workshop on Coconut Mite (*Aceria guerreronis*).Coconut Research Institute, Lunuwila, Sri Lanka, pp. 9–12.

Naseema Beevi, S., Mathew, T.B., Bai, H. and Saradamma, K. (2003). Status of eriophyid mite in Kerala-Resume of work done. In: Singh, H.P., Rethinam, P. (eds). Proc. Int. Workshop on Coconut eriophyid mite: Issues and strategies. Bangalore, pp. 64–75.

Nair, C.P.R. (2002). Status of coconut eriophyid mite *Aceria guerreronis* Keifer in India. In: Fernando, L.C.P., Moraes, G.J. de, Wickramanada, I.R. (eds). *Proc. Int. Workshop on Coconut Mite* (*Aceria guerreronis*). Coconut Research Institute, Lunuwila, Sri Lanka, pp. 9–12.

Navia, D., Gondim, Jr, M.G.C., Aratchige, N.S. and Moraes, G.J. de (2013). A review of the status of the coconut mite *Aceria guerreronis* (Acari: Eriophyidae), a major tropical mite pest. *Exp. Appl. Acarol.* 59(1-2): 67-94.

Nirula, K.K. (1957). Observations on the green muscardine fungus in populations of *Oryctes rhinoceros*. *J Econ. Entomol.* 50(6).

Norman, K. and Mohd Basri W. (2004). Immigration and activity of *Oryctes rhinoceros* within a small oil palm replanting area *J. Oil Palm Res.* 16(2): 64-77.

Oehlschlager, A.C. (2007). Optimizing trapping of palm weevils and beetles.Proc. 3rd Int. Date Palm Conference, Abu Dhabi, United Arab Emirates.

Perera, P.A.C.R. (1987). Studies on *Opisina arenosella* Walker and its natural enemies in Sri Lanka. *Ph.D. Thesis*, Univ. London.

Perera, P.A.C.R., Hassell, M.P. and Godfray, H.C.J. (1988). Population dynamics of the coconut caterpillar, *Opisina arenosella* Walker (Lepidoptera: Xyloryctidae) in Sri Lanka. *Bull. Entomol. Res.* 78: 479–492.

Perera, L., Sarathchandra, S.R. and Wickramananda I.R. (2013). Screening Coconut Cultivars for Tolerance to Infestation by the Coconut Mite, *Aceria guerreronis* (Keifer) in Sri Lanka. *CORD* 29(1): 46-51.

Rajapakse, C.N.K., Gunawardena, N.E. and Perera, K.F.G. (1998). Pheromone baited trap for the management of red palm weevil, *Rhynchophorus ferrugineus* F. (Coleoptera: Curculionidae) population in coconut plantations. *COCOS.* 13: 54-65.

Ramaraju, K., Natarajan, K., SundaraBabu, P.C., Palaniswamy, S. and Rabindra, R.J. (2002). Studies on coconut eriophyid mite, *Aceria guerreronis* Keifer in Tamil Nadu, India. In: Fernando, L.C.P., Moraes, G.J. de, Wickramanada, I.R. (eds). Proc. Int. *Workshop on Coconut Mite (Aceria guerreronis)*. Coconut Research Institute, Lunuwila, Sri Lanka, pp. 13–31.

Ramaraju, K., Palaniswamy, S., Annakodi, P., Varadarajan, M.K., Muthukumar, M. and Bhaskaran, V. (2005). Impact of coconut eriophyid mite, *Aceria guerreronis* K. (Acari: Eriophyidae) on the yield parameters of coconut. *Indian Coconut Journal* 25: 12–15.

Ranasinghe, S., Nainanayake, A. and Kumarasinghe, D. (2016). Coconut Yield Forecast. Volume 2, Issue 1. Coconut Research Institute of Sri Lanka.

Sadakathulla, S. and Ramachandran, T.K.A. (1990). A novel method to control rhinoceros beetle, *Oryctes rhinoceros* L. in coconut. *Indian Coconut Journal* 21,7-8, 10-12.

Sathiamma, B., Nair, C.P.R. and Koshi, P.K. (1998).Outbreak of a nut infesting eriophyid mite, *Eriophyes guerreronis* (K.) in coconut plantations in India. *Indian Coconut Journal* 29: 1–3.

Schliesske, J. (1988). On the gall mite fauna (Acari: Eriophyidae) of *Cocos nucifera* L. in Costa Rica. *Nachrichtenblatt Duetschen Pflanzenschutzdienstes* 40: 124–127.

Singh, S.P. and Arancon, Jr. R.N. (2007). Production and use of green muscardine fungus (*Metarhizium anisopliae* (Metschnikoff) in the IPM of rhinoceros beetle (*Oryctes rhinoceros* (Linnaeus). APCC – IPM Publication No.2.

Siriwardena, K.A.P., Fernando, L.C.P., Nanayakkara, N., Perera. K.F.G., Kumara.A.D.N.T. and Nanayakkara, T. (2010). Portable acoustic device for detection of coconut palms infested by *Rhynchophorus ferrugineus* (Coleoptera: Curculionidae). *Crop Prot.* 29: 25-29.

Sivapragasam, A. and Hong, L.W. (2007). *Plesispa reichei*: a pest of increasing importance in Malasia. (RAP Publication, Bangkok, Thailand, pp. 55-62.

Suarez, A. (1990). Influencia de diferentes intensidades de ataque del ácaro del fruto del cocotero Eriophyes guerreronisen los rendimientos en masa, copra y aceite. *CiencTecAgric* 13: 61–65.

Subhathma, W.G.R., Suwandharathna, N.I., Fernando, L.C.P. and Ahangama, D. (2012). Effect of temperature on local and Philippine isolates of *Metarhizium anisopliae* and their virulence on larvae of *Oryctes rhinoceros*, a pest of coconut in Sri Lanka *COCOS.* 20: 49-58.

Suwandarathna, N.I. and Kumara, A.D.N.T. (2007) On Farm Production of Green Muscardine Fungus to Combat Rhinoceros Beetle. In: COCOINFO Int. Bull. Asian and Pacific Coconut Community (APCC), Volume 14 (January 2007). pp. 26-28.

Suwandarathna, N.I., Fernando, L.C.P. and Edirisinghe, J.P. (2011).Effect of temperature on the developmental stages of two colour leaf beetle (*Plesispa reichei*), a pest of coconut. In: Proc. 2nd Annual Research Session. University of Peradeniya. p. 144.

ThirumalaiThevan, P.S., Muthukrishman, N., Manoharan, T. and Thangaraj, T. (2004). Influence of phenotypic and biochemical factors of coconut on eriophyid, *Aceria guerreronis* Keifer and predatory, *Amblyseius* spp. mites. *J. Entomol. Res.* 28: 291–299.

Varadarajan, M.K. and David, P.M.M. (2003). Effect of ground vegetation and nut characteristics on the severity of infestation by *Aceria guerreronis* in coconut. *Entomon* 28: 361–365.

Wickramananda, I.R., Fernando, L.C.P. and Aratchige, N.S. (2003). Use of chemicals for the management of coconut mite: Sri Lankan experience. In: Singh, H.P., Rethinam, P. (Eds). Coconut Eriophyid mite-Issues and Strategies. Proc. Int. workshop on coconut mite held in Bangalore, India. Coconut Development Board, India. April, 2003. pp. 50-56.

Wickramananda, I.R., Peiris, T.S.G., Fernando, M.T., Fernando, L.C.P. and Edgington, S. (2007). Impact of the coconut mite (*Aceria guerreronis* Keifer) on the coconut industry in Sri Lanka. *CORD* 23: 1–16.

Wood, B.J. (1968). Studies on the effect of ground vegetation on infestations of *Oryctes rhinoceros* (L.) (Col., Dynastidae) in young oil palm replantings in Malaysia. *Bull. Entomol. Res.* 59: 85-96.

Zelazny, B. (1979). Loss in coconut yield due to *Oryctes rhinoceros* damage. *FAO Plant Protection Bulletin* 27(3): 65-70.

Zelazny, B., Alfiler, A.R. and Lolong, A. (1989). Possibility of resistance to a baculovirus in populations of the coconut rhinoceros beetle (*Oryctes rhinoceros*) *FAO Plant Prot. Bull.* 37(2): 77-82.

2018, Pests of Plantation Crops   *Pages* **79–96**
Editors: **P. Chowdappa, Chandrika Mohan & A. Josephrajkumar**
Published by: **ASTRAL INTERNATIONAL PVT. LTD., NEW DELHI**

Chapter 3

Arecanut

☆ *P.S. Prathibha and Shivaji H. Thube*

1. Introduction

Arecanut is perennial plantation crop attacked by more than 90 species of insect and non-insect pests round the year. The pests infest all parts of the palm *viz.*, stem, leaves, inflorescence, fruits and roots. Pest attacking on fruits and inflorescence causes direct crop losses, and those attacking on vegetative parts like leaves, stem and roots causes indirect crop loss. Insect pests, such as root grubs (*Leucopholis lepidophora* Blanch. and *L. burmeisteri* Brens.) (Melolonthinae: Scarabaeidae: Coleoptera), spindle bug (*Carvalhoia arecae*), inflorescence caterpillars (*Tirathaba mundella*) and mites were found more prominent on arecanut. Among minor pests the coccids (scales and mealybugs), occupy an important place with possibilities of attaining the status of major pests in future due to climate change. As many pests and diseases are lethal to areca palm, a systematic monitoring and timely adoption of integrated pest management approaches are very critical for up keeping the health status of the palm.

2. Root Grubs/White Grubs *Leucopholis lepidophora* Blanchard *Leucopholis burmeisteri* Brenske and *Leucopholis coneophora* Burmeister (Scarabaeidae: Coleoptera)

Root grubs/white grubs are larval stage of scarabaeid beetles which feed on subterranean parts of arecanut and intercrops. White grub complex associated with arecanut based cropping system in India comprised of closely related three species of genus *Leucopholis viz., Leucopholis burmeisteri* Brenske, *Leucopholis lepidophora* Blanchard and *Leucopholis coneophora* Burmeister (Veeresh *et al.*, 1982). *L.*

lepidophora was first reported to feed on arecanut in Karnataka by Puttarudraiah and Channabasavanna (1957) and later by Nair and Daniel, (1982). It is the predominant species present in arecanut cultivated in hill tracts along Western Ghats areas of Karnataka (Subaharan *et al.*, 2001), Shivamogga and Chikkamagaluru district Sringeri, Thirthahalli (Prathibha 2015; Kalleswaraswamy *et al.*, 2015), parts of Northern western ghats of Maharashtra (Therurkar *et al.*, 2012) and North Eastern states of India (Thakur *et al.*, 2012). Adults are medium sized beetle with greenish black elytra. Raster of third instar *L. lepidophora* larva bears two rows of pali, arranged in ladder like fashion. Each row is separated by half the length of the pali, leading to overlap of the roundish pali along the mid line of the raster (Veeresh *et al.*, 1982; Kumar, 1997).

L. burmeisteri is one of the principal pests of areca palm (Daniel and Kumar, 1976). The larvae cause severe damage to roots of areca palms (Kumar, 1974; Kumar and Daniel, 1981). It is distributed in Dakshina Kannada district of Karnataka in areas having altitude > 200 m above MSL where arecanut is cultivated (Kumar, 1997). It is present in Sullia and Belthangadi taluks of Dakishna Kannada district of Karnataka (Prathibha, 2015). Adults of this species are yellowish brown in colour covered by yellow scales all over the body and dorsal surface is strongly convex. The raster of third instar grubs bears relatively strongly sclerotized, flat and dagger like pali which are regularly arranged in a barrel shape. It was found feeding on subterranean parts of cocoa, banana and rubber in the field (Veeresh *et al.*, 1982). *L. burmeisteri* and *L. lepidophora* are known as arecanut white grubs and have biennial life cycle.

Leucopholis coneophora Burmeister was recorded as a pest on coconut by Nirula *et al.* (1952) and followed by Shekhar (1958), Abraham and Kurian (1970), Kurian *et al.* (1974), Abraham and Mohandas (1988a; 1988b) and Abraham (1993). It is distributed in coastal areas and plains and at an altitude < 200 m above MSL (Kumar, 1997). Eventhough, *L. coneophora* is known as coconut white grub, it feeds on roots of arecanut grown in plains and coastal areas. Veeresh (1981) furnished a key for the identification of the South Indian genera of root grubs and these included three species of genus *Leucopholis*. The description provided by Patil and Veeresh (1981) and Veeresh *et al.* (1982) clearly differentiated the larvae of three species of *Leucopholis viz.*, *L. coneophora*, *L. lepidophora* and *L. burmeisteri*. Adults differed in respect of shape, size, clypeal notch, and density of scales, shape of the scale and punctuations on the body and with respect to relative orientation of the parameres of male genetalia. All the three species exhibited sexual dimorphism with respect to antennae and meta tibial characters. In females antennal club was short and tibial spur was narrow. Kumar (1997) suggested population representing species *L. burmeisteri* in Dakshina Kannada district was morphologically similar to *L. coneophora*. Differentiating characters identified by Veeresh *et al.* (1982) were not strong enough to warrant delineation of populations into specific status. He described them as *L. coneophora* – coastal strain which occur at altitude of < 200 m above MSL and *L. coneophora* – hill strain for those occurring at altitudes of over 200 m above MSL to represent two geographical conditions. Suggesting further studies involving karyology and DNA finger printing might confirm these suppositions and also their specific status. Nagesha *et al.* (2011) partially amplified mitochondrial cytochrome oxidase sub unit I (*COI*) gene of *L. burmeisteri sulyareca* isolate and deposited sequence (accession number AEM 24341) in gene bank data base of NCBI. Study of phylogenic relation by partial amplification of 16s rRNA and *COI* gene of *L. burmeisteri*, *L. coneophora* and *L. lepidophora* (collected from Dharmasthala, Kasaragod and Sringeri, respectively) revealed 98, 83 and 89 per cent similarity, respectively with *L. burmeisteri sulyareca* isolate (Prathibha *et al.*, 2013a).

In addition to *Leucopholis*, a close relative of this genus, *Lepidiota* was also reported to feed on arecanut in Karnataka. *Lepidiota* sp. damaged arecanut seedling by feeding on roots and inflicted sort of similar symptoms *viz.*, drooping and drying of leaves as in case of *Leucopholis* infestation (Rao *et al.*, 1961).

2.1. Damage Symptoms

It induced seedling mortality due to damage of root system (Rajamani and Nambiar, 1970; Kumar, 1974; Daniel and Kumar, 1976; Kumar and Daniel, 1981; Padmanabhan and Daniel, 2003). Grub infestation on arecanut seedlings led to damage to the root system and the seedling could be pulled out easily (Anonymous, 1967). Second and third instar grubs were voracious feeder which fed about 2.3 g arecanut root tissue/grub/day (Padmanabhan and Daniel, 2003). The grubs preferred younger roots as compared to older roots (Rajamani and Nambiar, 1970). Prolonged damage to root induces gradual yellowing of leaves, immature nut fall there by reduction in nut yield, delayed flowering and lack of production of inflorescence. Grub feeding on apical tender region of roots affect the nutrient absorption and restrict the nutrient flow which in turn led to stem tapering and reduction in crown size (Figure 3.1). Root grub infested palms can be easily pulled out with a small jerk as the entire root system in eaten away by the grubs. In addition, it caused poor production of inflorescence in areca palms which eventually led to yield loss (Rajamani and Nambiar, 1970). Nair and Daniel (1982) observed that the young palms attacked by *L. burmeisteri* died fast while older palms continued to survive and the palms damaged by root grubs are susceptible to secondary infection (Veeresh *et al.*, 1982).

2.2. Bionomics

Adults of white grubs are known as cockchafer beetles which emerge with the setting of South West monsoon in Southern part of peninsular India. The emergence of *L. burmeisteri* occurs during the month of June after the receipt of 3 - 4 good rains in June in Dakshina Kannada district. *L. lepidophora* emerges in last week of July or in the first week of August in hilly tracts of Chikkamagaluru district of Karnataka. *L. lepidophora* is active flier when compared to *L. burmeisteri*. Adult emergence of *L. coneophora* commence with the summer shower in March/April after a pause in May it resumes with the setting of south west monsoon (Abraham, 1993; Prathibha *et al.*, 2013b; Prathibha, 2015). Adults remain active noticed in the evening hours between 6.30 - 7.30 PM in case of both the species during this time they feed and copulate then go back to soil (Figure 3.2). Eggs are laid singly in interspaces during June. Incubation period varies from 12-15 days and hatch into tiny grubs with a brown head and white body. Larval stage prolongs for 226 to 346 days and has

Figure 3.1: Root Grub.

(a) Yellowing of fronds, (b) Stem tapering, (c) Root damage, (d) Reduction in crown size, (e) Root grub, *L. burmeisteri.*

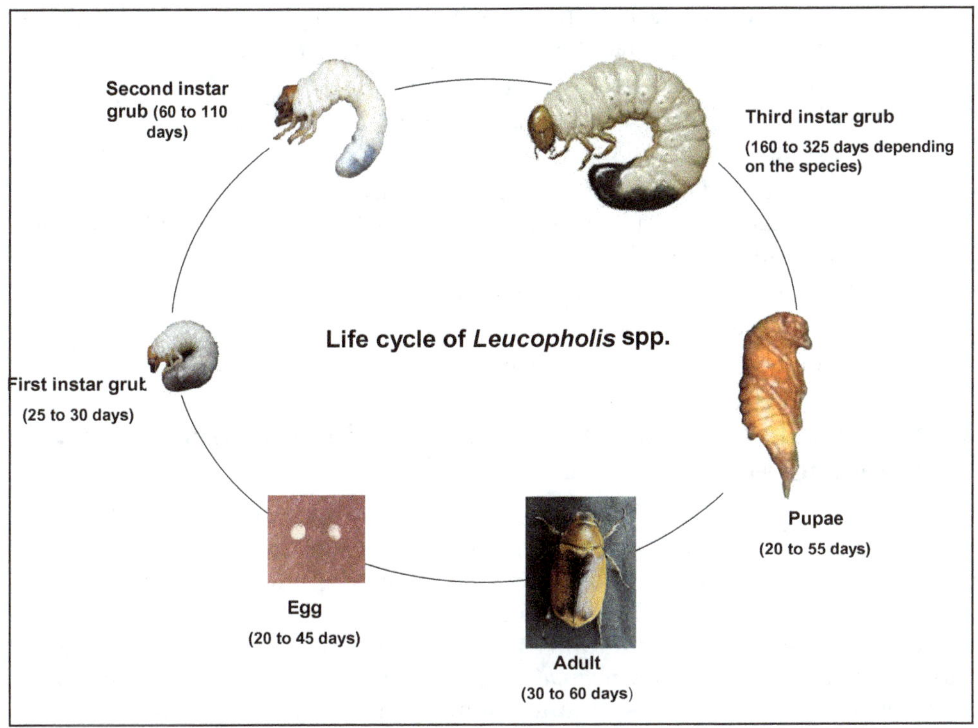

Figure 3.2: Life Cycle of White Grub.

three instars, with first instar larvae feed on organic matter present in the soil later it starts feed on roots of grasses and weeds present in the interspaces. Late second and third instar larvae feed on roots of arecanut and other intercrops. More number of third instar larvae can be noticed in root zone area during September to February. In some area overlapping generations are noticed where third instar larvae can be seen round the year. During summer months the grubs move deeper and deeper layers of soil as moisture in top layer deplete. From initial infestation, the spread of root grub infestation takes 4 - 6 years (Padmanabhan and Daniel, 2003). Therefore it is easy to manage the white grub infestation by means of integrated approach. Several natural enemies have been found infecting on the white grubs.

2.3. Management

- ☆ Hand picking of adults during peak emergence period of two weeks commencing from first day of South west monsoon daily in the evening 6.30–7.30 pm. It commence during May last to June first week in planes and during August in hilly tracts

- ☆ Application of neem cake @ 2 kg/palm/year during June to July in the basin which help in the rejuvenation of roots

- ☆ Patch application of chlorpyriphos or bifenthrin @ 2kg ai/ha covering interspaces and root zone area where grub infestation is noticed (*ie.,* chlorpyriphos 20 EC @ 10 litre/ha, bifenthrin 10 EC @ 20 litre/ha) coinciding with early instar stage in the field in planes. In hilly areas, first round application may be carried out during first week of September

☆ Application of Entomopathogenic nematode liquid suspension, *Steinernema carpocpasae* during September - October in plains and November - December in hills

☆ Second round need based root zone application of chlorpyriphos 20 EC @ 7 ml/palm or bifenthrin 10 EC @ 14ml/palm after 45 days of first round insecticide application

☆ Repeated ploughing (3- 4 ploughings) of gardens from October- December to expose the grubs to predators

☆ Provide good drainage in field

3. Spindle Bug, *Mircarvalhoia arecae* Miller and China (Miridae: Hemiptera)

Spindle bug/capsid bug, *Carvalhoia arecae* Miller and China (F. Miridae) is a major pest of areca palms which was first reported from DK district of Karnataka by Khandige in 1955. Miller and China (1957) recorded this as a new genus and species from *Areca catechu*. These are brightly coloured red and black bugs which are active during morning hours and nymphs are brownish red in colour. Kerzhner and Schuh (1998) proposed *Mircarvalhoia* as a replacement name for *"Carvalhoia"*, as latter has preoccupied as a junior homonym of *Carvalhoia* Kormaed, 1951 (Hemiptera : Colobathristidae). *Mircarvalhoia arecae* is considered as an endemic species of southern India, and are of a chronic problem in areca plantations of Kerala, Karnataka and parts of Tamil Nadu (Nair, 1964). At shimogga in Karnataka, it has assumed a serious pest status and spread extensively in areca gardens and causing economic loss to the farmers (Kantharaju *et al.*, 2009). As per survey conducted in Kerala during 1990, it is distributed in all states except middle zone *ie.*, Wadakkanchery and Palakkad districts (Stanley, 1990). Its limited occurrence outside the centre of origin of the areca palm was explained in terms of the origin and evolution of the insect on the native rattan palm *Calamus travancoricus* Bedd. ex Becc. and Hook. f. and subsequent host shift and adaptation to introduced palms such as *Areca catechu* L., *Areca concinna* Thwaites, *Areca lutescens* Bory, *Areca triandra* Roxb. ex Buch.-Ham., *Chrysalidocarpus madagascariensis* Becc., *Elaeis guineensis* Jacq., *Loxococcus* sp. and *Pinanga* sp. (Shameem and Prathapan, 2014). It is reported on areca palms of North eastern states of India (Thakur *et al.*, 2012). Most recently, it is reported from Sipighat and Mitha Khari in the South Andaman district and Diglipur in North Andaman and Nicobar islands for the first time on arecanut (Yaswant and Prathapan, 2014).

3.1. Damage Symptoms

Both nymphs and adults hiding in leaf axils and suck the sap from the emerging spindle, tender leaflets and leaf axil. While feeding, the bug injects toxic saliva, as a result fresh feeding marks appear as watery streaks on the infested leaflets and spindle. These linear lesions turn brown and become necrotic which appears

Figure 3.3:

**(a) Spindle bug Nymph, (b) Adult,
(c) Necrotic linear lesions on arecanut leaf caused by spindle bug.**

as elongated patch on the fronds while unfurling (Figure 3.3). Severely infested spindle leaf fails to unfurl. Such feeding damage results in stunted growth and reduction in yield. Infestation on seedlings leads to complete decay and drying of seedlings. Observation on gardens in Southern Kerala and parts of DK district revealed that feeding by these bugs caused 80 per cent damage to the spindle (Nair, 1964; Abraham *et al.*, 1976).

3.2. Bionomics

The bug completes life cycle in 24- 33 days (Nair and Das, 1962). Adults thrust eggs singly into the tender tissues as a result, oviposition site becomes dark in colour. Freshly laid eggs are white in colour, oval in shape and measures 1.36 mm x 0.34 mm. The anterior end is distinctly demarcated into a short neck, bearing at its tip thick and rigid convex oval operculum (Figure 3.3). The chorion is smooth and leathery. Two curved bristle which are unequal in length arise from the operculum. During the course of development egg turns to pink and then red and hatches out in a period of 10 days. Kantharaju *et al.* (2011a) reported five nymphal instars in the biology. The total nymphal duration ranged from 22-28 days. The preoviposition and oviposition period prolonged 2.45 ± 0.43 and 2.00 ± 0.40 days, respectively. The bug had the fecundity rate of 10 - 18 eggs. The adult male and female lived for 12 - 28 days and 14 - 35 days with a mean of 18.77± 6.39 and 25.66 ± 7.85 days, respectively. The adult bugs are black in colour and measures 6.0 mm long and 2.8 mm broad. The head is triangular with a pair of four segmented antennae. Three segmented rostrum reached up to hind coxa. The thoracic segments are equal in size. The six segmented legs bear a two segmented tarsus. The abdomen is oval with nine visible segments. The antennae, legs and rostrum are deep violet brown; thorax and boarder of abdomen are light violet brown, remaining part of abdomen is greenish yellow. The head is light yellow with scarlet red eyes. The fifth instar nymphs are 4.43 mm long and 2.15 mm broad. Wing pads are well developed reaching up to the third abdominal segment in the fifth instar nymph.

Though, the pest incidence occurs throughout the year, the peak incidence of the pest in Kerala is from June to October with maximum population in August and September (Nair, 1964). A peak in population density was noticed during

December, January and July (Anonymous, 1972). According to Koya *et al.* (1979) the pest population was high during the monsoon and post monsoon periods and low during summer months. There was a positive correlation between the rainfall and population of *C. arecae* and peak period of abundance varied at different places (Sathiamma *et al.*, 1980; 1985). Kantharaju *et al.* (2011b) reported the highest incidence during I fortnight of August, while the lowest incidence during II fortnight of March, 2007 and II fortnight of December, in arecanut growing areas of Shivamogga, Karnataka. Incidence of spindle bug had the significant negative correlation with maximum temperature and significant positive correlation with rainfall and relative humidity.

3.3. Management

Studies conducted for the past five decades revealed that majority of the insecticides were found to be effective in controlling spindle bug population. But most of the insecticides used are banned now. Menon *et al.* (1962) suggested spraying either of the chlorinated hydrocarbons/OP compound *viz.*, DDT, endrin, ethyl parathion or wettable BHC are effective in managing spindle bug population. Palms treated with BHC 0.2 per cent and endrin 0.025 per cent were completely free from damage (Nair and Das, 1962). Filling the innermost two leaf axils around the spindle with granular systemic insecticides like phorate 10 G or carabaryl 4G or methyl o demeton at the rate of 10 g/palm at an interval of three months was recommended (Abraham *et al.*, 1976). For this, a granular leaf axil filling applicator was also devised (Abraham, 1975). General recommendations included keeping sachets with granular insecticides into the innermost two or three leaf axils of arecanut palm (Sathiamma *et al.*, 1985; Jacob, 1990). Malathion 5 per cent dust, phorate 10 per cent granules, monocrotophos 0.15 per cent spray, lamda cyhalothrin 0.10 per cent spray showed cent per cent mortality in bug population on seventh day after application. Chlorpyriphos 0.20 per cent spray and profenophos spray exhibited highest efficacy in controlling the bug population (Kantharaju *et al.*, 2009). Spraying of Fish Oil Rosin Soap (FORS) (1 kg in 80 litres of water) or quinalphos @ 1 ml/litre or endrin 1 kg in 900 litres of water could control the pest or drenching with lindane (1.3 D) at 2.5 g/lt of water on the spindles or placing 2g of phorate at 10g/palm in perforated poly bags on the top most leaf axils is also recommended (Jayaraj and Balakrishnan, 2007).

   ☆ Regulation of shade in the garden

   ☆ Placement of thiamethoxam 25 WG of (2 g) in perforated poly-sachets in the inner most two leaf axils of areca palms during April- May

   ☆ Spraying with thiamethoxam 25 WG (0.25 g per litre water) in and around the spindle and inner whorl of leaves

4. Pentatomid Bug *Halyomorpha marmorea* Fab. (Pentatomidae : Hemiptera)

Pentatomid bug assumed status of serious pest and spreads extensively now a days, earlier it was a low density insect which is conspicuous only by the damage that it causes. It was first reported by Vidyasagar and Bhat (1986) that its infestation caused dropping of tender nuts. In parts of Karnataka and Kerala tender nut drop due to infestation of pentatomid bug was reported during April to July (Chandramohanan and Shantaram, 1986). *H. marmorea* are not easily found on the palm host, but can be seen on other host plants for most part of the year. The incidence could be recognized by the shed immature fruits, which is concentrated on palm basins. Cowpea, *Vigna sinensis* is reported as collateral host of *H. marmorea*. But so far, *H. marmrea* has not been reported as a pest on any crop in India except arecanut. According to Daniel (2010), intensity of attack varied widely and ranged from 8 – 78 per cent and it was is a low density pest. Even within a plantation, the activity of the bug was confined to a few palms. Wherever the intensity of attack was severe in a plot it was surrounded by secondary jungles of different trees, shrubs *etc.* Such locations were found in the south and southeast parts of Dakishna Kannada district of Karnataka bordering Kerala (Daniel, 2010).

4.1. Damage Symptoms

The bug pierces tender nuts with long proboscis and feeding continues for several hours. Due to continuous feeding the developing kernel was depleted of the most vital sap, leading to the nut shedding (Figure 3.4). Adult bug could feed on the sap of a single nut in a day, which causes drop in two days (Anonymous, 1985; 1990; Vidyasagar and Shama Bhat, 1986). Feeding activity is very intense during morning and evening hours in comparison to hot hours of the noon. It infests on fertilized flowers and immature fruits and do not feed on fruits once the endosperm is hardened. It is noticed that point of proboscis insertion on the fruit surface may be at the periath's attachment to rachillae, or at the end of perianth on the fruit body, or at the middle of the fruit body, or at the lower end of the fruit body. As a result microscopic pin prick black dot could be observed on the fruit. Later, tissues surrounding the puncture discolored later to light brown to black and this is more conspicuous on the inner surface of the husk when cut open. At the point of proboscis insertion on endosperm, the surface appears as thickened and rough. Presence of shriveled and sunken endosperm is the typical symptom of pentatomid bug attack. Developing kernel would be discoloured and blackened. In some cases, rotting of the kernel is also reported. Gummosis can be seen in some of the attacked fruits due to hyper sensitive reaction by the plant.

4.2. Bionomics

H. marmorea (Figure 3.4) completes the life cycle in a month. Biology was studies by Vidyasagar (1991) and reported five nymphal instars with an average duration of 3.7, 6.1, 5.1, 6.5 and 9.3 days respectively. The nature and extent of damage to arecanut and host range of this bug also described by Vidyasagar (1991). It is a seasonal pest occurs from February – March and continues up to July – August with a peak infestation in June depending on the availability of green nuts. The first incidence of the pentatomid bug on arecanut occurs in March. The percentage of shed nuts showing *H. marmorea* infestation was maximum during June (29.15) followed by July, Aug. and October (Anonymous, 1988). The incidence of the bug is found to become very serious during some years making it a major pest causing considerable loss to arecanut production.

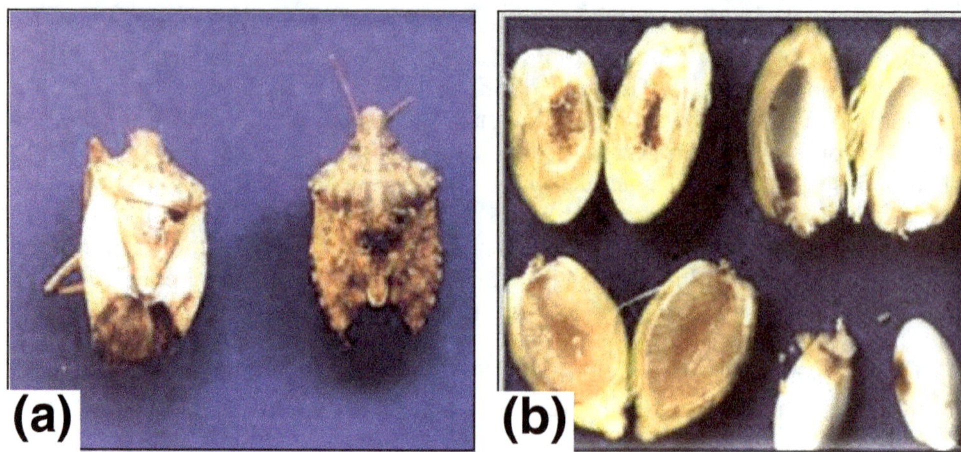

Figure 3.4
(a) *Halyomorpha marmorea*, (b) Damaged caused by pentatomid bug.

4.3. Management

Earlier, spraying with fenvalerate, monocrotophos, dimethoate, endosulfan and methyl parathion was followed, among which endosulfan 0.05 per cent and fenvalerate 0.02 per cent spraying were found to be cost effective (Anonymous, 1990). The bugs were collected and destroyed from alternate host plants such as chillies, ladies finger and bitter gourd. In case of severe infestation it was observed that spraying two rounds of endosulfan 0.05 per cent during April - May and the second round after an interval of 45 days was effective

    ☆ Adopt clean cultivation practices: Collection and destruction of the various stages of this insect seen on alternate hosts like cowpea, bhendi, bitter gourd, chillies *etc*, before these bugs shift to areca palms,

    ☆ Application of neem oil emulsion (0.5 per cent) two - times in fortnightly intervals only to the infested palms and surrounding palm. Care may be taken to avoid spraying of oil emulsion in freshly opened inflorescence as it affect the fruit set

    ☆ Spraying of 0.008 per cent imidacloprid (17.8 per cent ai) @ 0.5 ml/litre

5. Foliar Mites

Mites are widely distributed in all areca growing tracts of India and are polyphagous in nature. There are two species of foliar mites reported on arecenut *viz.*, false spider mite/palm mite/scarlet mite/red mite, which is scientifically known as *Raoiella indica* Hirst. Another one is a spider mite/jowar mite/white mite, *Oligonychus indicus* Hirst.

5.1. *Oligonychus indicus* Hirst. (F. Tetranychidae)

It was first reported in 1956 on arecanut seedlings in Bangalore. Both adults and nymphs colonise under webs on lower surface of leaves (Figure 3.5). The incubation period varies from 72 to 95 h. Total duration of immature stages (larval proto nymph, deuteron nymph) stage ranges form 6.5 – 9 days. Females mites lays on an average of 3 - 4 eggs per day and ovipostion period prolongs for 10 days.

Figure 3.5: White Mite, *Oligonychus indicus*.

5.2. *Raoiella indica* (Tenuipalpidae)

Among false spider mites genus (*Brevipalpuls, Raoiella* and *Tenuipalpus*), the genus *Raoiella* gained economic importance in the recent years and the species, *Raoiella indica*

R. indica was first described in the district of Coimbatore (India) by Hirst in 1924 on coconut leaflets [*Cocos nucifera*, whereas, on arecanut the first report was made in 1956 in Bangalore, Karnataka. The Red Mite/scarlet mite, *Raoiella indica* heavily infests palms of nursery stage and young palms and induces damage by sucking sap from the leaves of young palms during hot dry weather. The mite attained economic significance when it was first reported as an invasive species in the Caribbeans in 2004 (Flechtmann and Etienne, 2004), which was reported as important pest of coconut, and banana. (Nagesha and Channabasavanna, 1984; Welbourn, 2006). This polyphagous species of false spider mite rapidly spreads through the Neotropical region where the mite damages economically and ecologically important plants. Both the nymphs and adults are seen in large number on the ventral surface of arecanut leaves. In severe case of infestation they were seen on the upper surface and on the spindle too. It completes life cycle in 11.2 to 12.9 days. Other than arecanut it is infesting coconut, date palms and other ornamental palms also.

Population builds up immediately post monsoon. With the onset of hot weather from April–May, and becomes more active and virulent form. Study on population dynamics of *R. indica* in northern Kerala indicated that occurrence of the pest during March–September, with a peak incidence on March–April and more number located on bottom fronds (Prabheena and Ramani, 2014). Gardens under water stress and nurseries are more prone to mite infestation. The population declines with the onset of monsoon.

5.2.1. Damage Symptoms

Scarlet mite and Jower mite colonies are coexisting on the same leaf. They suck the sap from the green portion of the plant. Feeding by the mites leads to development of yellow speckles on the lamina (Figure 3.6). These speckles coalesce, become bronze coloured and the leaves wither away. In older palms, the infestation starts from the lower whorls. Severe infestation affects the photosynthetic capacity of the leaf. In case of infested seedling, yellowing and subsequent mortality occurs.

5.3. Perianth Mite: *Dolichotetranychus* sp. (Tenuipalpidae)

Mite infestation results in severe tender nut fall in affected palms. The infestation is noticed extensively in or around central Kerala. The mites are slender, orange coloured and seen colonized inside the perianth of tender nuts. As a result of feeding activity, the nuts shriveled and later on fall off resulting up to 10 per cent crop loss. The period of infestation is during November – May.

5.4. Management

- ☆ Cut and burn heavily infested and dried leaves to check the lateral spread of mites.

- ☆ Regulate shade in the field and provide adequate irrigation for seedlings

- ☆ Conserve indigenous natural enemies (predatory mites, coccinellid beetles (*Stethorus keralicus*) Neuropteran (*Chrysopa* sp.), predatory mites (*Amblyseius channabasavanni*) as they exert good check on the mite population

- ☆ Application of neem oil emulsion (0.5 per cent) two times at fortnightly intervals

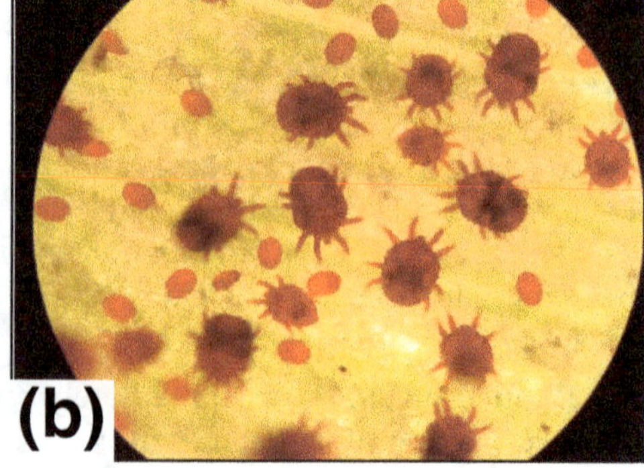

Figure 3.6

(a) Mite–induced damage symptom on areca palms, (b) Scarlet mite *Raoiella indica*.

6. Scales and Mealybugs

6.1. Damage

Many species of mealybugs are found to colonize almost all parts of areca palm *viz.*, spadices, inflorescence, developing fruit bunches, leaves and leaf sheaths. They colonise on the lower surface and in severe cases even the tender nuts also are affected. In the early stage of the crop, scales and mealybugs suck sap from leaves, which results in development of yellowish patch on leaf blade there by inhibit the growth (Figure 3.7). In the later stage they interfere with pollination and photosynthesis thereby severely affecting the yield of arecanut. Severe feeding leads to withering and shedding of buttons/fruits. Damage is very high during drought conditions. As of now, infestation by most of the mealybugs do not cause an economic yield loss to plantations. But there are chances for these sucking pests for major pest status due to huge shift in climate. The following mealybugs are associated with the leaves and leaf sheath.

Pseudococcus cryptus Hempel. Polyphagous in nature and are found to infest leaves inflorescence and developing fruit bunches.

Icerya seychellarum Westwood which is polyphagous in nature. In addition to outer surface of the leaf sheath they infest developing bunches and spadices.

Aonidiella orientalis Newstead is highly polyphagous hard diaspidid scale that feeds on inflorescence as well as on immature fruits

6.2. Management

☆ Collection and destruction of the various stages of this insect seen on alternate hosts like cowpea, bhendi, bitter gourd, chillies *etc*, before it shift to areca palms, must be done.

☆ Conserve and augmentation of lady bird beetle *Chilocorus nigrita*. These can be mass multiplied and released in affected areca garden to control the scale insects.

Figure 3.7. Scale Infested Arecanut.

☆ Regulation of shade in the garden

☆ Application of neem oil emulsion (0.5 per cent) two times in fortnightly intervals

7. Whitefly *Aleurocanthus* sp. (Aleyrodidae : Hemiptera)

It infests leaflets of areca palm from under surface. Population is high on young areca plantation in Dakishna Kannada district of Karnataka. It causes severe blotching and drying of leaves. The honey dew secreted by the insect promotes the colonization of sooty mould which interferes with photosysnthesis of palm.

8. Palm Aphid *Cerataphis brasiliensis* (Aphididae : Hemiptera)

This is commonly known as palm aphid and they colonized the leaves, spindle leaves and inflorescence. Natural enemies like *Paragnus yebuiensis* and *Pseudasphidimerus* sp. were recorded.

Sprayinng of 0.6 per cent neem oil emulsion or commercial neem formulation would check the whitefly and aphid population

9. Inflorescence Caterpillar

9.1. *Thirathaba mundella* (Pyralidae: Lepidoptera)

The pest causes damage to areca inflorescence, by boring on tender button, female flowers and spathe and feed on tender rachille and male flowers from the tip. It webs on the terminal portion of the inflorescence with silken thread, which delays the spadix opening. Mechanical injury on the palm pre-disposes the infestation (Figure 3.8). It is distributed in parts of Dakshina Kannada districts of Karnataka, Kerala, Andhra Pradesh, Bihar, Gujarat, Tamil Nadu, Madhya Pradesh and Orissa. Female moth deposits eggs into the spadix through punctures made on the spathe by slugs or snails. Incubation period is for 5 days. The fully grown larvae are greyish brown with a reddish brown head and measure 23- 25 mm in length. The larval period lasts for about 26 days covering five instars. Pupation is in silken cocoons with a wet mass of frass inside the spathe. Pupal period lasts for 9- 11 days.

Figure 3.8: Inflorescene Caterpillar and Infested Areca Inflorescence.

9.2. Inflorescence Caterpillar *Batrachedra arenosella* (Batrachedridae: Lepidoptera)

A new pest reported in the recent past causing significant economic loss was noticed in Southern plains of Karnataka during 1999. It is found to infest on coconut inflorescence also. Caterpillars feed on inflorescence and bore on female flower. Removal and burning of affected inflorescence. Spraying of chlorpyriphos (0.04 per cent) or lambda cyhalothrin (0.005 per cent) would keep the pest under check.

References

Abraham, V. A. (1975). An applicator for filling granular insecticides into the leaf axils of arecanut palms. *Pesticides.* 9(2): 26- 28.

Abraham, V. A. (1993). Behaviour of adult beetles of the coconut white grub, *Leucopholis coneophora* Burm. *Indian Coconut J.,* 24: 2-4.

Abraham, V. A and Kurian, C. (1970). Save your coconut palms from the ravages of white grubs. *Coconut Bull.,* 1: 3-4.

Abraham, V.A., Sathiamma, B., Abraham, K.J. and Kurian, C. (1976). Control of arecanut spindle bug, *Carvolhoia arecae* Miller and China using granular insecticides. *J. Plantn. Crops* 4: 24-2.

Anonymous (1967). Annual report Central and Regional Arecanut Researcch Station. Vittal, ICAR. p. 92.

Anonymous (1972). Annual report for 1971. Central plantation crops research Institute, Kasaragod India.

Anonymous (1985). *Annual Report for 1985.* Central Plantation Crops Research Institute, Kasaragod. pp 120.

Anonymous (1988). *Annual Report for 1988.* Central Plantation Crops Research Institute, Kasaragod. Pp 8.

Anonymous (1990). *Annual Report for 1989.* Central Plantation Crops Research Institute, Kasaragod. pp 262.

Chandramohanan, R. and Shantaram, M. (1986). *Indian Cocoa Arecanut and Spices J.,* 2: 27-28.

Daniel, M. (2010). Bionomics of the marmorated bug, *Halyomorpha marmorea* Fab. (Hemiptera: Pentatomidae) in arecanut plantation ecosystem. *J. Plantation Crops* 38(1): 78-81.

Daniel, M. and Kumar, T. P. (1976). Pests of arecanut. *J. Plantation Crops,* 4: 68-77.

Flechtmann, C.H.W. and Etienne, J. (2004). The red palm mite, Raoiella indica Hirst, a threat to palms in the Americas (Acari: Prostigmata: Tenuipalpidae). *Systematic and Applied Acarology* 9, 109.

Gupta, K.M. (1973). Neem leaves attract white grub beetles. *Indian J. Entomology,* 35(3): 276.

Jacob, S.A. (1990). Distribution of the spindle bug of arecanut, *Carvalhoia arecae* Miller and China in Kerala, its bio-ecology, suspected role as a vector of yellow leaf disease and control. Central Plantation Crops Research Institute, Palode, Trivendrum. Series/Report No: RNT VIII (131).

Jayaraj, J and Balakrishnan, S. (2007). Spindle bug menace in arecanut. Daily *The Hindu* dated 20th December 2007.

Kalleshwara Swamy, C.M., Adarsha,S.K., Naveena, N.L and Sharanabasappa. (2015). Incidence of arecanut white grubs (*Leucopholis* spp.) in hilly and coastal regions of Karnataka, India. *Current Biotica* 8(4): 423-424.

Kantharaju, N., Thippeswamy, C and Venkatesh Hosamani, (2009). Efficacy of Certain Insecticides Against Arecanut Spindle Bug, *Carvalhoia arecae* Miller and China (Heteroptera : Miridae) Under Field Conditions. *International Journal of Plant Protection* 2: 237-239.

Kantharaju, N., Thippeswamy, C., Siddalingappa, H. V. and Shivasharanappa, Y (2011). Incidence of arecanut spindle bug, *Carvalhoia arecae* Miller and China (Heteroptera : Miridae). *Environment and Ecology,* 29: 143-147.

Kantharaju, N.C., Thippeswamy, Venkatesh, H., Shivasharanappa, Y. and Siddalingappa (2011). Biology of arecanut spindle bug, *Carvalhoia arecae* Miller and China (Heteroptera: Miridae MKK Publication, Kolkata, India. *Environment and Ecology* 29: 148-151.

Kerzhner, I.M. and R.T. Schuh. (1998A). Replacement names for junior homonyms in the family Miridae (Heteroptera). *Zoosystematica Rossica* 7: 171-172.

Khandige, S.B. (1955). A Capsid bug on areca. *Arecanut Bull.,* 6: 120-121.

Koya, K.M. Rawther, T.S.S., Sathiamma, B. and Kurian, C. (1979). Evaluation of six granular insecticides for the control of arecanut spindle bug *Carvalhoia arecae* Miller and China in field. *Pesticides.* 13(8): 50- 51.

Kumar, A.R.V. (1997). Bio-ecology and management of arecanut white grubs. *Leucopholis* spp. (Coleoptera : Scarabaeidae) in Karnataka. Ph.D. thesis, UAS, Bangalore, p. 230.

Kumar, T.P. (1974). Chemical control of areccanut white grubs *Leucopholis burmeisteri* Bren. *Arecanut and Spices Bull.,* 6: 35.

Kumar, T.P. and Daniel, M. (1981), Studies on the control of soil grubs of arecanut palm. *Pesticides,* 15(9): 29-30.

Menon, R., Nair, R.B. and Abraham, K.J. (1962). A note on pests and diseases of arecanut seedlings. *Arecanut J.* 13: 26-29.

Miller, N.C.E. and China, W.E. (1957). A genus and species of Miridae from *Areca catechu* in South India (Hemiptera : Heteroptera).*Bull. Ent. Res.,* 47: 429-431.

Nagesh, M., Shylesha, A. N., Thippeswamy, R. and Ramesh, B. D. (2011). Cytochrome oxidase subunit I, partial (mitochondrion) *Leucopholis burmeisteri* Sulyaarecae isolate. NCBI data base, gene bank accession no: AEM24341.1

Nagesha, C.B.K. and Channabasavanna, G.P. (1984). Development and ecology of *Raoiella indica* Hirst (Acari : Tenuipalpidae) on coconut. In: D.A. Griffiths and C.E. Bowman [eds.], *Acarology* VI, 2(8) pp. 785–798.

Nair, C. P. R. and Daniel, M. (1982). Pests of areca nut. In: K.V.A. Bavappa, M. K. Nair and T. Premkumar (Eds). The Arecanut Palm (*Areca catechu* Linn.) CPCRI, Kasargod (India). pp: 162-184.

Nair, M.R.G.K and Das, N.M. (1962). On the biology and control of *Carvalhoia arecae* Miller and China. A pest of areca palms in Kerala. *India J. Ent.*, 24: 86-93.

Nair, R.B. (1964). *Carvalhoia arecae* Miller and China, a major pest of *Areca Catechu*. *Arecanut J.*, 15 : 57-61.

Nirula, K.K., Antony, J. and K.P.V. Menon (1952). A new pest of coconut palm in India. *Indian coconut J.* 12(1): 10-34

Patil, S.P. and Veeresh, G.K. (1981), Description of larvae *Leucopholis burmeisteri* and *L. lepidophora* (Scarabaeidae: Coleoptera) In: *Progress in Soil Biology and Ecology in India* (Ed. G.K. Veeresh), 1979, UAS Tech. Ser. No. 37, pp. 172-178.

Prabheena, P. and Ramani, N. (2014). Seasonal incidence and injurious status of *Raoiella indica* (Hirst) (Acari: Tenuipalpidae) on arecanut palms of Kozhikode district of Kerala. *International J. Plant, Animal and Environmental Sciences.*

Prathibha, P.S. (2015). Behavioural studies of palm white grubs, *Leucopholis* spp. (Coleoptera: Scarabaeidae) and evaluation of new insecticides for their management. Ph.D. Thesis submitted to University of Agricultural Science Bangalore. P. 119.

Prathibha, P.S., Kumar, A.R.V. and Subaharan, K. (2013b). Ethology of coconut root grub chafer, *Leucopholis coneophora* Burmeister (Melolonthinae: Scarabaeidae). *International J. Agric. Food Sci. Technology*, 4(2): 24-28.

Prathibha, P.S., Rachana, K.E., Shafeeq, R., Sabana, A. A., Subaharan, K. and Rajesh, M.K. (2013a). Molecular characterization of arecanut white grubs, *Leucopholis* spp. (Scarabaeidae: Coleoptera). P. 13 Abstract of paper: National Seminar on Application of Bioinformatics Tools in Agriculture Nov. 11th to 12th.

Puttarudraiah, M. and Channabasavanna, G.P. (1957). Some new insect and mite pests of areca palm in Mysore. III. *Arecanut*, 8(1): 10-11.

Rajamani, S. and Nambiar, K.K.N. (1970). Pests of arecanut palm. *Indian Fmg.*, 20(8): 23.

Rao, K.S., Naidu, G.V. and Bavappa, K.V.A. (1961). The white grub pest of arecaut. *Arecanut J.*, 12: 22-25.

Sathiamma, B., Abraham, V.A., Koya, K.M.A. and Abraham, K.J. (1980). Bionomics and control of spindle bug, *Carvalhoia arecae* Miller and China. *Final Report*, CPCRI, Kasaragod, pp. 27.

Sathiamma, B., Koya, K.M.A., Abraham, V.A., Rawther, T.S.S. and Kurian, C. (1985). Control of arecanut spindle bug, *Carvalhoia arecae* M and C using granular insecticides in the field. In : Arecanut Research and Development. (Eds. Shama Bhat, K. and Nair, C.P.R). Proc. SIJAR 1982. CPCRI, Kasaragoda. pp. 137-139.

Shameem, K. M. and Prathapan, K. D. (2014). Rattan cane *Calamus travancoricus* - a new host plant record for spindle bug of arecanut palm, *Mircarvalhoia arecae* (Miller and China). *Indian J. Ent.*, 76(3): 252-253.

Shekhar, P.S., 1958. Studies on the cock chafer *Leucopholis coneophora* Burm. A pest of coconut palm and other inter-cultivated crops. *Indian Coconut J.*, 11: 67-80.

Stanley, A. J. (1990). Distribution of spindle bug of arecanut *Carvalhoia arecae* Millar and China, its bioecology, suspected role as the vector of YLD and control. Ph. D. Thesis. Kerala Agricultural University.

Subaharan K., Vidyasagar P. S. P. V. and Mohammed Basheer B. M. (2001). Bioefficacy of Insecticides Against White Grub, *Leucopholis lepidophora* Blanch Infesting Arecanut alm. *Indian J. Plant Protection 29(1 and 2)*: 25-29.

Thakur, N.S.A., Firake, D.M., Behere, G.T., Firake, P.D. and Saikia, K. (2012). Biodiversity of Agriculturally Important Insects in North Eastern Himalaya: An Overview. *Indian Journal of Hill Farming* 25,37-40.

Theurkar S. V., Patil S. B., Ghadage M. K., Zaware Y. B. and Madan S. S. (2012). Distribution and abundance of white grubs (Coleoptera: Scarabaeidae) in Khed Taluka, part of Northern Western Ghats. *International Research Journal of Biological Sciences* 1: 1- 6.

Veeresh, G. K., Vijayendra M., Reddy, N. V. M. and Rajanna, C. (1982). Bio-ecology and management of areca nut white grubs (*Leucopholis* spp) (Coleoptera; Scarabaeidae: Melolonthinae). *J. Soil Biol. and Ecol.*, 2, 78-86.

Veeresh, G.K., 1981. Larval taxonomy of white grubs with special reference to Melolonthinae beetles. Final project report, ICAR Ad hoc Project on larval taxonomy of Melolonthinae beetles, 1977- 80, Dept. of Entomology, UAS, Bangalore.

Vidyasagar, P.S.P.V. 1991. Studies on *Halyomorpha marmorea* E. (Pentatomidae: Heteroptera), associated with tender nut drop in arecanut. *J. Plantn Crops* 18 (Suppl.): 305-311.

Vidyasagar, P.S.P.V. and Shyama Bhat, K. 1986. A pentatomid bug causes tender nut drop in arecanut. *Curr. Sci.*, 55: 1096-1097.

Welbourn, C. (2006). Red palm mite Raoiella indica (Acari:Tenuipalpidae). Pest Alert. DPI- FDACS. Available from http://www.doacs.state.fl.us/pi/enpp/ento/r.indica.html. (Accessed October 10, 2008).

Yeshwanth, H.M and Prathapan, K.D (2014). First report of the occurrence of the arecanut spindle bug, *Mircarvalhoia arecae* (Miller and China) in the Andaman and Nicobar Islands. *Journal of Tropical Agriculture* 52(2): 82-84.

2018, Pests of Plantation Crops   *Pages* **97–117**
Editors: **P. Chowdappa, Chandrika Mohan & A. Josephrajkumar**
Published by: **ASTRAL INTERNATIONAL PVT. LTD., NEW DELHI**

Chapter 4

Cocoa

☆ *M. Sujithra and M. Alagar*

1. Introduction

Cacao (*Theobroma cacao* L). is one of the most important cash crops grown worldwide. It is grown throughout the humid tropics with worldwide cultivated area of 50530790 ha in more than 75 countries. In India, it is cultivated in around 71000 with production of 15000 tonnes (FAO, 2014). Although cocoa has been cultivated for centuries in Central America, it is relatively new to Africa and even more recent in Asia. During 2013-14, the global production of cocoa is around 4370 thousand tonnes with the major producing countries are Cote d'Ivoire followed by Indonesia, Ghana, Nigeria and Cameroon (ICCO, 2015). It is mainly grown as a mixed crop in coconut and arecanut plantations and as under-storey crop in partially cleared forests. Over 1500 different insects are known to feed on cocoa, only about 2 per cent are of economic importance. These have been estimated to be responsible for up to 30 per cent losses in global production (Ploetz, 2007) and can play a major role in cacao boom and bust cycles (Clough *et al.,* 2009). In India, cocoa is known to be attacked by about 50 pests comprising mainly of mealybugs, tea mosquito bug, stem borer, aphids, stem girdler, leaf eating caterpillars and leaf eating beetles *etc.* (Rajagopal and Ananda, 2006).

2. Tea Mosquito Bug (TMB)/Mirids: *Helopeltis antonii* Signoret, *H. bradyi* Waterhouse and *H. theivora* Waterhouse

2.1. Occurrence and Distribution

Mirids are the major insects that affect cocoa worldwide. In Ghana, cocoa mirids have been recognized as a serious pest since 1908 due to their devastating effect. The most common species in Ghana and West African countries are *Distantiella theobroma* and *Sahlbergella singularis.* In South-East Asia, the *Helopeltis* spp. is responsible for the damage related to mirids while *Monalonion* sp. are present in South and Central America. Little is known about the genetic background of mirids (Latip *et al.*, 2010). Of the 41 species of *Helopeltis* recorded in the old world tropics, only 3 species are confined to India *viz. H. antonii, H. theivora* and *H. bradyi* (Stonedahl, 1991). Among these, *H. theivora* is a major pest of cocoa and tea in India and Asia followed by *H. bradyi* and *H. antonii* (Srikumar and Bhat, 2013). It also has been reported damaging other economically important plants such as black pepper, camphor, cashew and cinchona (Stonedahl, 1991). Additional host plants of this species were given by Miller (1941) and Das (1984). In peninsular India, the population build-up of tea mosquito bug on cocoa occurs in the month of October to November and it synchronizes with cessation of the monsoon rain. Peak abundance noticed during January to February and the insects remain active until the onset of the monsoon.

2.2. Nature and Symptoms of Damage

A summary of feeding damage and visible symptoms on cocoa, tea and cashews is given by Stonedahl (1991). Nymphs and adults suck the sap from the young leaves and shoots of the host plants (Karmawati, 2007) to cause discolored necrotic area or lesion around the point of entry of the labial stylets into the plant tissue (Stonedahl, 1991). *Helopeltis* is typically a low density pest with high damage. Plants are strongly affected by the toxin injected during feeding by nymphs and adults (Asokan *et al.*, 2012), results in plasmolysis of the cells. Mirids feeding on shoots often result in the death of terminal branches and leaves, causing dieback (Figure 4.2). Cherelles develop characteristic eruptive spots and finally shriveled and fall off prematurely (Figures 4.1 and 4.2). Das (1984) reported that a single late-instar nymph of *H. theivora* could make as many as 80 feeding lesions in 24 hours. Feeding activity is highest in the early morning and late afternoon. Heavy infestations can lead to substantial levels of pod malformation and drop (Tan, 1974). Economic Threshold Level for cocoa mirid is about the one mirid per tree (Asogwa *et al.*, 2006; CRIN 2011).The chemical nature, potency of the toxin and the extent of damage by different species of *Helopeltis* vary depending on the species. Similarly there is a need to monitor natural enemies of *Helopeltis*, which requires their proper identification to the species level (Asokan *et al.*, 2012).

2.3. Biology

Adult bug is reddish-brown, about 6-8 mm long with a black head, red thorax, black and white abdomen (Figure 4.2). Female bug lays eggs on the tender tissues of new shoots, and soft tissues cherelles and pods. Eggs are reniform in shape and creamy white and microscopic. Presence of chorionic threads projecting outside the tissues is indicative of the presence of eggs inside. Mated female mirids lay up to 60 eggs that are embedded in the bark of stems or inside the pod husk. On hatching, young nymphs feed on tender leaves, which later become necrotic. Nymphs are wingless, orange coloured and ant-like with long legs. Nymphal development takes about 10-15 days with 5 instars. Longevity of female bug is for about 7 days, whereas male longevity is 9-10 days. Total life cycle completes in 25-32 days.

Guidelines for the determination of mirid damage threshold as reported by CRIN (2011) are:

1. Take a random sample of 100 cocoa trees for damage symptoms, such as lesion on pods, twig die back and canker

2. If less than 5 per cent damage is observed, do not spray

3. If between 5-25 per cent is damaged, apply spot spray,

4. If higher than 25 per cent damage is observed, then carry out blanket spray,

5. However, do not spray if more than 70 per cent of pod are ripen, and do not spray if harvesting of pod is due in less than 13 days

6. After pod harvesting, carry out a re-assessment to know whether to spray or not.

Table 4.1: Morphological differences of TMB Complex in Cocoa Garden

Sl.No.	Morphological Characteristics	H. antonii	H. bradyi	H. theivora
1.	Thorax/Body colour	Bright red or brownish black	Bright red or brownish black	Whitish stripe on the anterior portion with black thorax
2.	Abdomen colour	Black and white banded	Black and white banded	Light green in colour
3.	Hind femur	White band on the distal end	White band on the basal end (proximal)	No such white band
4.	Length of the antennae	Longer than body length	Longer antennae than other species	Longer than body length
5.	Body size	–	–	Smaller than other species

Scoring of TMB in Cocoa Garden

TMB damage can be quantified by recording die-back symptoms on shoots (per cent) from all four sides of the cocoa tree and the no of pods and cherelles damaged.

Grading of TMB attack on cocoa pods is as follows:

Score	Damage Symptoms on Pods
0	No lesions found on pods
I	25 per cent of the pods and cherelles with lesions
II	50 per cent of the pods and cherelles with lesions
III	75 per cent of the pods and cherelles with lesions
IV	Almost all the pods with lesion symptoms

2.4. Pest Management

☆ Shade and canopy management should be designed to achieve a balance between mirid control, flowering and black pod management and proper pruning to get maximum sunlight inside the canopy.

☆ Carry out regular observation for infestation during the month of September to April.

Figure 4.1: Tea Mosquito Bug Damage.

☆ Alternative hosts should not be used as shade trees on cocoa farms. Check the buildup of the pest population on cocoa as well as on the alternate hosts like cashew, guava, neem, drumstick, mahogany and black pepper in the vicinity of cocoa.

☆ Observe for any oozing out, necrosis symptoms. If the damage is more than 1 per cent, go for control measures.

☆ Spray imidacloprid (0.004 per cent) 17.8 SL @ 0.25 ml/lit. or lambda cyhalothrin (0.003 per cent) 5 EC @ 0.3 ml/lit. or Bifenthrin (0.008 per cent) @ 10 EC 0.8 ml/lit. If infestation persists, one more spray can be given at 20 to 30 days interval after first spray.

☆ Nymphs and adults of TMB feed generally in the morning and late afternoon hours. Hence, spraying operation should be carried out either in the early morning or late in the afternoon. Avoid spraying during mid-day as the bugs may hide under dense canopy.

☆ Spray *Beauveria bassiana* at 400g/acre at an interval of fifteen days (three times).

☆ Release of *Chrysoperla zastrowi sillemi* at 25,000 eggs per acre once.

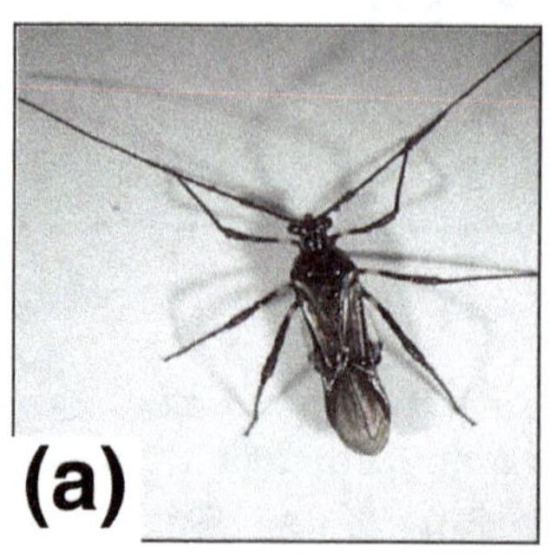

Figure 4.2

(a) Adult of *H. bradyi*, (b) Shoot with die back, (c) TMB infested cherelles.

3. Mealybugs: *Planococcus lilacinus* (Ckll.) and *P. citri* (Risso)

3.1. Occurrence and Distribution

P. lilacinus occurs mainly in tropical Asia and Oceania. Williams (1982) reported that the species was probably introduced into the South Pacific from Southern Asia. This pest is extremely polyphagous, feeding on tropical and sub-tropical fruit and shade trees within 35 families. Williams (1982), Cox (1989) and Ben-Dov (1994) provide comprehensive lists of hosts. Its chief hosts are cocoa, *Annona muricata*, *Psidium guajava* (guavas), *Ceiba pentandra* and species of *Bauhinia*, *Spondias* and *Erythrina* (Le Pelley, 1943). Other hosts include *Amaranthus gracilis*, bamboos, Citrus, Coffea, coconuts, *Ludwigia hyssopifolia*, *Mangifera indica*, *Mirabilis jalapa*, *Solanum nigrum*, *Solanum tuberosum*, *Sonchus arvensis*, *Spilanthes acmella* and *Vitis*. In India, it is reported as a serious pest causing damage to cocoa and is present in all the cocoa growing tracts of the country.

3.2. Nature and Symptoms of Damage

Both nymphs and adults of mealybug occur in colonies and infest growing shoots, terminal buds, flower stalks, foliage and pods and start sucking the sap (Figure 4.3). Symptoms on coconuts and cocoa are described as button nut shedding and drying up of inflorescence (Nair, 1981; Fernando and Kanagaratnam, 1987) and the death of tips of branches (Williams and Watson, 1988). Due to its damage, sunken patches in the developing pods result in the formation of scabs. Brown patches, irregular cracks and pits can be seen on mature pods. Dense colonies form conspicuous patches on fruits; copious honeydew excretion may result in sooty mould development near colonies and the attraction of attendant ants.

Figure 4.3: Mealybugs.
(a,b) *Crisicoccus hirsutus*, (c) *Planococcus lilacinus*.

3.3. Biology

Mealybugs are small, oval and light yellow. A female can lay around 200 eggs which hatch into nymphs within six hours. Nymphal period lasts for 20-25 days. Ants are attracted to the sugary substance excreted by the mealybugs and thus ants serve as protective and carrying agents for them. Though the pest occurs throughout the year, but peak population are noticed during April to May when the high temperature and low rainfall prevail in the Western Ghats region. With the advent of southwest monsoon, the population gradually declines (Nair, 1981).

3.4. Pest Management

☆ Timely pruning of the cocoa trees will reduce the colony build up.

☆ Neem oil suspension of 3 per cent may be sprayed on pods and foliage at the early stage of infestation.

☆ Spray Fenthion (0.04 per cent) 80 EC 0.5 ml/L or Dimethoate (0.06 per cent) 30 EC 2 ml/L. Give a second round of spray after 30 days if the incidence persists.

☆ Conserve natural enemies like coccinellid beetles (*Coccinella septempunctata, Scymnus coccivora, Chilocorus nigrita etc.*), syrphids (*Eristalis* spp., *Volucella* spp) and chrysophids (*Chrysoperla zastrowi sillemi*) which feed on the aphids and reduce the population.

☆ Release predatory lady bird beetles at 5-10 adults/tree.

☆ Destroy ant nests to minimise the spread of the pest.

4. Aphid: *Toxoptera aurantii* Boyer

4.1. Occurrence and Distribution

This is a polyphagous pest and its recorded hosts are chiefly shrubs and trees which comprises of 122 species belonging to 37 families in India. Alternate hosts includes camellia, coffee, *Ficus, Hibiscus, Ixora,* kamani, lime, macadamia, mango, mock orange, *Pittosporum,* pomelo and *Vanda* orchid. Occasional out breaks of *T. autantii* in citrus orchards and tea plantations made this insect pest as an economically important species in India. This insect is present throughout the year with a peak from August to January. Aphids, in general, are not considered as serious pests, but their severity results in leaf curling and flower wilt.

4.2. Nature and Symptoms of Damage

Both nymphs and adults colonize the tender leaves and shoots, cushions, cherelles, flower buds and feed on the tender plant parts. Severe feeding results in flower shedding and crinkling and twisting of leaves (Figure 4.4). Honey dew excretion attracts the sooty mould development, which hinder the photosynthetic rate.

4.3. Biology

Reproduction is mainly by parthenogenetic and viviparous and it occurs exclusively on the undersurface of flush leaves. The development of the aphid is temperature-dependent but at the optimum temperature, it takes 6 days to become adult passing through four instars. The incidence of alates in populations is influenced both by high aphid density and leaf age. Adult apterus, 1.60 ± 0.11

Figure 4.4: Colonies of *T. aurantii* on Cocoa Leaves and Flower Bud.

mm long, shiny black colour, oval in shape, antennae about two-thirds the length of the body, whereas alate are winged form, 1.65 ± 0.20 mm long, smaller in width than apterate and also relatively thinner (Firempong, 1997).

4.4. Pest Management

☆ Cultural practices like timely pruning of the cocoa trees will reduce the colony build up.

☆ Spraying Fenthion (0.04 per cent) 80 EC 0.5 ml/lit or Dimethoate (0.06 per cent) 30 EC 2 ml/lit. will reduce the incidence and spread. If re-occurrence of the pest is noticed, second spray may be given after an interval of 20 to 30 days.

☆ Natural enemies feed on the aphids and reduce the population. These include coccineld beetles (*Coccinella septempunctata, Scymnus coccivora, Chilocorus nigrita* etc.), syrphids (*Eristalis* spp., *Volucella* spp) and chrysophids (*Chrysoperla zastrowi sillemi*).

5. Thrips: *Selenothrips rubrocinctus* (Giard)

5.1. Occurrence and Distribution

The redbanded thrips, *Selenothrips rubrocinctus* (Giard), was first described from Guadeloupe, West Indies, where it was causing considerable damage to cocoa. As a result, it was referred to as the cacao or cocoa thrips. The earliest report relating to this thrips was a report by W.E. Broadway in 1898, when he called attention to the "blight" of cocoa. The redbanded thrips is a tropical-subtropical species thought to have originated in northern South America (Chin and Brown, 2008).

5.2. Nature and Symptoms of Damage

Symptoms of *S. rubrocinctus* attack on cocoa result from feeding by adults and/or larvae on the leaves and pods. On leaves, the feeding punctures cause the development of chlorotic spots and premature leaf drop, while on the pods, they cause brown patches that coalesce during severe infestations to form a dark brown, corky layer of dead cells that makes the estimation of pod ripeness virtually impossible. Necrotic lesions are produced in the leaves and pods by adults and nymphs, and in the flowers by adults.

5.3. Biology

The female is about 1.20 mm in length and has a dark brown to black body underlain by red pigment chiefly in the first 3 abdominal segments; the anal segments retain a reddish black color, and the wings are dark (Figure 4.5). The male is similar, but smaller and is seldom collected. The nymph and pupa are light yellow to orange with the first three and last segments of the abdomen bright red. After hatching, there are two nymphal stages lasting 9-10 days. Fully-grown second stage nymphs are about 1 mm long. The two nymphal stages are followed by two resting stages (pre-pupal and pupal stages). The resting stages last 3-5 days before adults emerge (Chin and Brown, 2008).

Figure 4.5: Adults and Nymphs of Redbanded Thrips.

5.4. Pest Management

☆ Redbanded thrips are preyed upon by a large assortment of natural predators such as spiders and mites, lacewings, predatory thrips, and predatory bugs, especially minute pirate bugs (Chin and Brown 2008; Funderburk *et al.*, 2007).

☆ Spray Fenthion (0.04 per cent) 80 EC 0.5 ml/lit or Dimethoate (0.06 per cent) 30 EC 2 ml/lit. Give a second round of spray after 30 days if the incidence persists.

6. The Red Branch Borer: *Zeuzera coffeae*

6.1. Occurrence and Distribution

This pest is found distributed in all the cocoa growing areas. The population is greater during post-monsoon months when up to 65 per cent of the plants may be damaged by the pest, in an endemic area. Entwistle (1985) reported this pest on cocoa from Sri Lanka, Malaysia, Java, West Irian and the territory of Papua New

Guinea, and a similar species on cocoa from Sabah and Peninsular Malaysia. *Zeuzera coffeae* is a pest of cocoa and is important as a branch and trunk borer of many trees.

6.2. Nature and Symptoms of Damage

This pest often flares up and causes conspicuous leaf necrosis and dieback of branches in mature cocoa, particularly in early maturity. The caterpillar bores into branches and the distal part subsequently dies. As the caterpillars grow older they will move down the stem, progressively attacking the older stem. The galleries formed are especially damaging to young cocoa, often causing the snapping-off of smaller branches (Figure 4.6). A characteristic yellowish or reddish mixture of fluid and frass particles is present on the attacked portion. The excreta are seen in lumps sticking on to the bark or in a heap on the ground below. The attacked part dies eventually, and in saplings, the whole plant may be killed from a single attack. This pest inside the branches is protected to a larger extent, direct application of insecticides into the bore holes leads to death of the caterpillar (Keane and Putter, 1992).

Figure 4.6

(a) Galleries on Stem made by *Z. coffeae*, (b) Larva boring into cocoa stem, (c) Adult of *Z. coffea*.

6.3. Biology

The eggs are yellowish and are laid in groups on the small stems and branches. They turn dark yellow prior to hatching. The young larvae remain together for some time, and then disperse by a silken thread which is caught by wind and carried to different branches. The full-grown larva has a brownish head and is predominantly reddish brown in colour, with an average length of 6.5 cm. The larval period lasts for 60-100 days. Before pupation the larva makes an exit hole covered by a loosely severed piece of bark. Pupation occurs inside the larval tunnels, and the life cycle takes about four to five months for completion. The adult is a leopard moth with wings which have dark spots on a white translucent background (Figure 4.6).

6.4. Pest Management

☆ Pruning of affected and dried branches and killing of larvae should be practiced. If the branches are not dried, killing larvae through a bore hole by inserting sharp iron needle can be tried.

☆ Applying insecticide solution (2 ml of Chloripyrifos/lit of water) by swabbing with cotton lint and keeping it in the bore hole will be effective.

7. Capsule Borer, *Conogethes punctiferalis* Guenee

7.1. Occurrence and Distribution

Conogethes punctiferalis is an important polyphagous pest attacking many economically important crops. Occurrence of this pest has been reported in cocoa during 1972 in Sri Lanka, Malaysia and Indonesia (Entwistle, 1985). The caterpillars cause damage up to 80 per cent of the flower cushions in about 40 per cent of the trees. During 1975, a study of cocoa pests in Thrissur district of Kerala revealed that *C. punctiferalis* was found to damage cocoa flower cushions severely and occasionally on cocoa bark (Chandramohanan and Harishkumar, 1975).

Figure 4.7. Capsule Borer.

(a) Larva of *C. punctiferalis*, (b) Damaged cocoa pods with larva, (c and d) Rotten cocoa beans due to *C. punctiferalis* damage.

7.2. Nature and Symptoms of Damage

Larvae feed on the rind of cocoa cherelles/pods, later bore into pods, feed the internal contents of the pods, the granular faecal pellets are seen outside the pods (Figure 4.7). When pods/cherelles touch each other, it is easy for the larvae to damage more than one pod/cherelle. Pods damaged by *Conogethes* are exposed to secondary infection by pathogens that lead to pod rot (Alagar *et al.*, 2013). The larvae sometimes feed on flower buds and flowers cushions. Damaged flower cushions may dry and shed prematurely. The damage of *C. punctiferalis* on cocoa is observed from December and peak incidence is noticed during March to May.

7.3. Biology

Adult moth is medium sized, wingspan of about 3 cm, with small black dots on pale yellowish wing (Figure 4.7). A single female moth laid pinkish oval flat eggs singly or in groups of 2 or 3 mostly in between wart or grooves of pods/cherelles/ flower cushions of cocoa. Incubation period of eggs is about 3- 4 days. The full grown larvae very active, 3 to 3.5 cm long, reddish brown, has brown marks on each segment with pinkish tinge, fine hairs on the body with dark head and prothoracic shield. Total larval period is about 24 days. Longevity of male and female moth is about 6 and 7 days respectively. Larvae were seen under a cover of silk and frass or excreta throughout their development. Pupation takes place inside the damaged pods or cherelles or in a thin silken cocoon outside the damaged pods/cherelles. Adult emerged in 7 to 10 days. Total life cycle completes in 25 to 33 days (Alagar *et al.*, 2013).

7.4. Pest Management

☆ Clearing off the damaged orchards and debris, scraping off the fruit tree bark in which larvae overwinter, and burning the crop after harvest may reduce the overwintering pest population.

☆ Setting light traps (some fumigant on a piece of cotton placed under a 60 W black light) and sugar-vinegar traps (containing a mixture of sugar, vinegar water and insecticide) in orchards and fields may reduce the adult population.

☆ In China, covering fruits with paper bags to keep away from adults from egg laying proven successful.

☆ Monitoring and mass trapping of male adults with different blends of E10-hexadecenal and Z10-hexadecenal.

8. Pod Borer, *Conopomorpha cramerella* Snellen

Cocoa pod borer (*Conopomorpha cramerella*) is a pest of cocoa in South-East Asia. Although the earliest report of damage caused by this pest was noted by Jansen (1860), the cocoa pod borer was first taxonomically described at the beginning of the 20th century and named *Acrocercops cramerella* (Snellen, 1904). Recently the generic placement of this species has been revised and it is now known as *Conopomorpha cramerella* (Bradley, 1986). In cocoa, damage by this pest makes the processing of cocoa pods difficult and reduces cocoa bean quality.

8.1. Occurrence and Distribution

The cocoa pod borer is known to occur in Saudi Arabia, China, Thailand, Brunei, Indonesia, Malaysia, Vietnam, Papua New Guinea, the Philippines, Samoa, Sri Lanka, Taiwan and Vanuatu. Currently, this pest is not present in India.

8.2. Nature and Symptoms of Damage

Cocoa pod borer (CPB) causes losses to cocoa by boring through the wall and into the pod, feeding on the pulp of bean and placenta of the pod (Figure 4.8). Damage to the funicles of pods results in malformed and undersized beans, in severe infestation it produce small flat beans that are often stuck together. It also causes the pod to yellow or ripen unevenly and prematurely. Young, green cocoa pods are particularly susceptible to attack by the pest. The beans from seriously infested pods are completely unusable. The fruit pulp becomes hard and the normal fermentation process used to produce the cocoa flavour precursors is made ineffective. Live pod borers are tough and can disperse over long distances. The CPB is also a pest of rambutan (Janny *et al.*, 2003).

Figure 4.8: Cocoa Pod Borer.
a) Larvae, b) Adult moth, c) Entry point, d) Tunnel made by CPB.

8.3. Biology

CPB lays oval shaped, disk-like shape eggs and yellow-orange in colour. Eggs are laid singly anywhere on the pod surface, although there appears to be some preference for the pod furrows. Eggs are laid on pods more than 5 cm in length. Egg stage lasts for 6- 9 days. On hatching, the first larva stage (instar) is translucent white in colour and about 1 mm long. Larvae tunnel out through the pod wall, leaving an easily identifiable exit hole. Entire larval stage takes 14-18 days to complete, with 4-6 instars. Pupation site could be in the furrow of the pod, or green or dried leaves and other debris. Larvas spins an oval-shaped cocoon over itself and enter into pupation. Pupal stage normally takes 6-8 days to complete. Moths are most active at night; A female can normally produce 50-100 eggs in its lifetime. During the day, adult moths normally rest underneath horizontal or near-horizontal cocoa branches. The adult has a protective coloration that blends with the resting place making them difficult to spot (Figure 4.8). Adult longevity is generally about one week, but they can live up to 30 days. In total, the entire life cycle takes about 1 month to complete.

8.4. Pest Management

Control of pod borer is difficult. However, it can be managed effectively using an integrated approach that includes good crop hygiene, early pod harvesting, insect trapping and chemical sprays.

- ☆ Sanitation practices involving the complete harvesting of ripe or damaged pods, burying of pod husk, placenta, rotten pods are recommended.

- ☆ Regular pruning of the cocoa canopy to less than 4 m in height is also good practice.

- ☆ Timely pruning and regular harvesting of pods minimises the pod borer attack.

- ☆ Destroy infected pods/infested crop residues.

- ☆ Follow early practice of "rampasan" involved removal of all pods longer than 5 cm during intercrop period, which breaks the life cycle of the pest (Entwistle, 1985).

- ☆ Grow clones with smooth pods are less susceptible than clones with rough pods (Wessel, 1983).

- ☆ Pod-sleeving with plastic bags also reduces attacks of CPB. Pods should be sleeved when they are about 8-10 cm long and the sleeves should be left throughout the pod maturation period (Entwistle, 1985).

- ☆ Ant species such as black ant (*Dolichoderus thoracicus*) and the weaver ant (*Oecophylla smaragdina*) are known to prey on larvae at emergence from the pods and on pupae, and disturb adults. Ants can be augmented and manipulated to colonise areas within a cocoa garden.

- ☆ Release of an egg parasitoid, *Trichogrammatoidea bactrae fumata* is found to be effective.

☆ The fungus, *Beauvaria bassiana* has also been found to infect larvae and pupae, causing a 100 per cent death rate.

☆ Sex pheromone for CPB identified as a blend of (E,Z,Z)- and (E,E,Z)-4,6,10-hexadecatrienyl acetates and corresponding alcohols. Trap densities of four and eight per ha were used to give economic control (Day, 1985; Beevor *et al.*, 1986; Beevor *et al.*, 1993).

9. Leaf Eating Insect Pests

Some of the leaf eating caterpillars occurs sporadically and seasonally in the cocoa garden. Generally, the damage caused by these caterpillars is simply loss of leaf, although in young cocoa, they can distort growth.

9.1. *Lymantria obfuscate* Walk. (Lepidoptera: Lymantriidae)

This is a serious pest of cocoa in all the cocoa growing tracts in India.

9.1.1. Nature and Symptoms of Damage

The brownish hairy caterpillars feed voraciously on tender leaves. During day time, the caterpillars congregate on the fallen leaves and twigs around the base of the plant or on the basal surface of the main stern.

9.1.2. Biology

A female moth, on an average, lays about 300 eggs. Eggs hatch in 8 days. Larval period lasts for 25 to 38 days and the pupal period is about 7 days. Longevity of female and male is for 3 - 4 days and 4 - 6 days, respectively. Males are winged, slender and brownish in colour with wavy markings on the fore-wings.

9.2. *Lymantria ampla* Walk. (Lepidoptera: Lymantriidae)

This is a closely related species of *L. obfuscata*, resembling it in biological features and damage caused.

9.2.1. Nature and Symptoms of Damage

The caterpillars are seen in the field after monsoon showers, and the damage is caused by feeding on the tender flushes.

9.2.2. Biology

A single female lays about 1000 eggs. The incubation period is about for 9 days. The larval period lasts for 27-38 days by having 4- 5 instars.

9.3. *Euproctis subnotata* Walk. (Lepidoptera: Lymantriidae)

9.3.1. Nature and Symptoms of Damage

The hairy caterpillars are found feeding on tender leaves as well as on the surface tissue of young pods. Pods are preferred over leaves for feeding, and the attacked young pods dry up. The yellowish moth lays eggs in masses of 8-10 on the lower surface of leaves and covered with hairs. The egg, larval and pupal stages last for 5, 4 to 7 and 8 to 10 days respectively.

9.4. *Euproctis guttata* Walk (Lymantriidae : Lepidoptera)

This is a related species of *Euproctis subnotata* which is observed to feed on cocoa leaves and pods causing similar damage. The moth is yellowish, lays eggs in groups on the lower surface of the leaf. The egg stage is from 6-8 days, the larva 18-27 days, and the pupa 9-14 days. During the survey it was found that the caterpillars of both these species together cause about 28 per cent damage to pods. The period of infestation is during June - July.

9.5. *Adoxophyes privatana* Walk (Tortricidae: Lepidoptera)

The caterpillar web young leaves together and feed from within, making irregular holes. The tiny moth lays eggs on tender twigs which hatch in about 6 days. Larval period is completed in 16-20 days and the pupal period is about 4 days. Pupation takes place within the webbed leaves. Total life cycle is completed in 26-30 days.

9.6. *Spodoptera litura* Biosd. (Noctuidae: Lepidoptera)

The caterpillars damage the nursery plants by voraciously defoliating the leaves. Affected seedlings show a tendency for premature branching. Eggs are laid on the leaves and covered with buff coloured hairs. On an average, 200- 300 eggs are laid by a female. Egg, larval and pupal period last for 5, 15-17 and 8-14 days respectively.

9.7. Other Caterpillar Pests

Other caterpillar pests of cocoa recorded are *Hyposidra talaca* Walk (Geometridae), *Spilosoma obligua* Walk (Arctiidae), *Dasychira mendosa* Hb. (Lymantriidae), *Pericallia ricini* F. (Arctiidae) Semilooper, *Achaea janata* L. (Noctuidae), *Pteroma plagiophelps* Hamps (Psychidae) and *Clania sp.* (Psychidae).

10. Vertebrate Pests

The vertebrate pests include rats, squirrels, palm civet and birds and they inflict loss of the crops by feeding and damaging the pods. A heavy damage (75 per cent) can be seen by rodents in any of the farmers' fields having cocoa plantations (Advani, 1982).

10.1. Rats and Squirrels

The Western Ghats squirrel, *Funambulus tristriatus* Waterhouse; the South Indian palm squirrel, *F. palmarum* Linnaeus and the black rat, *Rattus rattus* Linnaeus were observed to be causing much damage to cocoa in South India; the palm civet, *Paradoxurus hermaphroditus* Pallas and the bonnet monkey, *Macaca radiata* Geoffroy were causing minor damage (Bhat *et al.*, 1981). Regular trapping of rodents and bats in cocoa plots resulted in cost benefit ratio of about 1:500 in plantations near Kasaragod, increasing the productivity by more than ten times per ha (Advani, 1982).

10.1.1. Occurrence and Distribution

Rat damage is reported from most of the tropical countries including islands. In the Far East, palm rats and squirrels, in particular *Callosciurus notatus*, caused damage. The latter caused severe bark stripping of young cocoa plants (Hafidzi, 1982). Pod losses vary widely, but may reach 90 per cent at times (Han and Bose, 1980: Wood, 1984). Heavy losses noticed in India, e. g from the Western Ghats squirrel, *Funambulus tristriatus,* the South Indian palm squirrel (*F. palmarum*) and *Rattus wroughtoni* (Bhat, 1992; Baco *et al.,* 2010). Other territories where rodent damage to cocoa is recorded include West Indies, Pacific islands and South America. Han and Bose (1980) estimated 100-300 rats/ha. Without control measures, losses from small mammals over 7 weeks fluctuated in the range 75-90 per cent of pods, equivalent to 70 kg of dry beans per week. Rodents adapted themselves to this plant within a span of 8-10 years after the initial cultivation. The extent of rodents damage varies from 8 -51 per cent per cent in different cocoa growing tracts of Southern India (Mariamma, 2008). The amount of damage is very variable and depends upon the conditions under which the cocoa is grown. Cocoa grown under coconuts is susceptible to attack by both rats and squirrels. In the pacific islands, up to 60 per cent of pods may be lost (Williams, 1973).

10.1.2. Nature and Symptoms of Damage

The rats usually gnaw the pods near the stalk portion whereas squirrels gnaw the pods in the center (Figure 4.9). The rats are known to damage the mature as well as immature cocoa pods whereas the squirrels damage only the mature ones. They gnaw the pods and feed on the mucilage covering of the beans (Rajagopal and Ananda, 2006). The damage is compounded by ensuing fungal infection and affected pods are all lost. All rodent species while feeding on cocoa pods leave tooth marks on pods. Though the tooth marks made by different rodent species are not distinguishable, the same can be distinguished from the marks on the pods that had been attacked by monkeys, civet cats or birds. Squirrels are diurnal and rats are nocturnal in habits. The population of squirrels is found to be 4 to 5 numbers per hectare of cocoa garden. The home range of squirrels is 2 ha. Rat population is about 25-30 per hectare and home range is 0.5 hectare only. The population of *Rattus rattus* is about seven times that of Western Ghats squirrels, the intensity of squirrel damage to cocoa is nearly three times more than that caused by rats (Mariamma, 2008). Many species of rats

Figure 4.9: Vertebrate Pests.
(a) Rat damage, (b) Squirrel damage.

damage both young and mature pods. Often, squirrels are more important than rats. One squirrel can attack up to 4 cocoa pods per day. A rat on average takes one whole week to attack up to 4 pods. However, the rats alone have been reported to damage more than 9000 pods per hectare. Squirrels like heavily shaded farms. Both squirrels and rats prefer badly maintained farms.

10.1.3. Management

☆ An integrated approach on a locality basis is essential for the management of rodents.

☆ Clean cultivation should be followed to reduce the refuges of rodents.

☆ Habitat manipulation is yet another important step for rodent management. Crown cleaning of palms generally expose the nesting places of the rats to predators.

☆ Timely harvest of mature cocoa pods will reduce rodent damage to a certain extent.

☆ Squirrels are best controlled by trapping with wooden or wire mesh single catch 'live' trap with ripe coconut kernel as the bait. Trapping is more successful if carried out during the lean period (September-November) when the alternate foods such as paddy, cashew apples and jackfruit are not available.

☆ Single dose anticoagulant rodenticide 10g Bromadiolone (0.005 per cent) wax cakes twice at an interval of 10-12 days on the branches of one cocoa tree out of every five trees is recommended.

☆ Indiscriminate killing of predators such as owls, rat snakes, mongoose etc should be discouraged.

10.2. The Palm Civet, *Paradoxurus hermaphrodites*

10.2.1. Occurrence and Distribution

The palm civet, *Paradoxurus hermaphroditus* Pallas also known as toddy cat, is a nocturnal tree climber. It is present in many cocoa growing tracts of India and Malaysia. About 12.8 per cent damage is noticed in some areas by this pest.

10.2.2. Nature and Symptoms of Damage

The palm civet bites and breaks the husk of cocoa pods. The piece of broken chunk is 2.0 to 3.0 cm in diameter. There is no distinct pattern of damage. Two distantly placed markings caused by the canine teeth are very much prominent on the pod husk. Piles of defecated beans are seen scattered around the cocoa plantations.

10.2.3. Management

Trapping is best method for the management of palm civet.

11. Birds

The Golden backed woodpecker, *Dinopium sp.* is seen making small holes on cocoa pods and eats the mucilage inside. Extent of damage by this animal in cocoa plantation is very negligible.

12. Conclusion

Effective management of pest problems is frequently complex and usually involves agronomic, pest control, logistic, economic and sometimes even political factors. All these factors need to be taken into account to fully appreciate and accurately diagnose pest problems and make appropriate recommendations to farmers. It is also essential to understand current practice and how decisions are made before it is possible to improve the decision making process in pest management in cocoa farming systems. The relevant information should be gathered such as proper knowledge of the pests under cropping conditions, Monitor development of insect pests and their natural enemies, Identify critical time of pest incidence, Make decisions based on the problems and economic thresholds, Use correct dose and application technique and finally management strategy. Also, the closer the IPM system fits a farmer's needs then the greater is the likelihood of adoption. Thus, successful implementation of an IPM system in the protection of farm products should logically be preceded by the development of effective IPM programmes and the packaging of expert systems.

References

Advani, R. (1982). Ecology, Status and post natal development of the Black rat, *Rattus rattus* (Linnaeus) in the plantation crops in Sahyadri tract. *Proc. Plant. Crop. Syn.* (626-632).

Alagar. M., Rachana. K. E., Bhat. K. S., Rahman. S. and Rajesh. M. K. (2013). Biology, damage potential and molecular identification of *Conogethes punctiferalis* Guenee in cocoa (*Theobroma cacao* Linn.) *Journal of Plantation Crops* 41: 350-356.

Asogwa, E. U., Ojelade, K. T. M., Anikwe, J. C. and Ndubuaku, T. C. N. (2006). Insect pests of cocoa, kola, coffee, cashew, tea and their control. Answer Communication Concepts, Apapa, Lagos, Nigeria. pp. 80-86.

Asokan, R., Rebijith, K. B., Srikumar, K. K., Bhat, P. S. and Ramamurthy, V. V. (2012). Molecular identification and diversity of *Helopeltis antonii* and *Helopeltis theivora* (Hemiptera: Miridae) in India. *Florida Entomologist* 95: 350-358.

Baco, D., Nasruddin, R. and Juddawi, H. (2010). Rodent outbreaks in South Sulawesi, Indonesia: the importance of understanding cultural norms. In: Singleton, G. R., Belmain, S. R., Brown, P. R. and Hardy, B. (eds) *Rodent Outbreaks: Ecology and Impacts.* IRRI, Los Banos, Philippines, pp. 129-137.

Beevor. P. S., Mumford. J. D., Shah. S., Day. R. K., Hall.and D. R. (1993). Observations on pheromone-baited mass trapping for control of cocoa pod borer, *Conopomorpha cramerella*, in Sabah, East Malaysia. *Crop Protection* 12(2): 134-140.

Beevor. P.S., Cork. A., Hall. D. R., Nesbitt. B. F., Day. R. K. and Mumford. J. D. (1986). Components of female sex pheromone of cocoa pod borer moth, *Conopomorpha cramerella.* *Journal of Chemical Ecology* 12(1): 1-23.

Ben-Dov. Y. (Editor), (1994). A systematic catalogue of the mealybugs of the world (Insecta: Homoptera: Coccoidea: Pseudococcidae and Putoidae) with data on geographical distribution, host plants, biology and economic importance. Andover, UK; Intercept Limited, 686 pp.

Bhat, S. K. (1992). Plantation Crops. In: Prakash, I. and Ghosh, P. K. (eds) *Rodents in Indian Agriculture,* Vol 1. Scientific Publisheres, Jodhpur, India, pp. 279-288.

Bhat, S. K., Nair, C. P. R. and Mathew, D. N. (1981). Mammalian Pests of Cocoa in South India. *Tropical Pest Management* 27(3): 297-302.

Bradley. J. D, (1986). Identity of the South-East Asian cocoa moth, *Conopomorpha cramerella* (Snellen) (Lepidoptera: Gracillariidae), with descriptions of three allied new species. *Bulletin of Entomological Research* 76(1): 41-51.

Chandramohanan, R. and Harishukumar, P. (1975). The castor capsule borer, *Dichocrosis punctiferalis* Guen. as a pest of cocoa in India. *Agricultural Research Journal of Kerala* 14: 79-80.

Chin, D. and Brown, H. (2008). Red-banded thrips on fruit Trees. Agnote. (http://www.nt.gov.au/dpifm/Content/File/p/Plant_Pest/719.pdf) (19 August 2008).

Clough, Y., Faust, H. and Tscharntke, T. (2009). Cacao boom and bust: sustainability of agroforests and opportunities for biodiversity conservation. *Conservation Letters* 2: 197–205.

Cox. J. M. (1989). The mealybug genus *Planococcus* (Homoptera: Pseudococcidae). Bulletin of the British Museum (Natural History), *Entomology* 58(1): 1-78.

CRIN (Cocoa Research Institute of Nigeria). (2011). New cocoa varieties for Nigeria-Attributes and field management requirements. Library, Information and Documentation Department, CRIN, P.M.B. 5244, Ibadan. E-mail: directorcrin@yahoo.com; www.crin-ng.org.

Das, S.C. (1984). Resurgence of tea mosquito bug, *Helopeltis theivora* Waterh., a serious pest of tea. *Two and a Bud* 31: 36-39.

Day. R. K. (1985). Control of the cocoa pod borer. PhD. thesis. London, UK: Imperial College, University of London.

Entwistle, P. F. (1985). Insects and cocoa. In Cocoa. 4th Edition. G.A. R. woood and R. A. Lass, Longman. NewYork and London. pp. 366-443.

FAO (2013). Faostat3. fao. org/down;load/Q/QL/L.

FAO (2014). www.fao.org/faostat/en#data/ QC.

Fernando, L.C.P. and Kanagaratnam, P. (1987). New records of some pests of the coconut inflorescence and developing fruit and their natural enemies in Sri Lanka. COCOS 5: 39-42.

Firempong, S. (1997). Biology of *Toxoptera aurantii* (Homoptera: Aphididae) on cocoa in *Ghana. J. Nat. Hist.,* 11: 409-416.

Funderburk, J., Diffie, S., Sharma, J., Hodges, A. and Osborne L. (2007). Thrips of ornamentals in the southeastern U.S. EDIS. (19 August 2008).

Hafizdi, M. N. (1982). Some notes on bark stripping behaviour of *Callosciurus notatus*. *The Planter* 68: 501-505.

Han, K. J. and Bose, S. (1980). Some studies on mammalian pests in cocoa planted under coconuts. *The Planter* 56: 273–83.

ICCO (2015) Quarterly Bulletin of Cocoa Statistics, Vol. XLI, No. 3, Cocoa year 2014/15. http://www.icco.org/Published: 28-08-2015.

Janny G. M. V., Barbara, J. R. and Julie, F. (2003). Discovery learning about cocoa. An inspirational guide for training facilitators. CABI Bioscience, pp: 1-110. Jansen. A. J. F. 1860. Aantekingen betreffende de kakao cultuur in de residentie Manado. Natuurkundig Tijdschrift voor Nederlandsch Indie 20 vierde serie deel VI, 289-307.

Jansen AJF (1860). Aantekingen betreffende de kakao cultuur in de residentie Manado. Natuurkundig Tijdschrift voor Nederlandsch Indie 20 vierde serie deel VI, 289-307.

Karmawati, E. (2007). Pengendalian Hama Helopeltis spp. pada Jambu Mete Berdasarkan Ekologi: Strategi dan implementasi. Orasi Pengukuhan Peneliti Utama sebagai Profesor Riset Bidang Entomologi Pertanian. Badan Penelitian dan Pengembangan Pertanian, Jakarta. 61 hlm.

Keane, P. J and Putter, C. A. J. (1992). Cocoa pest and disease management in Southeast Asia and Australasia. *FAO Plant Production and Protection Paper*, 112: 213.

Latip, S. N. H., Muhamad, R., Manjeri, G. and Tan, S. G. (2010). Development of microsatellite markers for *Helopeltis theivora* Waterhouse (Hemiptera: Miridae). *African J. Biotechnol.* 9(28): 4478-4481.

Le Pelley, R. H. (1943). An Oriental mealybug (*Pseudococcus lilacinus* Ckll.) (Hemiptera) and its insect enemies. *Transactions of the Royal Entomological Society of London* 93(1): 73-93.

Mariamma, D. (2008). Pest Management in Cocoa. In: Model Training course on "Crop Management technologies for Coconut, Arecanut and Cocoa cultivation". CPCRI, pp. 283-288.

Miller, N.C.E. (1941). Insects associated with cocoa (*Theobroma cacao*) in Malaya. *Bulletin of Entomological Research,* 32: 1-15.

Nair, C.P.R. (1981) Investigations on insect pests of cocoa Theobroma cacao. L. in Kerala with special reference to the mealy bug *Planococcus lilacinus* (Ckll.) (Homoptera: Pseudococcidae). PhD thesis, Kerala Agric. University, pp. 150.

Ploetz, R.C. (2007). Cacao diseases: Important threats to chocolate production worldwide. *Phytopathology,* 97: 1634–1639.

Rajagopal, V. and Ananda, K.S. (2006). Technical bulletin on cocoa cultivation practices. Central Plantation Crops Research Institute, Regional Station, Vittal. pp. 1-14.

Snellen. P. C. T. (1904). Over de ontwikkelingstoestanden van eenige Microlepidoptera van Java. *Tijshrift voor Entomologie*, 46: 79-90.

Srikumar K. K. and Bhat, P. S. (2013). Biology of the tea mosquito bug (*Helopeltis theivora* Waterhouse) on *Chromolaena odorata* (L.) R.M. King and H. Rob. *Chilean Journal of Agricultural Research*, 73(3): 309-314.

Stonedahl, G. L. (1991). The oriental species of *Helopeltis* (Hemiptera: Miridae): A review of economic literature and guide to identification. *Bull. Entomol. Res.*, 81: 465-490.

Tan, G.S. (1974). *Helopeltis theivora* theobromae on cocoa in Malaysia II. Damage and Control. *Malayan Agricultural Research* 3: 204–212.

Wessel, P. C. (1983). The cocoa pod borer moth (*Acrocercaps camerella* Sa.). Review of Research in Indonesia 1900 -1918. In cocoa research in Indonesia 1900-1950. Ed. H. Toxoplus and P. C. Wessel. America Cocoa Research Institute. pp. 35- 88.

Williams, D. J. and Watson, G.W. (1988). The Scale Insects of the Tropical South Pacific Region. Part 2. The Mealybugs (Pseudococcidae). Wallingford, UK: CAB International, 260 pp.

Williams, J. M. (1973). Rat damage to coconuts and cocoa in FIJI. In: *Proceedings 2nd Regional conference of Directors of Agriculture and Livestock Production*. South Pacific Commission, Noumea, New Caledonia.

Williams. D. J, (1982). The distribution of the mealybug genus *Planococcus* (Hemiptera: Pseudococcidae) in Melanesia, Polynesia and Kiribati. *Bulletin of Entomological Research* 72(3): 441-455.

Wood, B. J. (1984). Rat pests of tropical crops - a review of practical aspects. In: *Proceedings of a conference on the Organization and Practice of Vertebrate Pest Control*, 30 August–3 September 1982, Elvetham Hall, England. ICI Plant Protection Division, Fernhurst, UK, pp. 265-287.

2018, Pests of Plantation Crops   *Pages* 119–142
Editors: **P. Chowdappa, Chandrika Mohan & A. Josephrajkumar**
Published by: **ASTRAL INTERNATIONAL PVT. LTD., NEW DELHI**

Chapter 5

Oil Palm

☆ *P. Kalidas and L. Saravanan*

1. Introduction

Oil palm (*Elaeis guineensis* Jacq.), is the richest source for vegetable oil production and is grown in large areas to overcome the edible oil shortage in the country. Its perennial nature with an economic life span of 25-30 years gives the advantage over its counterparts of other oil seeds crops. Oil palm is a tropical crop which is mainly cultivated in Asia, Africa and South America. In India, oil palm is grown in an extent of 2.68 lakh ha. of which around 1.20 lakh ha is in Andhra Pradesh alone. The productivity of the crop is often challenged by water and nutrient stress, drought, biotic stresses like pests and diseases *etc.* Since irrigation is given at regular intervals, the micro climate that is formed with high humidity and low temperature makes congenial conditions for the development of pests and diseases.

Although oil palm is a new crop to India, the pest incidence is found increasing with the age of the palms. The chances of introduction of foreign pests are remote due to the strict quarantine measures that are being followed at various ports of import of seed sprouts. The only chance for the pest outbreak is the migration of pests from other crops that are found in the vicinity of the oil palm plantations. This is mainly because of the continuous availability of food material (Kochu Babu and Kalidas, 2004; Kalidas, 2011). The perennial nature of palms and monocropping system as practiced in many locations provide ample opportunities for the buildup of the pest. The incidence of pests is slowly spreading to new areas.

In India around 60 species of insects are reported to be infesting oil palm (Dhileepan, 1991 and 1992; Kalidas *et al.,* 2002, 2011). Among the insect pests, rhinoceros beetle and defoliators are important ones throughout the world causing heavy yield losses (Norman and Basri,2007; Cheong *et al.,* 2010). The insect pests which are reported as minor and secondary pests on other crops are getting pest status in oil palm by causing damaging to both nursery and field palms. In this paper a brief account on pest scenario of Indian oil palm industry and the strategies for their management are presented.

2. Pests of Nursery

Prior to field planting, oil palm is generally pass through single or double stage nursery system. Oil palm seedlings are maintained for 12-14 months in nursery and culled and planted in the field. The nurseries are almost free from pest incidence as they are protected with preventive measures of insecticide application at frequent intervals. However, when in the neglected conditions, few pests are observed causing damage to the seedlings. Some of them are listed below.

2.1. Insect Pests

2.1.1. Tussock Caterpillar: *Dasychira mendosa* Hb. (Lepidoptera: Lymantridae)

It is a polyphagous insect, which is generally found on arecanut, citrus, guava, banana *etc.* Larvae occasionally cause significant damage to nursery seedlings by feeding voraciously on the leaves. They are found causing damage in oil palm plantations also (Wood, 1968). Occurrence of this pest was noticed in oil palm nurseries of Kerala, Karnataka, Andhra Pradesh and Tamil Nadu. The pest is noticed throughout the year with the highest incidence during June- July, coinciding with the onset of heavy rains (Dhileepan 1992). Incidence is seen only in those nurseries which are over grown and not lifted for transplantation at correct time.

2.1.1.1. Biology and Nature of Damage

A female moth lays about 302.9 eggs in its life time in confinement. The incubation period is about 5.5 days. After hatching, the larvae pass through 9-10 instars in 43.1 days. The larvae are found defoliating the oil palm severely at secondary nursery stage. Initially the young larvae scrap the leaves in congregation and disperse in later stages and start defoliating the tender leaves severely. The total period for developing from egg stage to adult is about 65.1 days. Caterpillars are characterized by the dense tufts of hair on the body. They are capable of causing minor irritation and sometimes rash on human body.

2.1.1.2. Management

Under field conditions, the larvae are parasitized by tachnid flies to the extent of 10.2 per cent and pupa are parasitized by *Brachymeria albotibialis* to the tune of 40.0 per cent. The pest can be managed with one or two sprays of quinalphos 0.05 per cent.

2.1.2. Spindle Bug: *Carvalhoia arecae* Miller (Hemiptera: Miridae)

It is primarily a serious pest of arecanut palms but also attained pest status on oil palm in Karnataka and Kerala states where arecanut is commercially grown. Pest is noticed in nursery seedlings and young plants with low levels of incidence and infestation. Infestation is noticed throughout the year with the highest incidence during June and the lowest in February. The per cent infestation declined with the increase in the age of the palms and no incidence is noticed at 30 months after field planting (Dhileepan, 1992). The incidence is not reported from other than the above mentioned states. Existence of low temperatures may be the critical factors for the pest incidence.

2.1.2.1. Nature and Symptoms of Damage

Both the nymphs and adults suck the sap from the tender spear leaves causing typical linear brown lesions. Spear leaves fail to open fully when the infestation is severe. The infested portions develop necrotic patches, which turn brown and subsequently dry up. The central portions of the necrotic patches drop off forming numerous holes on the leaves. Due to severe infestation the leaves are shredded and the palms become stunted (Nair and Daniel, 1982).

2.1.2.2. Management

The bugs are naturally suppressed by an entomopathogen, *Aspergillus candidus* Link. during the rainy season, coinciding with the peak period of its incidence (Dhileepan *et al.*, 1990). Placing of phorate 10 G granules @ 20g/sachet in perforated polythene sachets within the innermost two leaf axils is an effective management practice (Jacob, 1985). The sachets are transferred over and again to the innermost leaf axils as and when new spindles emerge. The longevity of the sachet is about eight to ten months.

2.1.3. Shoot Borer: *Sesamia inferens* Walker (Lepidoptera: Noctuidae)

The pest is found on oil palm seedlings in the primary and secondary nurseries in all the southern states of the country including Andhra Pradesh, Karnataka and Kerala.

2.1.3.1. Nature and Symptoms of Damage

Caterpillars tunnel into the stem through the spindle leaf rachis and reach the meristematic tissues and feed which cause dead heart and little leaf symptoms thereby arresting the growth (Figure 5.1). Caterpillars are pink in colour and pupate inside the leaf rachis. It is observed that severe infestation causes mortality to oil palm secondary nursery plants, there by warranting control measures. Poorly managed nurseries are found more prone to pest attack.

2.1.3.2. Management

Application of carbaryl 50 per cent WP 0.01 per cent at bimonthly intervals will bring down the pest incidence (Jacob and Kochu Babu,1995).

2.1.4. Cockchafer Beetles (Root Grub): *Apogonia/Adoretus* spp. (Coleoptera : Scarabaeidae)

The incidence of adults is observed during the onset of monsoon where as the grubs are observed all round the year, preferably during the first two to three months of planting.

Figure 5.1: Shoot Borer, *Sesamia inferens* Walker Infested Oilpalm Seedling.

2.1.4.1. Nature and Symptoms of Damage

Both the grubs and adults cause damage to oil palm. Grubs feed on the roots of young seedlings of 1-2 months old, which leads to the seedling mortality. Adults feed on the leaves causing defoliation. It is a problem in red soils of Karnataka and to a little extent in Andhra Pradesh.

2.1.4.2. Management

Application of 20 gms of phorate 10 G granules in to each nursery bag reduces the incidence of grub stage where as application of contact insecticides like quinalphos (0.05 per cent) will effectively reduce the adult incidence.

2.1.5. Psychid, *Metisa plana* Walker (Psychidae: Lepidoptera):

It is the important pest on the nursery seedlings. The pest is found feeding on the undersurface of the old leaves. Caterpillar is the damaging stage which nibbles the chlorophyll making brown patches. These patches ultimately form into holes. Severe incidence affects the growth rate. Nurseries that are raised near coconut and oil palm plantation surroundings are found to get more incidence. (See pests of adult palms for details)

2.1.6. Tobacco Caterpillar, *Spodoptera litura* Fb. (Lepidoptera: Noctuidae)

Stray incidences of the pest feeding on the oil palm nursery leaves are observed during the months of February-March. This is more common in nurseries which are raised adjacent to maize/tobacco fields. The pest might be migrating from the adjoining maize fields after the harvest of the later.

2.1.6.1. Nature and Symptoms of Damage

Defoliation of new leaves and feeding symptoms on the primordial region are the damaging symptoms.

2.1.6.2. Management

The pest can be brought under control with the application of contact insecticides like quinalphos (0.05 per cent).

2.1.7. Leaf Hopper, *Proutista moesta* Westwood (Derbidae: Homoptera)

Low incidence of leaf hopper during the months of September-October and again in the months of January-February is observed on 10-14 month old nursery plants in Andhra Pradesh. Though this has no damaging status on the seedlings, it has been proved as carrier of many viral/MLO diseases particularly the Spear rot (Kochu Babu, 1989).

2.1.8. Leaf Webworm, *Acria meyricki* (Lepidoptera: Depressaridae)

The larvae are found feeding on the undersurface of the old leaves inside the silken web. Caterpillar is the damaging stage which nibbles the chlorophyll content making brown patches. These patches ultimately coalesce and form into holes. Severe incidence cause drying of leaves and affect the growth of the seedlings. (See pests of adult palms for details).

2.1.9. Aphids (Homoptera: Aphididae)

Schizaphis rotundiventris (Signoret) and *Mysteropneura setariae* (Thomas) infest the oil palm seedlings in Karnataka especially in areas where the nurseries are surrounded by sugarcane fields. *Astegopteryx rhaphides* (Van der Goot) has been found encrusting the oil palm leaves in Little Andamans. Due to feeding on nursery seedlings, twisting and distortions of the spears are caused. Sooty mould is developed due to honey dew secretion.

2.1.9.1. Management

Spraying of dimethoate 0.04 per cent or monocrotophos 0.05 per cent or malathion 0.05 per cent on the under surface of the leaves is recommended for the control of aphids (Nair *et al.*, 1997).

2.2. Vertebrate Pests

Apart from above insects, the vertebrate following pests are also reported in the Oil palm nurseries causing considerable damage.

2.2.1. Wild Boar, *Sus scrofa*

Wild boar is reported as an important mammal causing heavy damage to the nursery in both Karnataka and Kerala states. The damage is found on the boll region of 18-24 months old seedlings. They come in groups during the dusk period and feed on the seedlings. The damage is found more due to their disturbance compared to their eating. Wild boar scaring devices like putting the electric fencing and other smoking devices are found futile.

2.2.2. Black Rat, *Rattus rattus* Wroughtoni

Rat incidence on oil palm seedlings is recorded in all the nurseries. Its more serious in poorly managed nurseries. They are found feeding on the kernel portion of the nuts causing mortality to the seedlings (Figures 5.2 and 5.3). Incidence is so severe when spreading of primary seedlings is not done even after 6 months and where the soil is drained and the roots are exposed due to poor irrigation practices. In Mizoram, *boi* (bamboo rat *Cannomys badius*) is a common pest feeding on roots of the newly planted oil palms (Figure 5.2).

Figure 5.2

(a) Rat damage on oil palm seedlings, (b) Bamboo rat (Boi) *Cannomys badius.*

2.3. Molluscan Pests

2.3.1. Black Slug, *Laevicaulis alte*

It is an occasional pest feeding on oil palm one day old nursery plants. The

Figure 5.3: Rat Damage on Juvenile Palms and Oil Palm Bunch.

pest is found feeding on the primordial portion of nursery plants that are planted in the primary bags (Figure 5.4). It is found migrating from the nearby leftovers and rubbles of removed crops like banana. Plenty of pests are found hiding under the banana stumps that are thrown out after removal of the crop. Its incidence is severe during rainy season. During the period, it is found migrating long with the stream of water. During night time they are coming to the nursery and feeding on the just planted sprouts. This pest can be managed by plugging of entry points for the pest in the green house application of salt pellets on raised bunds on all sides of the green house keeping the lights on during night in the green house *etc.* (Kalidas *et al.*, 2006).

Figure 5.4: Black Slug, *Laevicaulis alte* on Seed Sprouts.

3. Pests of Adult Palms

3.1. Rhinoceros Beetle, *Oryctes rhinoceros* L. (Coleoptera : Scarabaeidae)

It is a common pest of many palms including coconut, oil palm, areca nut and palmyrah palms in all the states of India. It has attained the major pest status in all the oil palm growing states of India (Dhileepan, 1992). It is found throughout South-East Asia and many oil palm growing countries. Pest incidence is found throughout the year. Peak period of adult emergence is during South-West Monsoon (June to September). This indicates that pest emergence synchronizes with the monsoon showers. Infestation is severe in plantations where field hygiene and sanitation are neglected. Incidence is more in oil palm plantations adjoining coconut gardens. Palmyrah palm (*Borasses flabellifer*), coconut, arecanut are collateral hosts of this pest (Pillai *et al.*, 1993; Kalidas, 2002, 2006 and 2012).

The pest is found migrated from other arecaceae palms like coconut and palmyra during the initial periods of crop establishment. But in the recent period it is found dejecting the oil palm. This is particularly seen when coconut plants are adjacent to oil palm. Presence of more lignin at the place of feeding (leaf petiole) in oil palm compared to leaf tips in case of other arecaceae palms is the prime reason for low pest incidence of the pest (Kalidas, 2011). This also indicates the pest preference to particular place of the crop. Costa Rican material, Deli X Ghana is found to be the least susceptible to rhinoceros beetle attack followed by Deli X Nigerian cross. The Malaysian variety Gutherie is found more susceptible to the beetle followed by Palode material (Kalidas, 2004a,b,c).

3.1.1. Biology and Nature of Damage

The adult beetle lays oval, white, seed like eggs in 5-10 cm below the soil surface in decaying organic matter. The early stages of the beetle are generally passed in manure pits, decomposing organic matter such as cattle dung, compost, rubbish heaps, rotting palm logs and stumps. The beetles also breed in the leaf axils of oil palm, rotting inflorescences and on mesocarp heaps dumped in the plantation (Dhileepan, 1988; Ponnamma *et al.*, 2001). A female lays about 100-150 eggs which hatch in 8-18 days and the grubs start feeding on the decaying matter. The larvae pass through 3 instars to complete their development in 100-180 days depending on weather factors. Pupation takes place in a chamber at a depth of about 30 cm and the beetle emerge after 10-25 days. They remain in pupal cell for about 10-20 days before coming out of the soil. They lay eggs after 20-60 days. Beetles are active at night and attracted to a source of light. Adults can live for more than 200 days. Generally, one generation is completed in a year. Adult beetles bore into the palms at the base of the spear cluster (Figure 5.5) to consume the sap and tender parts of the leaves when it is not opened. Through the outermost petiole of the spear cluster, the beetles penetrate to the interior, leaving it permanently marked with a hole (Figure 5.5). The wedge-shaped gap in the leaf silhouette and a hole in the petiole are the common characteristic symptoms (Figure 5.5). Young palms exhibit much more severe damage at the base of the spears as compared to mature

Figure 5.5: Rhinoceros Beetle and its Damage Symptoms on Oilpalm.

palms. The damaged spindle may collapse or expanded fronds may snap off or be truncated. Adult rhinoceros beetles are found boring and chewing the male and female inflorescences even before anthesis, while the inflorescences are inside the spathe (CPCRI, 1993). The entry holes of beetles can be recognized by the presence of chewed up fibrous tissues (Wood, 1968). Secondary rotting to the bud is commonly seen due to the entry of fungi and bacteria through the injuries made to the heart of the palm by the pest. The injuries made on the petioles and female inflorescences serve as sites for egg laying of red palm weevil.

3.1.2. Management

An integrated pest management approach by incorporating mechanical, sanitational, chemical and biological aspects are required to combat the pest menace. Detection of all possible breeding sites of the pest and monitoring the beetle population on the crown of the palm are essential components of the pest management technology.

1. The beetles, which burrow deep into the crowns of young palms, are to be extracted by means of a hooked pointed metal rod (beetle hook). After extraction of the beetle, the leaf axils around the injured spindle/leaf are to be filled with mixture of Mancozeb and sterilized fine sand at a ratio of 3 g: 1 kg.

2. Prophylactic leaf axil filling is to be done to protect young palms from beetle attack. The innermost 2-3 leaf axils may be filled with a mixture of Sevidol 8 G (25 g) + fine sand (200 g) per palm during April-May, September-October and December-January. However while using the sand, care should be taken as it may carry disease multiplying organism.

3. All potential breeding sites are to be eliminated from the plantation. Those breeding sites which cannot be eliminated/destroyed are to be sprayed with Carbaryl 50 per cent WP 0.01 per cent (Nair *et al.,* 1997).

4. Since oil palm produces more than two leaves per month, the granules that are applied in the crown portion proved ineffective unless it is removed and placed in the new spindle every month. As the leaves are having spines at their bases, its found laborious and time taking. In the recent years, application of second generation synthetic pyrethroid, lambda cyhalothirn is found very effective in reducing the pest population throughout the world.

5. Indigenous predators such as *Santalus parallelus, Harpalus* sp., *Scarites* sp., *Pteropsophus occipitalis, Agrypnus* sp. nr. *bifoveatus etc.* suppress the immature stages of rhinoceros beetle. The entomopathogen, *Metarhizium anisopliae* produces epizootics in the natural population of rhinoceros beetle when the moisture levels of the breeding medium are 50 per cent and temperature is 29°C and below (conditions ideal for mycosis) (Sundarababu *et al.,* 1993).

6. Application of *Metarhizium anisopliae* (Figure 5.6) on the dead coconut logs is also proved effective in controlling the pest. The spores of the microbial organism are found viable till 9 months after preparation at room temperatures causing same mortality to grub stages (Kalidas, 2004).

7. The *Baculovirus* of *Oryctes* is one of the most successful microbial control agents employed for the biosuppression of rhinoceros beetle infesting coconut. The viral infection causes reduction in the longevity of the beetles by 40 per cent and total reduction in the fecundity. Wherever,

Figure 5.6: *Metarhizium anisopliae* **Infection on Grubs and Adults of Rhinoceros Beetle.**

the virus was introduced into the habitat of the pest, an initial epizootic decimated the larval and beetle populations resulting in drastic reduction in the pest incidence and crop damage (Pillai *et al.*, 1993). Release of the infected beetles is the most economical, effective and easy method for dissemination of the viral inoculum into the natural population of the beetles (Dileepan, 1994).

8. Pheromone traps: Rhinolure/oryctalurte sachets placed in bucket vane trap's and kept in the oil palm plantation at a height of 10 ft. @ one trap/2 ha is very effective in trapping the floating population. However high temperatures and low humidity are detrimental for the pest attraction to the pheromone (Kalidas, 2004a).

3.2. Bag Worm, *Metisa plana* (Lepidoptera: Psychidae)

Bag worm, *Metista plana* is observed in moderate levels with more than six case worms per leaflet in Andhra Pradesh state all the year round. However in Karnataka the incidence is heavy during monsoon and winter months. In Tamil Nadu psychid infestation is not reported so far. Cocoa (*Theobroma cacao*), coconut (*Cocos nucifera*), areca nut (*Areca catechu*) and areca palm (*Dypsis lutescens*) are observed as collateral hosts for the pest. Nine species of caseworms infesting oil palm have been recorded in India. *Metisa plana, Manatha albipes, Crematopsyche pendula* are the common species observed. The pest is observed endemic in most of the gardens with the presence of all the stages at any date of observation. Psychid incidence is observed during of July to March leaving the summer months with a per cent incidence varying 0.39 to 8.8 in different age group of palms. Young gardens of less than 5 years recorded less incidence compared to middle aged ones. Lack of penetration of sunlight into the aged gardens due to overlapping/intermingling of leaves of adjacent palms is observed as the main reason for the increased incidence compared to young gardens where such conditions are lacking. Natural mortality of psychid that was observed during summer and rainy period could be correlated to uncongenial abiotic conditions like high temperatures during pre monsoon period, late starting of monsoon and heavy rains during monsoon period. Although the pest maintains congenial conditions suited for its growth and development in the bag by regulating the required temperature and humidity, during rainy season it could not maintain dryness and during hot summers, it is unable to maintain the required humidity leading to mortality of the pest.

3.2.1. Biology and Nature of Damage

Female adults are wingless and found to lay 60-80 eggs in the pupal cocoons concealed in bags. Sometimes, the fecundity is relatively high. The egg incubation period is 16 days. The larvae normally takes 100-125 days before pupating. Seven larval instars in case of females and five in case of males are observed. Both larvae and adults are black in colour with pectinate antennae. As the caterpillars mature, they turn around in the bag, changing from a feeding position with their heads oriented toward the plant surface. The pupal stage lasts for 26 days. The first instar larvae after hatching from eggs are naked without any bag. Immediately after hatching they start feeding on the mother's bag and thereby form its own bag.

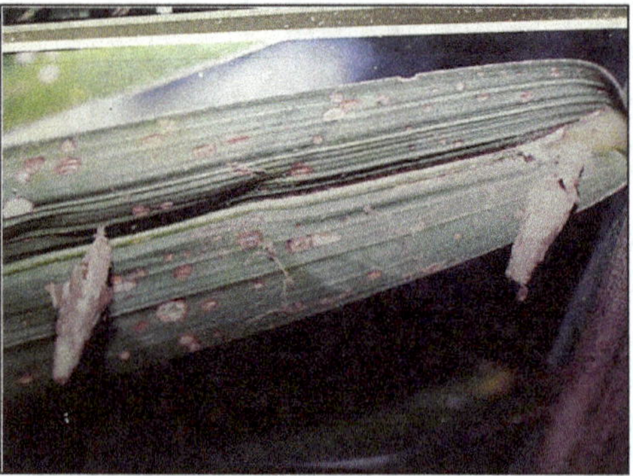

Figure 5.7: Psychid Damage on Oil Palm Leaves.

Caterpillars inhabit in a case throughout their development. The case/bags have a shape and appearance characteristic for each species. The males are winged and capable of flying. The pupae are simple cocoons and are found inside the bags.

The early stage caterpillars scarify the abaxial surfaces of fronds. Caterpillars of later instars chew the entire leaf tissue, making holes, finally feed at the leaf margins, causing notches. In addition to the holes and notches caused by bagworm feeding, dried necrotic areas are formed where they fed on the surface. Bagworms attack the middle and older fronds of the palms, younger fronds usually remaining free of damage. Damage by psychids is severe only when the population is very high. There is progressive necrosis and eventual skeletonisation, which adversely affect the yield of the palm. The pest is found causing a loss of 3.88 per cent to photosynthetic area per palm by means of defoliation.

3.2.2. Management

1. Cutting and burning the badly affected and dried leaves having insect stages at regular interval bring down the population.

2. Three species of parasitoids namely *Goriphous bunoh* infesting pupal stages and *Brachymeria* spp. and *Dolichogenidea metesae* infesting larval stages of the pest are recorded in majority of the plantations (Kalidas, 2012 and Ricardo *et al.*, 2012). Impact of parasitism on the pest population is observed more in the later stages of the pest. *Brachymeria* spp. is observed as the main parasitoid causing more than 40 per cent parasitism to larval stages. The average per cent parasitism is observed at 33.15 per cent with maximum parasitisation (65.23 per cent) during the month of November while minimum (15.83 per cent) during December.

3. In severe case of infestation, application of lambda cyhalothirn 0.05 per cent is recommended as aerial spraying. Spraying should be timed to coincide with the maximum occurrence of young larvae, which are more sensitive to insecticides. If the palms attain too tall and not accessible for spraying, stem injection or root feeding may be carried out using systemic insecticides like monocrotophos or imidacloprid. Application of granular

insecticides like phorate or carbofuran @ 100-150 g per palm in basin may also be effective if properly applied and irrigated.

3.3. Leaf Web Worm, *Acria meyricki* (Lepidoptera: Depressaridae)

The pest was first recorded during the winter months of 1995-96 (Kalidas and Rethinam, 1998). The occurrence was observed erratic for a decade confining to few gardens in alternate years. Since 2005 onwards, it has become a regular pest occurring every year in Krishna, East and West Godavari districts of Andhra Pradesh. The pest has become endemic in some of the areas where the palms attained tallness and the leaves of adjacent palms are intermingled creating congenial climate for the pest development. The infestation was further aggravated in those orchards where the palms were given basin as well as flood irrigation with excess quantity than the required. A yield loss of up to 34 per cent is reported for leaf web worm, *A. meyricki*. It is observed as occasional and sporadic pest only during the winter months. As the temperature increases the pest disappears. The incidence is generally observed from October to April. It is found that larval population was significantly negatively correlated with weather parameters like maximum temperature, minimum temperature, mean temperature and non significantly with relative humidity and rainfall (Kalidas, 2004a; Shashank *et al.*, 2015).

3.3.1. Biology and Nature of Damage

A female moth lays about 62.5 eggs in groups (in confined condition). The incubation period ranges from 4 -6 days with an average period of 4.7 days. The percentage of egg hatchability is 95.6 with ranging from 92.8-100 per cent. The larval stage pass through 6-7 instars in a period of 20.7 days. When the larva is full grown, it stops feeding and reduce in size and enter to pre-pupal stage, which lasts for 1 day. The pupal stage lasts for 5.8 days. Adult lives for about 5.4 days. The total life period from egg to adult stage is ranging from 30 -44 days (Saravanan *et al.*, 2012). The caterpillar is the damaging stage which stays inside a web on the underside of the leaves. Early instars scrap the leaves and later instars cause defoliation. Initially they are found feeding on the older leaves causing heavy defoliation. After complete defoliation of the lower leaves the caterpillars migrate to the next upper leaves. Due to severe infestation, the leaflets were dried and give burnt up appearance (Figure 5.8). Upon disturbance, the larvae hang on silken threads and spread to either nearby palms or retrieved back and feed on the same leaves. Intermingling of palm fronds in the garden paves way for easy spread of infestation. During severe infestation,

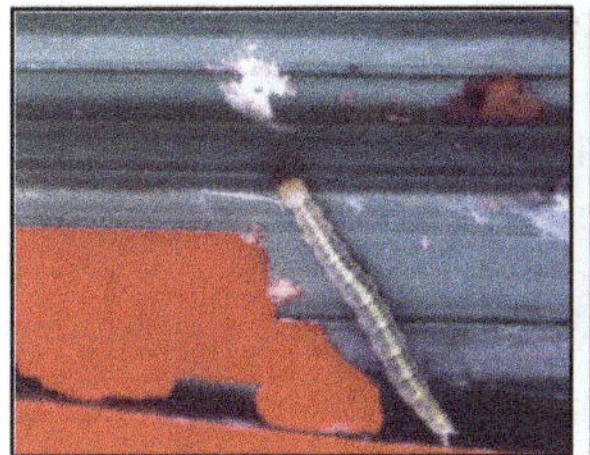

Figure 5.8: Leaf Webworm, *Acria meyerkii* and Infestation Symptoms on Oil Palm.

the larvae are found feeding on cocoa, *Theobroma cocoa*, coconut, *Cocos nucifera* and ornamental arecanut, *Areca catechu*, banana and some grasses in the gardens. The infestation on cocoa is normally seen when it is intercropped with oil palm.

3.3.2. Management

1. The lower fronds having pest stages are to pruned at the beginning of the pest activity period and burn them.

2. Under field conditions, larvae are found parasitized by two biocontrol agents *viz., Apanteles hyposidrae* Wilkinson (Hymenoptera: Ichneumonoidea : Braconidae) and *Elasmus brevicornis* Gahan (Hymenoptera: Chalcidoidea: Eulophidae). The per cent parasitism on larvae by Elasmids and Braconids are in the range of 21.41 to 36.58. whereas pupa are parasitized by *Brachymeria albotibialis* (Ashmead) (Hymenoptera: Chalcidoidea: Chalcididae) and per cent parasitism ranging between 21.21 to 79.75.

3. Pest is effectively controlled with the application of the microbial organisms namely *Beauveria bassiana*, *Metarhizium anisopliae* and *Lecanicillium lecanii*. Of these *Beauveria bassiana* check the pest population effectively.

4. Aerial spraying of quinalphos 0.05 per cent or lamda cyhalothrin 0.02 per cent twice during pest activity period at 15 days interval is proved effective in controlling the pest.

5. Stem injection with monocrotophos 25 ml/100 ml of water per palm is also effective in case of tall palms.

3.4. Slug Caterpillar *Darna catenatus* Snellen, (Lepidoptera: Limacodidae)

Incidence of slug caterpillar was first reported on oil palm in 2002 in few plantations of West Godavari and Krishna districts of Andhra Pradesh. Sporadic and erratic occurrence is noticed in oil palm gardens. The pest was reported as major problem in both Malaysia and Indonesia where the large hectares of Oil palm were infested with it causing heavy defoliation (Syed and Saleh, 1998). The pest was also observed feeding on the adjacent coconut plants but at low levels and probably migrated to the oil palm due to the availability of lush green leaves. High summer temperatures were however not found affecting the pest build up but high relative humidity was found to be important factor for natural suppression.

3.4.1. Biology and Nature of Damage

Each female lays a total of about 250-350 eggs. The eggs are laid in row on the abaxial surfaces of the more mature fronds, often near the tips of leaflets. These eggs hatch in about a week. The caterpillars are green with urticating spines. The first instar caterpillars feed only on the epidermis, forming translucent window-like areas. Caterpillars of later instars feed from the margin of the lamina inward leaving the mid vein. The larva takes 3-7 weeks for development. Caterpillars crawl down to the base of the trunk or among herbaceous vegetation to spin cocoons and pupate. Pupal period lasts for 2 to 4 weeks. Caterpillars are the damaging stage and found feeding on the leaf lamina causing heavy defoliation leaving only midribs.

Figure 5.9: Slug Caterpillar, *Darna* spp. and Infestation Symptom on Oil Palm.

The incidence is very heavy on the lower whorl leaves making them completely dried (Figure 5.9). Intensity of the pest could be assessed based on the presence of fecal droppings on the ground level/cover and gnawing sound during severe outbreak. The larvae are found to have setae on the dorsal side of the body causing irritation to handle. Presence of bluish tinge on the dorsal side of the caterpillar is the conspicuous identification mark of the pest. Apart from oil palm, incidence was also observed on cocoa and maize that are grown within or adjacent to oil palm. The caterpillars are characterized by their stinging spines which can cause nettle-rash on contact with the skin. Young caterpillars scrape strips of epidermis and as they become mature, generally commence feeding near the tips of pinnae, leaving only the midrib (Wood, 1968). During severe outbreak, the fronds are severely defoliated leaving midrib alone.

3.4.2. Management

1. The lower fronds having pest stages are to be pruned at the beginning of the pest activity period and burn them.

2. Natural suppression of the pest is observed in few plantations due to virus infections. Viral infection to the caterpillars arresting the moulting process is observed in severely infested gardens.

3. Application of microorganism, *Beauveria bassiana* (10^{-1}) that caused white muscardine disease to the caterpillar proved effective but time taking compared to chemical insecticides.

4. Under natural conditions, large number of parasitoids of Eulophidae, Chalcididae, Braconidae, Tachinidae, and Bombyliidae and predators of Hemiptera and Pentatomidae families regulating the host populations (Mariau, 1999).

5. Spraying of carbaryl 50 per cent WP at 0.1 per cent is recommended for the control (Nair *et al.*, 1997).

6. Aerial spraying of quinolphos 0.05 per cent or lamda cyhalothrin 0.05 per cent twice during pest activity period at 15 days interval can effectively control the pest.

7. Stem injection with monocrotophos 25 ml/100 ml of water per palm is also effective when the palms attained height.

3.5. Coccoids (Scales and mealybugs)

It is to mention that *Palmicultor palmarum* (Ehrhorn) and *Nipaecoccus nipae* (Maskell) are observed on *Elaeis* in Ecuador. *Ferrisia virgata* (Cockerell), *Dysmicoccus cocotis* (Maskell), *Pseudococcus cryptus* Hempel in the neotropical region. Two species of mealybugs, *Dysmicoccus brevipes* Cockerell and *Rhizoecus americanus* Hambleton, are found causing damage on oil palm in Colombian nurseries. Some species live on the roots of *Elaeis,* such as *Dysmicoccus brevipes* Cockerell in Ecuador (Mariau, 2001). In India, *Pseudococus citricutus* Green, *Palmicultor* sp. and Margarodids, *Icerya aegyptiaca* (Douglas) are reported infesting the spear leaves of oil palm seedlings in the nursery and field planted young oil palm seedlings. *Dysmicoccus brevipes* infests the pre anthesising male and female inflorescences and also unripe and ripe oil palm fruits (Dhileepan and Jacob, 1996; Ponnamma, 1999; Kalidas, 2016). Mealybugs on oil palm plants are migrants from other palmaeceae palms. Mealybugs are found in moist, warm climates. The mealybugs are occasional pests and are commonly seen in the plantations but outbreaks do not usually occur which could be due to the action of natural enemies like coccinellid beetles. Whereas poor hygienic conditions/sanitation practices that are being followed in the gardens are the major criteria for endemic infestation.

3.5.1. Nature and Symptoms of Damage

Coccids (soft scales) are found on the leaves of oil palms of all ages. Diaspids (armoured scales) are commonly found on the fruit bunches and leaves of oil palm. Both nymphs and adult females suck sap from the tender spear leaves, inflorescence and fruits. Attack by Diaspids results in appearance of chlorotic spots on leaf tissues. These pests secrete a waxy cover, which hardens to form a tough armour. The shape of the armour varies, being circular as in *Aspidiotus,* elongated/coma shaped/thread-like as in *Pinnaspis*. The females live in one place for all their lives and are often protected by ants. *Pseudococus citricutus* Green, *Palrnicultor* sp. and Margarodids, *Icerya aegyptiaca* (Douglas) infest the spear leaves of oil palm seedlings in the nursery and field planted young oil palm seedlings. Both *Pseudococus* and *Palmicultor* species attack on spindle leaves of young plants results into the yellowing of unfolding leaves and stunted growth of the palm. Although continuous feeding causes mottling appearance on leaves, ring spots were not observed so far. *Dysmicoccus* spp. infests the oil palm fruits of Fresh Fruit Bunches (FFB) by sucking the sap from mesocarp. These pests feed on plant sap and excrete honey dew which will attract ants and sooty mould development. Since the pest is found feeding only on ripe FFB when the harvest is delayed, quantification of loss by means of these pests has not been done so far. It is also observed that the pest attack leads to loosening of the fruits that lead to premature fruit drop.

3.5.2. Management

1. Since mealybugs are often carried by ants, elimination of the pest can easily be done by control of ants and keeping hygienic conditions in the garden.

Poor hygienic conditions/sanitation practices are the major criteria for endemic infestation. Leaf pruning and weeding at regular intervals are found to keep the plantation free from the pest attack.

2. Ladybird beetles are the most important predators of mealybugs.

3. Infestation of mealybugs can be controlled by spraying with phosphamidon or dimethoate at 0.05 per cent or methyldemeton at 0.025 per cent. Scale insects can be controlled by spraying with malathion 0.1 per cent (Nair *et al.*, 1997).

3.6.. Termites (Termitidae : Isoptera)

Termite infestation is noticed on seedlings maintained in polybags and also on spear leaves, male and female inflorescences and on field planted young seedlings. As majority of the oil palm growing areas are comprised of red soils, termite incidence is also found more in all the oil palm growing area. However they are found to be good decomposers causing fast decomposition of oil palm leftovers like pruned leaves, empty and rotten fruit bunches and male inflorescences.

In Karnataka, two species of termites, *Pericapritermes* sp. and *Hypotermes* sp. feed on the roots of the seedlings maintained in polybags, resulting in stunted growth of the seedlings, Infestation is mainly noticed in oil palm plantations without adequate irrigation (Dhileepan, 1992). This is noticed as severe in plantations of Tamil Nadu where most of the palms were showing symptoms of earthen sheathing over the stem position. In Malaysia, termite infestation in oil palm, especially in peat soil, has become a serious problem due to rapid expansion of the industry. In India *Odontotermes* spp. is the commonly found pest in all the oil palm growing areas (Kalidas *et al.*, 2002; Kochubabu and Kalidas, 2004; Kalidas and Lavanya, 2014).

3.6.1. Nature and Symptoms of Damage

The pest was found observed feeding on the trunk as well as on the other dried material like leaf butts, male inflorescence and dried/bunch rot and bunch failure and diseased FFBs (Figure 5.10). Infestation was observed at initial levels during the winter period and reached to peak by the end of March-April. Though they are found feeding only on the dried and useless things and causing no deathblow to the palms, however it makes the palms to appear ugly and indicates the poor maintenance of the orchards. The earthen sheathings that are formed over the foraging areas are made up of subsoil containing high amounts of potash are beneficial in enhancing the soil fertility (Kalidas, 1986; Kalidas and Veeresh,1989 and 1990). Termites are found to cause mortality to the young nursery seedlings by feeding on the collar portion of the plant with characteristic earthen sheathings on the stem portion. Poorly managed nursery and yielding plants are more prone to termite attack. Adult yielding palms are also found dead due to termite incidence in completely neglected plantations. Though termite activity in irrigated oil palm plantations is not causing any yield losses, it gives ugly look to the plantation. Apart from this, the termite mounds present in the plantation may become nests for reptiles causing panic to the working labour.

Figure 5.10: Termite Incidence on Adult Oil Palms.

3.6.2. Management

1. Application of Entomopathogenic fungi such as *Beauveria bassiana* and *Metarhizium anisopliae* have shown great potential for the management of various insect pests (Inglish *et al.*, 2001).

2. Drenching with chlorpyriphos 0.05 per cent is recommended (Nair *et al.*, 1997).

4. Vertebrate Pests

4.1. Avian Pests

Birds and rodents are the major pests of oil palm in the oil palm growing countries of the world. Both of them feed on the mesocarp of Fresh Fruit Bunches (FFB) and cause direct losses on yield. Several species of birds cause extensive damage to oil palm fruits. Birds such as crows (*Convus splendens protegatus; Corvus macrorhynchus cuiminatus*), Mynah (*Acridotheres tristis*), Babbler (*Turdoides affinis affinis*), Parrots (*Psitticula krameri manillensis*) feed on the mesocarp of fruits causing an estimated fruit loss up to 2.8 t/ha/year. Of these, Indian myna, Jungle crow, house crow and parakeets cause heavy fruit loss. The attack is observed throughout the year round and no seasonal variation in damage intensity is evident (Dhileepan, 1989). Crow pheasant, pariah kite, white-headed babbler and the large pied wagtail occasionally feed on oil palm fruits. Infestation is more (76 per cent) in ripe bunches compared to unripe (5.6 per cent) bunches. Among the ripe bunches, Duras are more susceptible (84 per cent) compared to Teneras (63 per cent). However damage to Pisifera bunches is uncommon which is due to low population of palms and poor fruit setting due to sterility. In Tamil Nadu, observations of high incidence (>20-30 per cent) of avian pests were recorded in Trichy and Karur districts where oil palm is cultivated in isolated patches. Incidence of myna was observed heavy (20-30 per cent) compared to crow. It is observed heavy incidence of bird damage in the oil palm gardens of isolated areas with 100 per cent damage even after the bunches are covered with oil palm leaves. The bird activity is more in Oil palm gardens during the rainy and summer seasons since no other food material is available to them during the above periods.

4.1.2. Nature and Symptoms of Damage

Feeding damage by birds is specific feeding exclusively on mesocarp leaving only fibers on the seeds. Damage by birds is either partial or complete. In partially damaged fruit, 40-50 per cent total weight of individual fruits is eaten away by birds. In bunches with total fruit damage, weight loss of 68-73 per cent can be observed. In many ripe fruit bunches, all the fruits are lost resulting 100 per cent loss of fruit weight. Partial fruit damage is more common during the initial stages of the fruit ripening (130-150 days old) whereas the complete fruit damage is seen with the progress in ripening of fruits progressed. Fruit loss is very high in fully ripened bunches of 160-180 days.

The damage due to birds is higher palms in border area (24.8 per cent) compared to interior of plantations (11.4 per cent). The average weight of fruits lost per bunch due to bird damage was reported at 2.3 kg in the border palms and 1.3 kg in the interior palms. In each harvested bunch an average of 1.8 kg corresponding to 4 per cent mesocarp was lost due to bird damage. The loss estimate due to these pests is 30 per cent in Malaysia. In India it is estimated to be a loss of 2.8 tonnes of Fresh Fruit Bunch (FFB) per ha per year, which is on par with 420 kg of palm oil (Dhileepan,1990).

4.1.3. Management

1. The ripe fruit bunches after 150 days of fruit set are to be covered with wire net of 1.25 cm mesh (60 x 90 cm size), reed baskets, plaited coconut leaf baskets or oil palm leaves to avoid bird damage (Dhileepan and Jacob, 1996).

2. Covering the bunches with oil palm leaf tips and tying with a piece of rope to keep them firm and impenetrable by the bird beak is found to be effective and cheap.

3. Tying nylon fishnets of 9 X 1 mtr size in between two palms is found to be most effective to manage the menace (Figure 5.11). Nets having 5 sq cm size holes are best fit to trap all the birds. Nylon. An average of five nets per ha could give maximum benefit and are found optimum to make the plantation free from bird infestation. Green and violet coloured fishnets are more effective in trapping more number of birds and thereby reducing the per cent infestation to zero within one month of implementation (Kalidas and Vasudevarao, 2005).

4. Use of sticky glue and glue traps are not found effective in reducing the incidence (Kalidas, 2006).

5. Tying the dead eagles in the periphery of the orchard to scare the birds entering inside the garden is commonly seen practice which is unique. However it is not effective.

4.2. Mammalian Pests

Black rat (*Ratus rattus wroughtoni*), House rat (*Rattus rattus rufescens*) Lesser bandicoot (*Bandicota bengalensis*), Larger bandicoot (*Bandicota indica*), Indian gerbil

Figure 5.11: Management Practices for Avian Pest Menace.
(a) Bio acoustic play, (b) Use of aterror, (c) Use of wire mesh and (d) Use of fishnet.

(*Tatera indica cuvieri*), Western Ghat squirrel (*Funambulus tristriatus*) and Procupines (*Hystrix indica*) attack oil palm at various stages of its development.

4.2.1. Nature and Symptoms of Damage

Burrowing rat, *Tatera indica* (Hardwicke) was found attacking the young oil palm plants by migrating from the adjacent maize fields as well as the forest plants. They burrow down to the bole region by making cavities to feed the cabbage tissue resulting into the wilting of leaves and mortality of the palms. Black rat, *Rattus rattus* Linn. was mainly observed feeding on the immature fruit bunches of 2-3 months old. They were found attracted to feed on the apical part of mesocarp and kernel portion of the fruit which is semi solid. The symptoms of attack depict as half cut of the fruits.

In mature palms rats eat the ripe and unripe bunches and gnaws the exposed pericarp of unripe and ripe fruits (Kalidas, 2002). Rodents feed on the pericarp with their incisors leaving characteristic gnawing marks on fruits. Rats also destroy spikelets of the male inflorescences while feeding on the grub and pupae of pollinating weevils. Wild boar (*Sus scrofa*) digs up newly planted seedlings and chew them up. They also eat away the fruits from the bunches on the tree when they are accessible.

4.2.2. Management

1. Damage to young seedlings can be prevented by placing barriers consisting of 1.25 cm mesh (Chickenwire mesh) collars around their base. They must be tightened around the palm and well fastened down to prevent the rats getting inside or underneath the ground.

2. Traps such as iron live traps, deathfall trap, bow trap *etc.* may be used as an integrated approach to minimise the rodent damage to oil palm.

3. Baiting with zinc phosphide and bromadiolone are found effective against rat menace. Baiting with zinc phosphide using banana leaves as packets with hand gloves proved 33 per cent more effective in managing the rat problem compared to used news paper without hand gloves.

4. A local wild boar scaring device has been developed to scare away wild boar from entering nurseries and young oil palm plantations. The plantation border is fenced with 18-gauge G.I. wire at 20 cm height on two lines parallel to the ground, supported on poles and kept in position with the help of guide hooks. The poles are positioned at 3 to 10 m spacing depending upon the terrain of the land. Junction boxes are made with the help of 4 poles, two crushing slabs, the two oval plays and cracker. This may be spaced at 5 to 15 meters apart depending on the landscape boundaries, roads, *etc.* The two fencing lines arriving at the junction boxes from opposite sides are joined on to the oval plays and pulled closer and held in position with the help of a crushing slab hung from a third play kept on the first two plays. Underneath this crushing slab a cracker is kept. When the animal hits the fence, it will cause the first plays to pull apart resulting in the fall of the crushing slab with the cracker on to the second crushing slab kept directly underneath, making the cracker burst. The method has been found very effective in scaring away the animals (Jacob, 1993; Nair *et al.*, 1997).

References

CPCRI. (1993). *Annual Report for 1992-93.* Central Planta-tion Crops Research Institute, Kasaragod, India.

Cheong., Y.L., Sajap, A.S., Hafidzi, M.N., Omar, D. and Abood, F. (2010). Out breaks of bagworms and their natural enemies in oil palm, *Elaeis guineensis* plantations at Huang Melintang, Perak, Malaysia. *Journal of Entomology.* 7: 141-151.

Dhileepan, K. (1988). Incidence and intensity of rhinoceros beetle infestation in the oil palm plantations in India. *Journal of Plantation Crops,* 16: 126-129.

Dhileepan, K. (1989). Investigations on avian pests of oil palm, *Elaeis guineensis* Jacq. in India. *Tropical Pest Management* 35(3): 273-377.

Dhileepan, K. (1990). Trials on the protection of Oil palm fruit bunch from bird damage in India. *The Planter* 66: 171-177.

Dhileepan, K. (1991). Insects associated with Oil palm (*Elaeis guineensis*) in India. *FAO Plant Protection Bulletin.* 39: 94-97.

Dhileepan, K. (1992). Insect Pests of oil palm (*Elaeis guineensis*) in India. *The Planter,* Kuala Lumpur 68: 183-191.

Dhileepan, K. (1994). Impact of release of Baculovirus oryctes into a population of *Oryctes rhinoceros* in an oil palm plantation in India. *The Planter* 70: 255-266.

Dhileepan, K. and Jacob, S.A. (1996). Pests. In: *Oil palm production technology.* pp. 49-58. CPCRI, RC, Palode.

Dhileepan, K., Nair, R.R. and Leena, S. (1990). *Aspergillus candidus* Link, as an entomopathogen of spindle bug, *Carvalhoia arecae* M and C (Miriidae; Heteroptera). *The Planter.* Kuala Lumpur 66: 519-521.

Inglis, G. D., Goettel, M.S., Butt, T.M and Strasser, H. (2001). Use of Hyphomycetous fungi for managing insect pests. In. 'Fungi as biocontrol agents: progress, problems and potential' (eds. T.M. Butt, C. Jachson, N. Magan). CABI Publishing, Wallingford, UK. pp. 23–69

Jacob, S.A. (1985). *Spindle Bug of Arecanut Extension folder No 10,* Central Plantation Crops Research Institute, Kasaragod, Kerala, India.

Jacob, S.A. (1993). A simple device for scaring away wild boar (*Sus scrofa*) in newly planted oil palm fields. *The Planter* 69(811): 475-477.

Jacob, S.A. and Kochu Babu, M. (1995). *Sesamia inferens,* a new pest of oil palm seedlings. *The Planter,* Kuala Lumpur 71(831): 265-266.

Kalidas, P. (1986). Studies on termites, *Odontotermes* spp. with special reference to their role in the fertility of soil. Ph.D. thesis, UAS, Bangalore. p.123

Kalidas, P. and Veeresh, G.K (1989). Plant growth in mound soils of *Odontotermes* species. *Indian Journal of Agricultural Sciences* 59(1): 8-10.

Kalidas, P. and Veeresh, G.K. (1990). Effect of Termite foraging on soil fertility. In. *'Social Insects and the Environment'.* Proceedings of the 11th International Congress of International Union for the Study of the Social Insects (IUSSI), 5-11th August, 1990, Bangalore, India. Oxford and IBH Publications, pp. 608-609.

Kalidas P. (2002), Incidence of leaf eating caterpillars on oil palm and strategies to be adopted for their management. PLACROSYM XV, 10-13th December, 2002, Mysore. Programme and Abstracts p. 100.

Kalidas, P. (2004). Effect of abiotic factors on the efficiency of rhinoceros beetle pheromone oryctalure in the oil palm growing areas of Andhra Pradesh. *The Planter* 80(935): 103-115.

Kalidas, P. (2004). Stress management of insect pests on Oil palm, *Elaeis guineensis* Jacq. *Journal of Oilseeds Research* 21(1): 220-223.

Kalidas, P. (2004). Susceptibility of Oil palm (Tenera hybrid) planting material for rhinoceros beetle, *Oryctes rhinoceros* L. *Journal of Plantation Crops* 32(Suppl.): 385-387.

Kalidas, P. (2006). Avian pest menace in oil palm and the impact of novel technologies on its management. *Journal of Plantation Crops,* 34: 172-178.

Kalidas, P. (2011). Strategies on pest management in oil palm. In. 'Recent trends in Integrated Pest Management' (Eds. A.K.Dhawan, Balwinder Singh, Ramesh Singh and Manmeet Brar Bhuller). Indian Society for the Advancement of Insect Science. pp. 177-185.

Kalidas, P. (2012). Pest Problems of Oil palm and Management Strategies for Sustainability. *Agrotechnology*. S11:001. doi:10.4172/2168-9881.S11-001Meilke, 2012.

Kalidas, P. (2016). Oil palm. In: Mealybugs and their management in agricultural and horticultural crops (eds) M. Mani and C. Shivaraju. Springer India, 569-571 pp.

Kalidas, P. and P. Lavanya, P.(2014). Termite management of in oil palm ecosystem using microbial organisms. *Indian Journal of Soil Biology and Ecology* 35(1 and 2): 257-265.

Kalidas P., Saravanan, L. and Deepthi, K.P. (2011). Incidence of *Silvanus* sp. (Silvanidae: Coleoptera) in oil palm plantations in Andhra Pradesh, India: A study. *The Planter*, Kuala Lumper. 87: 15-20.

Kalidas, P. and Rethinam, P. (1998). Incidence of leaf web formers on Oil palm, *Programme and Abstracts, National Seminar on Oil palm Research and Development in 21st Century*, pp. 31.

Kalidas, P. and Vasudevarao, V. (2005). New technology for the management of bird menace in Oil palm. Paper presented in the *National Seminar on Research and development of Oil palm in India*, 19-20th Feb.,05 held at NRC for Oil palm, Pedavegi. p. 173.

Kalidas, P., Ramprasad, K.V. and Rammohan, K. (2002). Pest status in irrigated oil palm orchards of coastal areas of India. *Journal of Indian Society for Coastal Agricultural Research*. 20(1): 41-50

Kalidas, P., Ch.Venkateswara Rao, Nasim Ali and Kochu Babu, M. (2006). New pest incidence on oil palm seedlings in India- A study of black slug (*Laevicaulus alte*). *The Planter*, 82(960): 181-186.

Kochu Babu, M. (1989). Spear rot of oil palm (*Elaeis guineensis* Jacq.) in India. *Journal of Plantation Crops*, 16 (suppl.): 281-286.

Kochu Babu, M. and Kalidas, P. (2004). Key pests and diseases of oil palm in India: Their biology, epidemiology and method of control. *Proceedings of the International Conference on the pests and diseases of importance to the Oil palm Industry* held in Malaysia during 18-19th May, 2004. pp. 184-206.

Mariau, D. (2001). The fauna of oil palm and coconut - Insect and mite pests and their natural enemies. CIRAD, France, p. 213.

Nair, C.P.R. and Daniel, M. (1982). Pests. In: *The Arecanut Palm, A monograph* (Eds. Bavappa, K. V.A., Nair, M.K. and Premkumar, T.) Central Plantation Crops Re-search Institute, Kasaragod, India.

Nair, C.P.R., Daniel, M. and Ponnamma, K.N. (1997). Integrated Pest Management in Palms (Eds. Nambiar, K, K. N. and Nair, M. K.), Coconut Development Board; Kochi, 30p.

Norman, K. and Basri, M.W. (2007). Status of common insect pests in relation to technology adoption. *The Planter*, 83: 371-385.

Pillai, G.B., Sathiamma, B. and Danger, T.K. (1993). Integrated control of Rhinoceros beetle. In: *Advances in Coconut Research and Development* (Eds Nair, M.K., Khan.H.H., Gopalasundaram, P. and Rao, E.V.V.B.) Central Plantation Crops Research Institute, Kasaragod, pp. 58-65.

Ponnamma, K. N. (1999). Coccoids associated with oil palm in India – A review. *The Planter* 75(882): 445–451.

Ponnamma, K.N., Lalitha, N and Sajeeb Khan, A. (2001). Oil palm mesocarp waste a potential breeding medium for Rhinoceros beetle, *Oryctes rhinoceros* L, *International Journal of Oilpalm*, 2(1): 37-40

Ricardo, S. T.Rafel, C.R., Marcelo, T.T., Evaldo, F.V., Walkymario, D.P.L. and Jose, C.Z. (2012). *Brachymeria* spp. (Hymenoptera: Chalcididae) parasitizing pupae of Hersperidae and Nymphalidae (Lepidopera) pests of oil palm in the Brazilian Amazonian region. *Florida Entomologist*, 95(3): 788-789.

Saravanan, L., Kalidas, P and Phani Kumar, T. (2012). Investigations on oil palm leaf web worm, *Acria* sp.: An emerging serious pest of oil palm. Abstracts: "Global conference on Horticulture for food, nutrition and livelihood options". 28-31st May, 2012, Bhubaneswar, Orissa, pp. 233-234.

Shashank, P.R., Saravanan,L., Kalidas,P., Phanikumar,P., Ramamurthy, V.V and Chandra Bose, N.S (2015). A new species of the genus *Acria* Stephens, 1834 (Lepidoptera: Depressariidae: Acriinae) from India. *Zootaxa*. 3957(2): 226-230.

Sundarababu, P.C., Balasubramanian, M. and Jayaraj, S. (1993). Studies on the pathogenicity of *Metarhizium anisopliae* (metschnikoff) Sorokin var. *Major* Tulloch on *Oryctes rhinoceros* (L) Research Publication. Tamil Nadu Agricultural University, 1: 32.

Syed, R.A., and Saleh, H.A. (1998). Integrated Pest management of bagworms in Oil palm plantations of PTPP London Sumatra Indonesia TBK (with particular reference to *Mahasena corbetti* Tams) in North Sumatra. *Proceedings of the 1998 International Oil palm Conference, Nusa Dua, Bali, Indonesia. p.7.*

Wood, B.J. (1968). *Pests of oil palms in Malaysia and their control.* The Incorporated Society of Planters, Kuala Lumpur, Malaysia, p. 222.

2018, Pests of Plantation Crops   *Pages* **143–161**
Editors: **P. Chowdappa, Chandrika Mohan & A. Josephrajkumar**
Published by: **ASTRAL INTERNATIONAL PVT. LTD., NEW DELHI**

Chapter 6

Spice Crops

☆ *S. Devasahayam, T.K. Jacob and
C.M. Senthil Kumar*

1. Introduction

Spices are high-value products obtained from over 100 plant species belonging to diverse families and are extensively used to preserve and add flavour and taste to human food and also in cosmetics, perfumes and indigenous medicine. In recent years spices are increasingly becoming an important source of high-value compounds possessing many health benefits. Spices are also major sources of foreign exchange among agriculture commodities in the country and play a crucial role in the economy of various states where they are produced. In India, over 50 species of plants yield spices and among them black pepper (*Piper nigrum* L.) (Piperaceae), cardamom (*Elettaria cardamomum* Maton), ginger (*Zingiber officinale* Rosc.) and turmeric (*Curcuma longa* L.) (Zingiberaceae) account for considerable share of export earnings. These crops are also ideal for cultivation as intercrops in perennial cropping systems and increasing the productivity per unit area of land and adding additional income to farmers.

Various species of insects have been recorded on spice crops in India and some of them cause severe economic losses. Excessive and indiscriminate use of pesticides for the management of these pests could result in pesticide residues, mortality of pollinators, natural enemies and other non-target organisms, development of resistance and environmental pollution. It is also important that the levels of pesticide residues are kept below tolerance limits in view of the harm caused to human health. There has been a renewed interest in developing environment–friendly pest management schedules in spice crops, reflecting the global concern over pesticide misuse. The distribution, nature of damage, life cycle, alternate hosts, natural enemies and management of insect pests of black pepper, cardamom, ginger and turmeric are highlighted in this chapter.

2. Black pepper

Black pepper, one of the oldest spice known to mankind and termed as the 'king of spices', is obtained from the perennial vine *P. nigrum*. The crop is grown in about 1,23,810 ha in India mainly in Kerala, Karnataka and Tamil Nadu with a production of about 50,870 tonnes annually. Around 60 species of insects have been recorded on black pepper in India, and among them *pollu* beetle, scale insects, root mealybugs, top shoot borer and leaf gall thrips are important (Devasahayam, 2000a).

2.1. Major Insect Pests

2.1.1. *Pollu* Beetle

The *pollu* beetle (*Lanka ramakrishnai* Prathapan and Viraktamath=*Longitarsus nigripennis* Mots.) (Coleoptera, Chrysomelidae) is the most destructive insect pest of black pepper in India, especially in Kerala. The pest infestation is higher in the plains and midlands in Kerala and up to 32 per cent of the crop is lost due to the pest infestation in some plantations (Premkumar and Nair 1988). Stray infestations of the pest were noticed in Wayanad, Idukki and Kodagu districts in Kerala and Karnataka at higher altitudes during later years (Devasahayam, 1992; Josephrajkumar *et al.*, 2000).

The adult beetle feeds on tender shoots, leaves and spikes resulting in necrotic patches on tender shoots and spikes and small irregular circular holes on tender leaves (Figure 6.1). The damage caused by the larva (grub) is more serious and it bores into tender spikes and berries and feed on the internal tissues (Figure 6.1). The infested spikes develop necrotic patches and the berries turn black and crumble when pressed (the term *pollu* in Malayalam indicates the hollow nature of the infested berries) (Devasahayam, 2000a).

Figure 6.1: *Pollu* Beetle Infested Leaf and Berries of Pepper.

2.1.1.1. Bioecology

The adult beetle measures about 2.5 x 1.5 mm in size, the head and thorax being yellowish-brown and the abdomen brown; the elytra (forewings) is black. A method for identification of sex of adult beetles and the morphology of internal reproductive structures of male and female beetles have been described (Vidyasagar *et al.,* 1988; Devasahayam *et al.,* 1998). The adults are often found on the undersides of tender leaves and jump away when disturbed. Vines trailed on standards (support trees) producing heavy shade are generally heavily infested by the pest. Fully-grown grubs are creamy yellow and measure about 5.5 mm in length. The eggs are laid on the rind of tender berries and also on tender shoots and spikes. The eggs hatch into creamy white larvae in 3-8 days. There are three larval instars lasting for 20-40 days. Pupation occurs in the soil in earthen cocoons and the pupal period lasts for 6-8 days. There are many overlapping generations in a year and the population of the pest is higher in the field during September-October (Premkumar 1980; Babu 1994). Apart from an entomophagous nematode (Mermithidae), spider (Araneae) and weaver ant (Fomicidae), no other natural enemies have been recorded (Devasahayam and Koya, 1994a).

2.1.1.2. Management

Regulation of shade in the plantation by pruning branches of support and shade trees before the onset of the monsoon helps in reducing the build-up of pest population in the field. Later spraying the vines during July and October with quinalphos (0.05 per cent) or spraying quinalphos (0.05 per cent) during July followed by Neemgold (0.6 per cent) (neem-based insecticide) during August, September and October is effective for the management of the pest (IISR, 2015a). Leaf extracts of *Chromolaena odorata* L., *Azadirachta indica* A. Juss and *Strychnos nux-vomica* L., root bark extracts of *Uvaria narum* Blume and *U. hookeri* King, neem oil, neem seed kernel extract and custard apple (*Annona squamosa* L.) seed extract posses appreciable antifeedant activity against *pollu* beetle, indicating their potential in utilizing them in IPM schedules (Babu *et al.,* 1996a, b; Devasahayam and Anandaraj, 1997; Devasahayam and Leela, 1998; Devasahayam and Koya, 1999).

2.1.2. Scale Insects

Scale insects such as mussel scale (*Lepidosaphes piperis* Gr.) and coconut scale (*Aspidiotus destructor* Sign.) (Hemiptera, Diaspididae) are major insect pests of black pepper at higher altitudes in Kerala, Karnataka and Tamil Nadu, especially in Thiruvananthapuram, Idukki, Wayanad (Kerala), Kodagu (Karnataka) and Nilgiris (Tamil Nadu) districts. The mussel scale encrusts main stems, lateral branches, mature leaves and berries of black pepper vines resulting in chlorotic patches, yellowing and drying of leaves (Figure 6.2). Younger vines may succumb to the pest infestation during 2-3 years when the infestation occurs on the main stem. On older vines, the infested lateral branches wilt and dry resulting in vacant spaces in the canopy. The coconut scale generally infests mature leaves and berries resulting in chlorotic patches in the infested tissues (Koya *et al.,* 1996).

Figure 6.2: Pepper Leaf Infested by Coconut Scale and Mussel Scale.

2.1.2.1. Bioecology

The adult females of *L. piperis* are elongated and dark brown measuring 3-4 mm in length. The I and II larval stages last for 9–12 and 9–10 days, respectively. In males, the pre-pupal and pupal stages last for 2–3 days each. Adult females of *A. destructor* are circular and light yellow measuring 1.5-2.0 mm in diameter. Males are weak-bodied with a pair of wings and do not live long and die after mating. The I and II larval stages last for 6–7 and 4–5 days, respectively. In males, the pre-pupal and pupal stages lasts for 6–7 days and 4–5 days, respectively. Both the species also reproduce parthenogenetically. Scale insects are observed in the field on black pepper vines throughout the year but their population is higher during the post monsoon and summer months. *A. destructor* is polyphagous and is known to infest more than 20 economically important plants in India. *L. piperis* occasionally infest cassava (*Manihot esculenta* Crantz) stems in the field and storage. Various predators and parasites have been recorded on scale insects in the field among which *Chilocorus* spp., *Pseudoscymnus* spp. (Coccinellidae), *Aphytis* sp. and *Encarsia* spp. (Aphelinidae) are common and widespread. Methods have been standardized for mass rearing of *C. nigrita* (Fab.) and *C. circumdatus* (Gyllen.) in the laboratory. Evaluation of these predators in the field indicated that their release brought down the population of *L. piperis* and *A. destructor* (Selvakumaran *et al.*, 1996; Devasahayam, 1998; Devasahayam 2000a).

2.1.2.2. Management

Scale insects can be managed effectively when the control measures are undertaken during early stages of infestation. Spraying of dimethoate 0.1 per cent may be undertaken on affected vines after clipping off severely infested branches after harvest of berries; the spraying has to be repeated after 21 days if the infestation persists (Devasahayam and Koya, 1994b; Josephrajkumar *et al.*, 2001; IISR, 2015). Natural products such as neem oil (0.3 per cent), Neemgold (0.3 per cent) and fish oil rosin (3 per cent) are promising for the management of scale insects especially during early stages of infestation and are safer to the coccinellid predators (Devasahayam, 1998).

2.1.3. Root Mealybugs

Root mealybugs (*Planococcus* spp. and *Dysmicoccus* sp.) (Hemiptera, Pseudococcidae) are serious insect pests of black pepper at higher altitudes in Wayanad (Kerala) and Kodagu (Karnataka) districts. Colonies of root mealybugs are observed at the basal portion of the stem under the soil and also on the roots causing yellowing and wilting of leaves and lateral branches and mortality of younger vines (Figure 6.3). The pest infestation is generally seen on vines affected with *Phytophthora* sp. and nematodes (Devasahayam *et al.*, 2009).

Figure 6.3: Root Mealybugs of Blackpepper.

2.1.3.1. Bioecology

Root mealybugs are oval and soft bodied, covered with white waxy filaments and measure about 1.5 mm x 1.0 mm in size. There are three larval instars in females of *D. brevipes* and males have not been recorded. Females are viviparous and the I, II and III instars last for 7-12, 4-6 and 5-9 days, respectively. In *P. citri* there are three larval instars in females lasting for 5–11, 4–13 and 4–10, respectively. In males there are two instars lasting for 5–13 and 5–12, days, respectively. The pre-pupal and pupal stages last for 4–10 days. Root mealybugs also infests banana (*Musa* sp.) corm, colocasia (*Colocasia* sp.), turmeric (*Curcuma* sp.) and cardamom (*E. cardamomum*) rhizomes and base of stems of coffee (*Coffea* spp.) and many species of weed plants commonly growing in and around black pepper gardens. Ants are generally associated with root mealybug colonies and they play an important role in the spread of the pest in the field. (Devasahayam 2006; Devasahayam *et al.*, 2009).

2.1.3.2. Management

An integrated pest management strategy involving planting root mealybug-free rooted cuttings in the field, removal of weeds in interspaces of black pepper vines during summer, drenching tobacco extract (3 per cent) on mildly affected vines or chlorpyriphos (0.075 per cent) on affected vines and adoption of control measures against *Phytophthora* and nematode infections (Devasahayam, 2006).

2.1.4. Top Shoot Borer

The top shoot borer (*Cydia hemidoxa* Meyr.) (Lepidoptera, Tortricidae) is a serious insect pest in young vines in all black pepper areas. The larvae bore into tender terminal shoots and feed on internal tissues resulting in decay and blackening of infested shoots (Figure 6.4). Repeated infestation of tender terminal shoots affects the growth of the vine and also the establishment of newly planted vines. In some young plantations even up to 100 per cent of new shoots are infested by the pest (Devasahayam and Koya, 1993a).

2.1.4.1. Bioecology

The adult is a tiny moth with a wingspan of 10-15 mm with crimson and yellow fore wings and grey hind wings. Fully-grown larvae are greyish green and measure 12-15 mm in length. There are five larval instars. The larval period lasts for about 14 days. The pupal period lasts for 8–10 days. The pest infestation is higher during the monsoon season (June-October) when numerous new shoots are available on the vines (Visalakshi and Joseph 1965). *Hexamermis* sp. (Mermithide) and *Apanteles cypris* Nixon (Braconidae) are major natural enemies parasitizing larvae of top shoot borer in the field (Devasahayam and Koya, 1993b; 1994c).

Figure 6.4: Top Shoot Borer.

2.1.4.2. Management

Spraying quinalphos (0.05 per cent) on tender terminal shoots of young vines is effective in controlling top shoot borer infestation. The spraying has to be repeated at monthly intervals to protect emerging new shoots especially during July to October (IISR, 2015a).

2.1.5. Leaf Gall Thrips

The leaf gall thrips (*Liothrips karnyi* Bagn.) (Thysanoptera, Phlaeothripidae) is a serious insect pest of young black pepper vines at higher altitudes and also on rooted cuttings in nurseries in the plains. The pest infestation is more serious in Wayanad, Idukki, Thiruvananthapuram (Kerala), Shimoga, Kodagu (Karnataka) and Nilgiris (Tamil Nadu) districts. Leaf gall thrips infest tender leaves causing the leaf margins to curl downwards and inwards resulting in the formation of marginal, tubular galls (Figure 6.5). The infested leaves later become crinkled, malformed and reduced in size and may also turn chlorotic at the margins. Severe infestations affect the growth of younger vines and rooted cuttings in the nursery (Devasahayam, 2000b).

2.1.5.1. Bioecology

The adults are black with the distal segments of the antenna and legs light lemon yellow, and measure about 2.5-3.0 mm in length; the larva the pupae are creamy white. The two larval stages, pre-pupal stage and two pupal stages last for 4–7, 4–7, 2, 2–3 and 2–3 days, respectively. The pest population is higher in the field during June-September when numerous tender leaves are produced on the vines. Among the predators recorded on leaf gall thrips, *Montandoniola* spp. (Anthocordiae) and *Andothrips flavipes* (Phlaeothripidae) are more common and widely distributed (Devasahayam, 2000b; Yamada *et al.*, 2011).

2.1.5.2. Management

Spraying dimethoate (0.05 per cent) during emergence of new flushes on younger vines and rooted cuttings in nurseries is effective for managing the pest infestation (Devasahayam, 1990; IISR, 2015a).

2.2. Minor Insect Pests

2.2.1. Mealybugs

Mealybugs such as *Planococcus* spp., *Pseudococcus* spp. and *Ferrisia virgata* Ckll. (Hemiptera, Pseudococcidae) infest tender shoots and leaves especially on rooted cuttings

Figure 6.5: Leaf Gall Thrips.

in the nursery resulting in wilting of affected plants. The pest infestation can be controlled by spraying dimethoate (0.05 per cent) on the affected plants (IISR, 2015a)

2.2.2. Scale Insects

Scale insects such as *Marsipococcus marsupiale* Green (Hemiptera, Diaspididae) infest leaves of black pepper vines especially at higher altitudes. Adult females are brown and circular, whereas the juveniles are white and oval. *Protopulvinaria longivalvata* Green (Hemiptera, Diaspididae) is observed on mature leaves of older cuttings in the nursery. Adult females are pyriform to oval shaped and reddish brown. The pest infestation results in yellowing and wilting of affected plants and the subsequent attack by sooty mould accelerates the deterioration of infested plants. Scale insects can be controlled by spraying dimethoate (0.05 per cent) (IISR, 2015a).

2.2.3. Gall Midge

Maggots of the gall midge *Cecidomyia malabarensis* (Felt) (Diptera, Cecidomyiidae) infest tender leaf petioles, leaf veins and shoots of young black pepper vines resulting in swelling of affected tissues. The pest infestation is more common in the nursery and on young vines in the field during the monsoon season. The adults are soft-bodied flies and the maggots are reddish pink. The pest infestation can be controlled by spraying dimethoate (0.05 per cent) (IISR 2015a).

2.2.4. Semilooper

The semilooper *Synegia* sp. (Lepidoptera, Geometridae) is commonly seen in the field on younger vines during June-October. The larvae are olive green and feed on tender shoots, leaves and spikes of younger vines especially during the flushing season. The pest can be managed by hand-picking and destroying the larvae and spraying quinalphos (0.05 per cent) if the infestation is severe (IISR, 2015a).

3. Cardamom

Cardamom, known as the 'queen of spices', is the dried fruit of a perennial rhizomatous plant *E. cardamomum* and is mainly cultivated in Kerala, Karnataka and Tamil Nadu at higher altitudes. The total annual production of cardamom in the country is about 15,370 tonnes from an area of about 63,570 ha. More than 60 species of insects have been recorded on the crop in India, among which cardamom thrips, shoot and capsule borer and root grub are important.

3.1. Major Insect Pests

3.1.1. Cardamom Thrips

The cardamom thrips (*Sciothrips cardamomi* (Ramk.) (Thysanoptera, Thripidae) is the most widespread and destructive insect pest occurring in all cardamom areas and causing up to 47 per cent crop loss in unprotected plantations. The adults and larvae lacerate the tissues of shoots, panicles, flowers and immature capsules and feed on the exuding sap resulting in shedding of flowers and immature capsules and scab formation on mature capsules (Figure 6.6). The infested capsules loose their aroma and the formation of seeds is also affected (Gopakumar and Chandrasekhar, 2002).

Figure 6.6: Thrips Damage on Cardamom.

3.1.1.1. Bioecology

The adults are minute and greyish brown measuring 1.5–2.0 mm in length and also reproduce parthenogenetically. Molecular characterization of 45 populations from various regions indicated that there were no significant variations among them (Asokan *et al.*, 2013). The I and II stage larva lasts for 2–5 days, and 7–9 days, respectively. The pre-pupal and the pupal stages last for 2 days and 4-6 days respectively. The life cycle from egg to adult is completed in 27–33 days. Populations of cardamom thrips rapidly increase during the post monsoon and summer months and decline with the onset of rains. The pest population is higher in thickly shaded areas in the plantation and a number of zingiberaceous plants growing in and around cardamom plantations serve as alternate host plants. The pest infestation is higher on Mysore and Vazhukka types of cardamom (Gopakumar and Chandrasekhar 2002). The bacterial endosymbiont Wolbachia has been recorded to infect adults and larvae (Jacob *et al.*, 2015). Recently the enotmogenous fungus *Lecanicillium psalliote* (Treschew) Zare and W. Gams (Ascomycota: Hypocreales) has been identified as a potential biocontrol agent of cardamom thrips (Senthil Kumar *et al.*, 2015).

3.1.1.2. Management

Regulation of shade and removal of alternate host plants in and around the plantations helps in reducing the buildup of cardamom thrips population in the field. Since the pest generally breeds within leaf sheaths especially during summer, pruning of leaf sheaths during February-March before spraying operations also reduces the pest population and increase the efficacy of insecticides. Later five to seven rounds of spraying quinalphos (0.025 per cent), during March, April, May, August and September is recommended for controlling the pest infestation (IISR, 2015b). Recently spraying of spinosad 0.0135 per cent, (natural product) during February-March, March-April, April-May, August and September was found effective for the management of the pest under Karnataka conditions (Jacob *et al.*, 2015).

3.1.2. Shoot and Capsule Borer

The shoot and capsule borer (*Conogethes punctiferalis* Guen.) (Lepidoptera, Cossidae) infests cardamom plants in nurseries and plantations and yield losses of 70-80 per cent have been recorded in severely infested plantations. The early instar larvae bore into panicles and immature capsules, whereas the later instars bore into pseudostems and feed on the internal tissues (Figure 6.7) The presence of bore holes with extruding frass and the withered central shoot are characteristic symptoms of the pest infestation (Gopakumar and Chandrasekhar, 2002).

Figure 6.7: Shoot and Capsule Borer.

3.1.2.1. Bioecology

The adult is a medium-sized moth with a wing span of about 20 mm; the wings and body are orange-yellow with black spots. There are five larval instars; fully grown larvae are pale pinkish brown and measure 25–30 mm in length. At Thadiyankudisai (Tamil Nadu), the life cycle from egg to adult is completed in 41–57 days during summer (March–May) and 84–123 days during winter (November–February). The pest infestation is noticed throughout the year in the field but its population is higher in the field coinciding with shoot, panicle, and capsule formation stages (Varadarasan, 1991). The shoot borer is highly polyphagous and has been recorded on more than 35 host plants including several economically important plants in India (Devasahayam and Koya, 2005). Twenty species of hymenopterous parasitoids have been recorded on the shoot and capsule borer among which, *Eriborus trocheanteratus* (Morely), *Xanthopimpla australis* Kr., *Friona* sp. and *Agrypone* sp. (Icheumonidae) are important (Varadarasan,1995; Devasahayam, 1996).

3.1.2.2. Management

Removal and destruction of alternate host plants in and around the plantation, removal of infested suckers (as indicated by extrusion of frass), during September-October when the infestation is less than 10 per cent, and collection and destruction of adults which are generally observed on the undersurface of leaves, helps in reducing the population of shoot and capsule borer in the field. Later, spraying quinalphos (0.075 per cent) twice, during January-February and September-October is to be undertaken for controlling the pest infestation (IISR, 2015b).

3.1.3. Root Grub

The root grub (*Basilepta fulvicorne* Jacoby) (Coleoptera, Chrysomelidae) is a serious insect pest of cardamom in nurseries and plantations resulting in crop losses of up to 66 per cent in various regions. The larvae feed on roots and tender rhizomes and in severe cases of infestation, the entire root system is eaten away resulting in yellowing of leaves and stunting in growth of plants (Figure 6.8). The pest damage leads to secondary infection by pathogens resulting in rotting. Severely infested plants, especially seedlings, succumb to the pest attack (Gopakumar and Chandrasekhar, 2002).

3.1.3.1. Bioecology

The adults are small beetles measuring about 5.0 x 2.5 mm in size and are metallic blue, bluish green or greenish brown. The grubs are stout and 'C' shaped, pale white and measure about 1 cm in length when fully grown. The adults emerge in large numbers after receipt of

Figure 6.8: Root Grub Infestation.

pre-monsoon showers (April-May and September-October). At Idukki (Kerala), the egg, larval and pupal stages last for 8–10, 45–60 and 10–17 days, respectively. At Mudigere (Karnataka), the egg, larval and pupal stages last for 10–15, 78–80 and 20–25 days, respectively. The adults are polyphagous and feed on the leaves of a number of trees in and around the plantation such as jack, Indian almond, mango, guava, cocoa, *Ficus* spp. and *Erythrina* sp. The natural enemies recorded on the root grub include the green muscardine fungus *Metarrhizium anisopliae* (Metsch) Sorokin (Hypocreales, Clavicipitaceae) and the entomogeneous nematode *Heterorhabditis* sp. (Secementea, Rhabditida) infesting grubs, and the white muscardine fungus *Beauvaria bassiana* (Bals.) Vuill. (Hypocreales, Clavicipitaceae) infesting adults in the field (Thyagaraj *et al.*, 1991; Vardarasan, 1995; Varadarasan and Naidu, 1993).

3.1.3.2. Management

Collection and destruction of adult beetles during peak periods of emergence is effective in reducing the pest population in the field. Along with collection of adults, application of phorate 10 G @ 20–40 g per clump or chlorpyriphos (0.075 per cent) twice a year during May-June and September-October synchronizing with emergence of adults and egg laying periods is effective for the management of the pest (IISR, 2015b).

3.2. Minor Insect Pests

3.2.1. Aphids

Colonies of aphid (*Pentalonia caladii* van der Goot) (Hemiptera, Aphididae) occur within leaf sheaths feeding on plant sap. Though the direct damage caused by them is not serious, they are vectors or mosaic or *katte* disease, a major debilitating disease of cardamom. The adults are dark brown and reproduce by sexual and parthenogenetic means. Removal of partly dried and decayed pseudostems and alternate reduces the pest population in field; a spray with dimethotae (0.05 per cent) may be given if the infestation is severe (NRCS, 1989).

3.2.2. Whitefly

The whitefly (*Kanakarajiella cardamomi* David and Subramaniam) (Hemiptera, Aleyrodidae) sucks the sap from leaves especially during the dry summer months, resulting in chlorotic patches; later the leaves turn yellow and dry up. The pest infestation is more common in Idukki District and Lower Palani Hills. The adult is a small soft-bodied insect with two pairs of white wings and reproduce by sexual and parthenogenetic means; the nymphs are flat and pale greenish yellow. The pest can be controlled by spraying neem oil (0.5 per cent) (when the infestation is less) or dimethoate (0.05 per cent) (NRCS, 1989).

3.2.3. Shoot Fly

Shoot fly (*Formosina flavipes* Mall) (Diptera, Chloropidae) infestations are seen in nurseries and new plantations especially in exposed areas. The larvae feed on growing shoots of young suckers resulting in dead heart symptoms and affect the growth and establishment of young plants. The adults are small black flies with yellow patches. Removal of affected shoots along with the larvae and spraying dimethoate (0.05 per cent) or quinalphos (0.05 per cent) is effective for the management of the pest (IISR, 2015b).

3.2.4. Hairy Caterpillars

Hairy caterpillars (*Euproctis* sp. (Lepidoptera, Lymantridae), *Lenodera* sp. (Lepidoptera, Lasiocampidae), and *Eupterote* spp. (Lepidoptera, Euproctidae) belonging to various families occur sporadically in enormous numbers especially during the post monsoon season and defoliate cardamom plants. Hairy caterpillars are gregarious in habit and congregate on trunks of shade trees during day time and migrate to the cardamom plants during the night. Collection and destruction of the congregated caterpillars, luring of adults to light traps and killing them and spraying insecticides like quinalphos (0.1 per cent) are effective in controlling the pest infestation (IISR, 2015b).

3.2.5. Capsule Borer

The larvae of early capsule borer (*Jamides* sp.) (Lepidoptera, Lycaenidae) bore into flower buds, flowers and young capsules especially during the early monsoon season. The infested capsules exhibit a characteristic circular hole, turn yellowish brown and eventually decay and drop off. The adult is a swift flying butterfly with metallic blue wings bordered with a white line and black shade on the dorsal surface of wings; the hind wings have a pair of delicate tail-like prolongations. The fully grown larva is flat with dense hairs. The spraying schedule undertaken with quinalphos (0.025 per cent) for the control of cardamom thrips is effective in controlling the pest infestation (IISR, 2015b).

4. Ginger and Turmeric

Ginger (*Zingiber officinale*) and turmeric (*Curcuma longa*) are herbaeceous crops and their dried rhizomes yield the spice of commerce. India produces about 6,55,060 and 11,89,890 tonnes of ginger and turmeric respectively, annually, from an area of about 1,32,620 ha and 2,32,670 ha, respectively. Kerala, Meghalaya and Arunachal Pradesh are the major ginger producing states whereas, Andhra Pradesh, Tamil Nadu and Orissa are the major turmeric producing states in India. More than 45 species of insects have been recorded on ginger and turmeric in India and many of them are common on both the crops. Among them, shoot borer, rhizome scale and white grubs are important (Devasahayam and Koya, 2005, 2007).

4.1. Major Insect Pests

4.1.1. Shoot Borer

The shoot borer (*C. punctiferalis*) (Lepidoptera, Crambidae) is the most widespread and serious insect pest of ginger and turmeric. The larvae bore into pseudostems and feed on the internal shoot resulting in yellowing and drying of infested pseudostems. The presence of a bore hole on the pseudostem through which frass is extruded (Figure 6.9) and the withered central shoot is a characteristic symptom of pest infestation (Devasahayam and Koya, 2005, 2007).

Figure 6.9: Shoot Borer.

4.1.1.1. Bioecology

The adult is a medium sized moth with a wingspan of 18–24 mm; the wings and body are pale straw yellow with minute black spots. There are five larval instars and they last for 3–4, 5–7, 5, 3–8 and 7–14 days, respectively. Fully grown larvae are light brown with sparse hairs and measure 16–26 mm in length. The pre-pupal and pupal periods lasted for 3–4 and 9–10 days, respectively (Jacob, 1981). The pest is observed in the field throughout the crop season and its population is higher during September-October in Kerala. Various natural enemies have been recorded on the shoot borer among which mermithid nematode (*Hexamermis* sp.) and hymenopterous parasitoids (*Xanthopimpla australis* Kr. (Ichneumonidae), *Apanteles taragamae* Viereck (Braconidae) are important. The shoot borer is highly polyphagous and has been recorded on more than 35 host plants including several economically important plants in India (Devasahayam, 1996; Devasahayam and Koya, 2005, 2007).

4.1.1.2. Management

Spraying malathion (0.1 per cent) at 21 day intervals during July to October or pruning and destroying freshly infested shoots during June to August and spraying malathion (0.1 per cent) during September to October has been suggested for controlling the pest infestation on ginger (IISR, 2015c). A sequential sampling strategy for monitoring the pest infestation in a field as guidance for undertaking control measures has also been formulated (Koya *et al.*, 1986). On turmeric, spraying malathion (0.1 per cent) or lamda cyhalothrin (0.0125 per cent) during July to October at 21 day intervals is effective in controlling the pest infestation (IISR, 2001).

4.1.2. Rhizome Scale

The rhizome scale (*Aspidiella hartii* Sign.) (Hemiptera, Diaspididae) infests rhizomes of ginger and turmeric both in the field and in storage. The pest infestation is generally seen during the end of the crop season in the field and severely infested plants wither and dry. The pest infestation results in shriveling of buds and rhizomes in storage (Figure 6.10) and when the infestation is severe, it adversely affects the sprouting of rhizomes (Devasahayam and Koya, 2005, 2007).

Figure 6.10: Scale Infested Rhizome.

4.1.2.1. Bioecology

The adult female scales are minute, circular and light brown to grey and measure about 1.5 mm in diameter. Females are ovo-viviparous and also reproduce parthenogenetically. No information is available on the life history of the pest. Two species of hymenopterous parasitoids namely, *Cocobius* (*Physcus*) *comperei* (Hayat) (Aphelinidae) and *Adelencyrtus moderatus* (Howard) (Encyrtidae) are important natural enemies of the pest. The rhizome scale also infest yams, tannia and taro (Devasahayam, 1996; Devasahayam and Koya, 2005; 2007).

4.1.2.2. Management

Timely harvest and discarding of severely infested rhizomes during storage reduces further spread of the pest infestation in storage. Dipping of seed rhizome in quinalphos (0.075 per cent) after harvest and storage in dry leaves of *Strychnos nux-vomica* + saw dust in 1 : 1 proportion is effective in controlling the pest infestation (IISR, 2008; 2015c).

4.1.3. White Grubs

White grubs (*Holotrichia* spp.) cause serious damage to ginger in certain regions of North East India. The grubs feed on roots and newly formed rhizomes resulting in yellowing of leaves. The pseudostems may be cut at the basal region when the infestation is severe. The entire crop may be lost in severely infested plantations (Varadarasan *et al.*, 2000).

4.1.3.1. Bioecology

The adult beetles of *H. setticollis* that commonly occur in Sikkim, are dark brown measuring about 2.5 x 1.5 cm in size. The grubs are creamy white. The adults emerge in large numbers with the receipt of summer showers during April-May (Varadarasan *et al.*, 2000).

4.1.3.2. Management

Collection and destruction of adults during peak periods of emergence and application of the entomophagous fungus *M. anisopliae* mixed with fine cowdung is effective for the management of white grubs. However, in severely affected areas drenching with chlorpyriphos (0.075 per cent) may be necessary along with collection and destruction of beetles (IISR, 2015c; Varadarasan *et al.*, 2000).

4.2. Minor Insect Pests

4.2.1. Leaf Thrips

The turmeric thrips (*Panchaetothrips indicus* Bagn.) (Thysanoptera, Thripidae) infest leaves of turmeric and is more common in drier regions. The infested leaves roll up, turn pale and dry up. The thrips can be controlled by spraying dimethoate (0.05 per cent) (IISR, 2008).

4.2.2. Lacewing Bug

The lacewing bug (*Stephanitis typicus* Dist.) (Hemiptera, Tingidae) is more common in the drier regions and occurs in colonies and sucks sap from turmeric leaves. The infested leaves turn yellow and dry up. The adults are pale white with transparent lace-like wings; the nymphs have a large black patch on their body. Spraying dimethoate (0.05 per cent) is effective in controlling the pest infestation (IISR, 2008).

4.2.3. Leaf Beetles

Adults and larvae of leaf beetles (*Lema* spp.) (Coleoptera, Chrysomelidae) feed on turmeric leaves especially during the monsoon season and form elongated parallel feeding marks on them. The larva is pale yellow with a brown head and carries its excrement dorsally in the posterior end of its body. The control measures undertaken for the management of shoot borer is sufficient to manage this pest.

4.2.4. Leaf Roller

The leaf roller (*Udaspes folus*) (Lepidoptera, Pieridae) is more common during the monsoon season and the larvae cut and fold leaves of ginger and turmeric and feed from within. The adult is a medium-sized butterfly with black wings with a large irregular white spot on the upper side of the hind wing and several small irregular white spots on the forewing. The larva is dark green with a black head when fully grown. The control measures undertaken for the management of shoot borer is sufficient to manage the pest along with hand picking and destruction of larvae.

References

Asokan, R., Rebijith, K. B., Krishna, V., Krishna Kumar, N. K., Jacob, T. K., Devasahayam, S., Kaomud Tyagi and Sujeesh, E. S. (2013). Molecular diversity of cardamom thrips *Sciothrips cardamomi* Ramakrishna) (Thripidae: Thysanoptera). *Oriental Insects* 47: 55-64.

Babu, S. (1994). Some aspects of biology of *Longitarsus nigripennis* Mots. (Coleoptera: Chrysomelidae), a serious pest on black pepper *Piper nigrum* L. *Entomon* 19: 159–161.

Babu, P. B. S., Padmaja, V. and Hishan, A. (1996a). Isodesacetyluvaricin-insect antifeedant against *Longitarsus nigripennis* Mots. *Indian Journal of Experimental Biology* 34: 377–379.

Babu, P. B. S., Rao, J. M., Joy, B. and Sumathykutty, M. A. (1996b). Evaluation of some plant extracts as feeding deterrents against adult *Longitarsus nigripennis* Mots. (Coleoptera: Chrysomelidae). *Entomon* 21: 291-294.

Devasahayam, S. (1990). Field evaluation of insecticides for the control of leaf gall thrips (*Liothrips karnyi* Bagnall) on black pepper. *Entomon* 15: 137-138.

Devasahayam, S. (1992). Management of insect pests in black pepper: Problems and prospects. In: Y. R. Sarma, S. Devasahayam and M. Anandaraj (Eds.) *Black Pepper and Cardamom-Problems and Prospects* (pp. 48-50). Calicut: Indian Society for Spices.

Devasahayam, S. (1996). Biological control of insect pests of spices. In: M. Anandaraj and K. V. Peter (Eds.). *Biological Control in Spices* (pp. 33-45). Calicut: Indian Institute of Spices Research.

Devasahayam, S. (1998). Biological control of scale insects infesting black pepper. Final Report of ICAR Ad-Hoc Research Scheme. Calicut: Indian Institute of Spices Research.

Devasahayam, S. (2000a). Insect pests of black pepper. In: P. N. Ravindran (Ed.). *Black pepper (Piper nigrum)* (pp. 309-334). Amsterdam: Harwood Academic Publishers.

Devasahayam, S. (2000b). Bioecology of leaf gall thrips (*Liothrips karnyi* Bagnall) infesting black pepper. PhD Thesis. Calicut: University of Calicut.

Devasahayam, S. (2006). Bio-ecology and integrated management of root mealybugs (*Planococcus* sp.) infesting black pepper. Final Report of Ad-hoc Research Scheme. Calicut: Indian Institute of Spices Research.

Devasahayam, S. and Anandaraj, M. (1997). Black pepper (*Piper nigrum* L.) research. In: S. S. Narwal, P. Pauro and S. S. Bisla (Eds.) *Neem in Sustainable Agriculture* (pp. 117-122). Jodhpur: Scientific Publishers.

Devasahayam, S. and Koya, K. M. A. (1993a). Effect of top shoot borer (*Cydia hemidoxa*) infestation on young vines of black pepper (*Piper nigrum*). *Indian Journal of Agricultural Sciences* 63: 762–763.

Devasahayam, S. and Koya, K. M. A. (1993b). Seasonal incidence of hymenopterous parasites of top shoot borer (*Cydia hemidoxa*) infesting black pepper. *Journal of Entomological Research* 17: 205-208.

Devasahayam, S. and Koya, K. M. A. (1994a). Natural enemies of major insect pests of black pepper (*Piper nigrum* L) in India. *Journal of Spices and Aromatic Crops* 3: 50–55.

Devasahayam, S. and Koya, K. M. A. (1994b). Field evaluation of insecticides for the control of mussel scale (*Lepidosaphes piperis* Gr.) on black pepper. *Journal of Entomological Research* 18: 213–215.

Devasahayam, S. and Koya, K. M. A. (1994c). Seasonal incidence of *Hexamermis* sp. (Dor., Mermithidae) parasitising larvae of top shoot borer *Cydia hemidoxa* Meyr. (Lep., Tortricidae) on black pepper. *Journal of Applied Entomology* 117: 31-34.

Devasahayam, S. and Koya, K. M. A. (1999). Integrated management of insect pests of spices. *Indian Journal of Arecanut, Spices and Medicinal Plants* 1: 19–23.

Devasahayam, S and Koya, K. M. A. (2005). Insect pests of ginger. In: P. N. Ravindran, and K. N. Babu, (Eds.). *Ginger-The Genus* Zingiber (pp. 367–389). Washington: CRC Press.

Devasahayam, S. and Koya, K. M. A. (2007). Insect pests of turmeric. In: P. N. Ravindran, K. N. Babu, and K. Sivaraman (Eds). *Ginger-The Genus* Zingiber, *Turmeric-The Genus* Cucurma (pp. 169-192). Washington: CRC Press.

Devasahayam, S., Koya, K.M.A., Anandaraj, M., Thomas, T. and Preethi, N. (2009). Distribution and ecology of root mealybugs associated with black pepper (*Piper nigrum* Linnaeus) in Karnataka and Kerala, India. *Entomon* 34: 147-154.

Devasahayam, S. and Leela, N. K. (1998). Evaluation of plant products for antifeedant activity against pollu beetle (*Longitarsus nigripennis* Motschulsky), a major of black pepper. In: P. P. Reddy, N. K. Krishna Kumar and A. Verghese (Eds.). *Advances in I P M for Horticultural Crops.* (pp. 172-174). Bangalore: Association for Advancement of Pest Management in Horticultural Ecosystems.

Devasahayam, S., Vidyasagar, P. S. P. V. and Koya, K. M. A. (1998). Reproductive system of pollu beetle (*Longitarsus nigripennis* Mots.), a major pest of black pepper. *Journal of Entomological Research* 22: 77-82.

Gopakumar, B. and Chandrasekhar, S. (2002). Insect pests of cardamom. In: P. N. Ravindran and K. J. Madhusoodanan (Eds.). *Cardamom-The Genus* Elettaria (pp. 179–206). London: Taylor and Francis.

IISR (2001). Annual Report 2000, ICAR-Indian Institute of Spices Research, Calicut, Kerala.

IISR (2005). Annual Report 2004-05. ICAR- Indian Institute of Spices Research, Calicut, Kerala.

IISR (2008). Turmeric (Extension Pamphlet), ICAR-Indian Institute of Spices Research, Calicut, Kerala.

IISR (2015a). Black Pepper (Extension Pamphlet), ICAR-Indian Institute of Spices Research, Kozhikode, Kerala.

IISR (2015b). Cardamom (Extension Pamphlet), ICAR-Indian Institute of Spices Research, Kozhikode, Kerala.

IISR (2015c). Ginger (Extension Pamphlet), ICAR-Indian Institute of Spices Research Kozhikode, Kerala.

Jacob, S. A. (1981). Biology of *Dichocrocis punctiferalis* Guen. on turmeric. *Journal of Plantation Crops* 9: 119–123.

Jacob, T.K., Sharon D'Silva, Senthil Kumar, C. M., Devasahayam, S., Rajalakshmi, V., Sujeesh, E.S., Bhat, A. I. and Abraham, S. (2015). Single strain infection of adult and larval cardamom thrips (*Sciothrips cardamomi*) by *Wolbachia* subgroup Con belonging to supergroup B in India. *Invertebrate Reproduction and Development* 59: 1-8.

Jacob, T. K., Senthil Kumar, C. M., Sharon D'Silva, Devasahayam, S., Ranganath, H. R., Sujeesh, E. S., Biju, C. N., Praveena, R. and Ankegowda, S. J. (2015). Evaluation of insecticides and natural products for their efficacy against cardamom thrips (*Sciothrips cardamomi* Ramk.) (Thysanoptera: Thripidae) in the field. *Journal of Spices and Aromatic Crops* 24: 133-136

Josephrajkumar, A. J., Kurien, P. S., Backiyarani, S. and Murugan, M. (2000). Occurrence of insect pests on black pepper at high altitudes of Idukki District in Kerala. In: *Proceedings, 12th Kerala Science Congress 2000, Kumily* (pp. 417-420). Thiruvananthapuram: State Committee on Science, Technology and Environment.

Josephrajkumar, A. J., Kurien, P. S., Backiyarani, S.and Murugan, M. (2001). Efficacy of conventional insecticides for suppression of black pepper mussel scale, *Lepidosaphes piperis* Gr. (Diaspididae: Hemiptera). *Entomon* 26 (Special Issue): 113 –114.

Koya, K. M. A., Balakrishnan, R., Devasahayam, S. and Banerjee, S. K. (1986). A sequential sampling strategy for the control of shoot borer (*Dichocrocis punctiferalis* Guen.) on ginger (*Zingiber officinale* Rosc.) in India. *Tropical Pest Management* 32: 343–346.

Koya, K. M. A., Devasahayam, S., Selvakumaran, S. and Kallil, M. (1996). Distribution and damage caused by scale insects and mealybugs associated with black pepper (*Piper nigrum* Linn.) in India. *Journal of Entomological Research* 20: 129-136.

National Research Centre for Spices (1989). Cardamom Package of Practices. Calicut: National Research Centre for Spices.

Premkumar, T. (1980). Ecology and control of pepper pollu beetle *Longitarsus nigripennis* Motschulsky (Chrysomelidae: Coleoptera). PhD Thesis, Kerala Agricultural University, Vellayani.

Premkumar, T. and Nair, M. R. G. K. (1988). Effect of some planting conditions on infestation of black pepper by *Longitarsus nigripennis* Mots. *Indian Cocoa Arecanut Spices Journal* 10: 83-84.

Senthil Kumar, C.M., Jacob, T.K., Devasahayam, S., Sharon D'Silva and Krishna Kumar, N.K. (2015). Isolation and characterization of a *Lecanicillium psalliotae* isolate infecting cardamom thrips (*Sciothrips cardamomi*) in India. *BioControl* 60: 363-373.

Selvakumaran, S., Mini, K. and Devasahayam, S. (1996). Natural enemies of two major species of scale insects infesting black pepper (*Piper nigrum* L.) in India. *Pest Management in Horticultural Ecosystems* 2: 79-83.

Thyagaraj, N. E., Chakravarthy, A. K., Rajagopal, D. and Sudarshan, M. R. (1991). Bioecology of cardamom root grub *Basilepta fulvicorne* Jacoby (Eumolpinae: Chrysomelidae: Coleoptera). *Journal of Plantation Crops* 18 (Suppl.): 316–319.

Varadarasan, S. (1991). Dynamics of life cycle of cardamom shoot borer, *Conogethes punctiferalis* Guen. *Journal of Plantation Crops* 18 (Suppl.): 302–304.

Varadarasan, S. (1995). Biological control of insect pests of cardamom. In: T. N. Ananthakrishnan (Ed.). *Biological Control of Social Forest and Plantation Crops Insects* (pp. 109–119). New Delhi: Oxford and IBH Publishing Company Private Limited.

Varadarasan, S. and Naidu, R. (1993). Studies on Root Grubs. Consolidated Report on Research Project Financed by NABARD (1989–93). Myladumpara: Indian Cardamom Research Institute.

Varadarasan, S., Singh, J., Pradhan, L.N., Gurung, N. and Gupta, S. R. (2000). Bioecology and management of whitegrub *Holotrichia seticollis* Mosher (Melolonthinae: Coleoptera), a major pest on ginger in Sikkim. In: N. Muraleedharan, and R. R. Kumar (Eds.) *Recent Advances in Plantation Crops Research* (pp. 323-326). New Delhi: Allied Publishers Limited.

Vidyasagar, P. S. P. V., Devasahayam, S. and Koya, K. M. A. (1988). Sexing live adults of the pollu bettle *Longitarsus nigripennis* Mots. (Chrysomelidae: Coleoptera). *Current Science* 57: 869.

Visalakshi, A. and Joseph, K. V. (1965). The biology of the pepper shoot borer *Laspeyresia hemidoxa* Meyr. (Eucosmidae: Lepidoptera). *Agricultural Research Journal of Kerala* 3: 48–50.

Yamada, K., Bindu, K., Nasreem, A. and Nasser, M. (2011). A new flower bug of the genus *Montandoniola* (Hemiptera: Heteroptera: Anthocoridae), a predator of gall-forming thrips on black pepper in southern India. *Acta Entomologica Musei Nationalis Pragae* 51: 1–10.

Figure 1.1: Rhinoceros Beetle Infestation.

a) Damage symptom on unfurled leaves, b) Spear leaf damage, c) Tusk like symptom due to beetle feeding, d) Twisted growth of juvenile palm due to beetle attack. (p. 2)

Figure 1.2
(a) Drying of spathe, (b) damage on the nut due to rhinoceros beetle. (p. 3)

Figure 1.3a-e: (a) Egg and neonate grub, (b) Fully grown grubs, (c) Pupae, (d) Male beetle and (e) Female beetle. (p. 5)

Figure 1.4

(a) Rhinoceros beetle breeding on decaying palm trunk, (b) *C. infortunatum* induced malformation in rhinoceros beetle, (c) Hooking out of beetle from infestested site. (p. 6)

Figure 1.5

(a) Leaf axil filling with oilcake sand mixture, (b) Chlorantraniliprole sachet for leaf axil filling, (c) Botanical cake and (d) Paste developed by ICAR-CPCRI for prophylactic leaf axil filling. (p. 7)

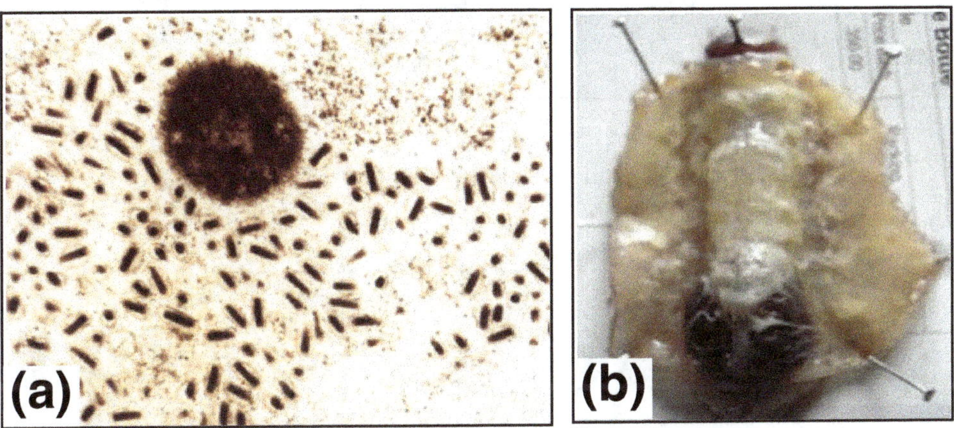

Figure 1.6

(a) OrNV particles under EM, (b) OrNV infected (fluid-filled) gut of *Oryctes* grub. (p. 8)

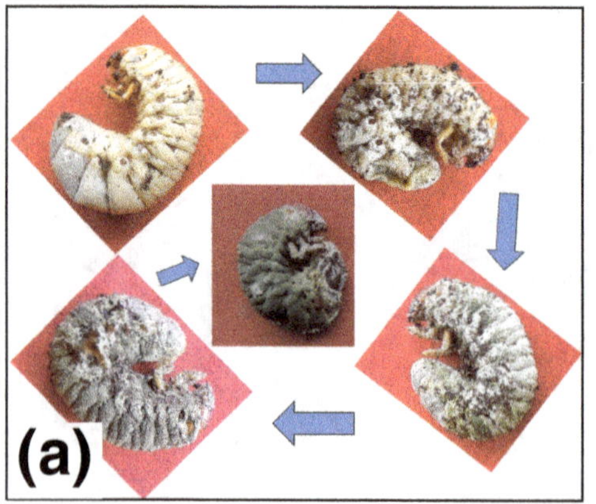

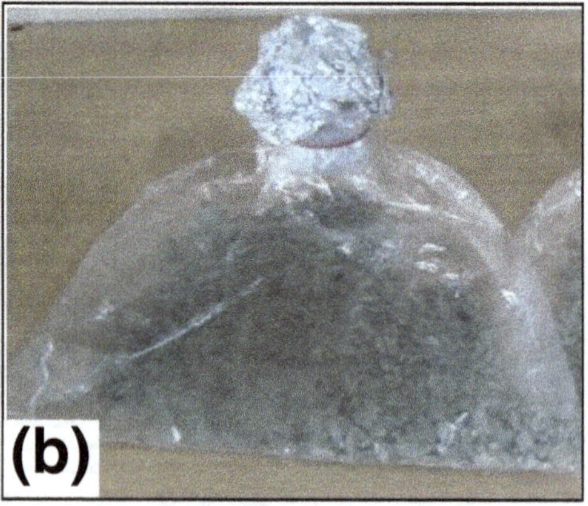

Figure 1.7
(a) Stages of GMF infection in *O.rhinoceros* grubs, (b) *M. anisopliae* culture in rice media. (p. 10)

Figure 1.8
(a) Red palm weevil and (b) Colour variant from NEH region. (p. 11)

Figure 1.9: Size and Colour Morphs of Red Palm Weevil form different Coconut Growing Tracts of India. (p. 12)

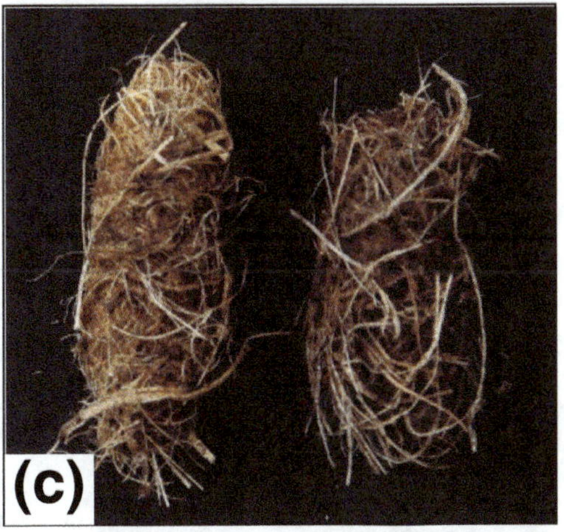

Figure 1.10: Life Stages of Red Palm Weevil.

(a) Egg of red palm weevil, (b) Grubs, (c) Pupal cocoons, (d) Various stages of grubs.
(p. 13)

Figure 1.11a-d: Symptoms of Red Palm Weevil Infested Palms.

(a) Holes on palm trunk, (b) Oozing of brown viscous fluid, (c) Splitting of petiole (d) Wilting of spear leaf, (e) Crown toppled palm. (p. 15)

Figure 1.11e: Symptoms of Red Palm Weevil Infested Palms (Crown Toppling). (p. 16)

Figure 1.12: Crop Habitat Diversification for Pest Regression. (p. 18)

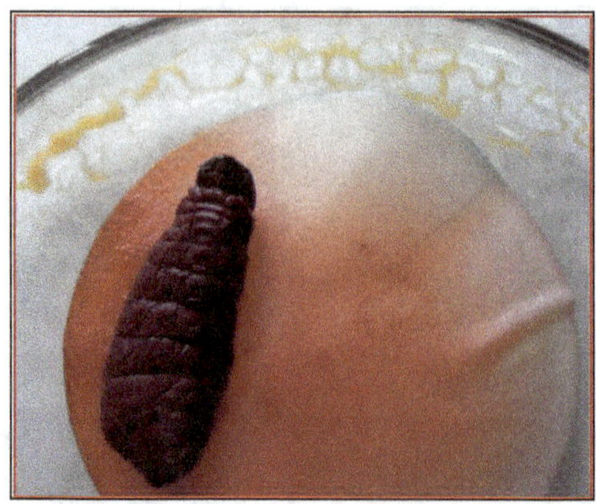

Figure 1.13: EPN Infected RPW. (p. 19)

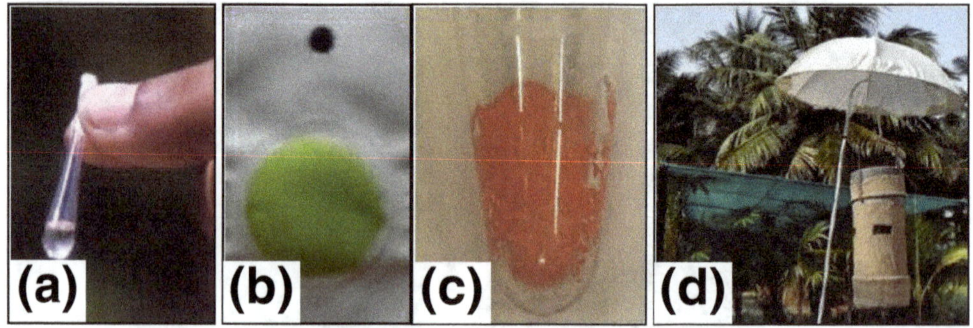

Figure 1.14: Pheromone Dispensation Strategies.
(a) Capillary vial, (b) Polymembrane matrix, (c) Nanoporus matrix and (d) Modified PVC pipe based pheromone trap. (p. 19)

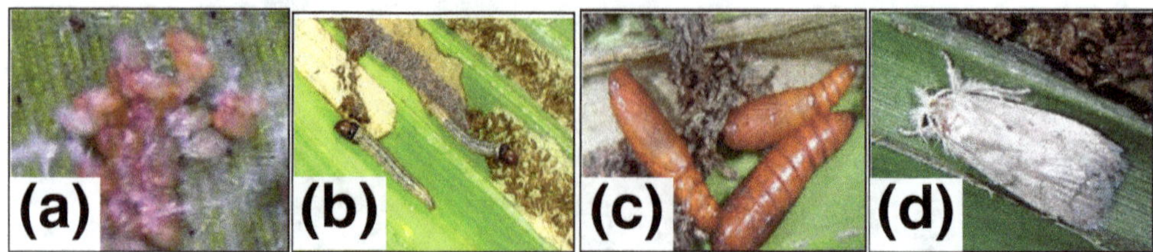

Figure 1.15: Life Stages of Coconut Black Headed Caterpillar.
a) Eggs, b) Larvae, c) Pupae and d) Adult moth. (p. 20)

Figure 1.16: Coconut Palms Infested by *O. arenosella*. (p. 21)

G.nephantidis on host larva   **Egg laying on host**   Progeny developing on host

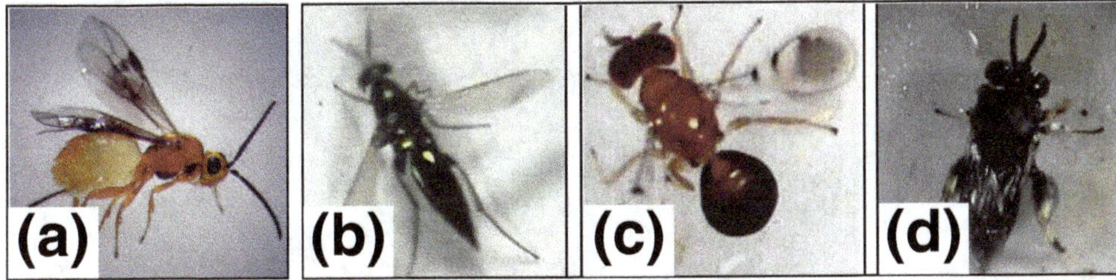

Figure 1.17: Parasitoids of *O.arenosella*
(a) *B. brevicornis*, (b) *E. nephantidis*, (c) *T. pupivora* and (d) *B. nephantidis*. (p. 23)

Figure 1.18: Black Headed Caterpillar Infested Area–Jajur Village, Arsikere during 2013. (p. 25)

Figure 1.19: Palms Completely Recovered from Black Headed Caterpillar Infestation. (p. 25)

Figure 1.20: Eriophyid Mite Damage Symptoms on Nuts of Varying Maturity. (p. 26)

Figure 1.21: Progression of Eriophyid Mite Infestation. (p. 27)

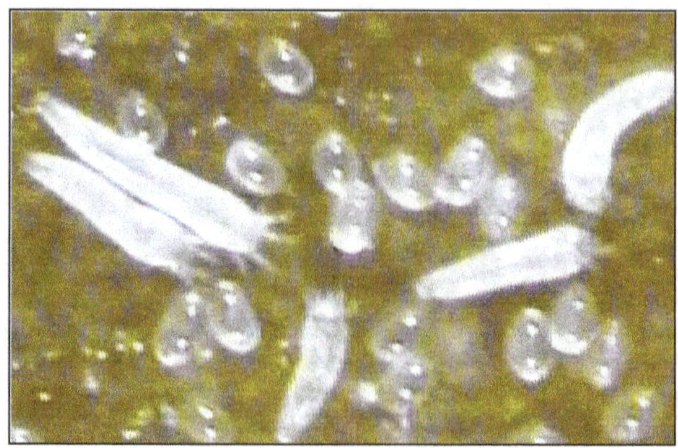

Figure 1.22: Eriophyid Mite Colony under Microscope. (p. 27)

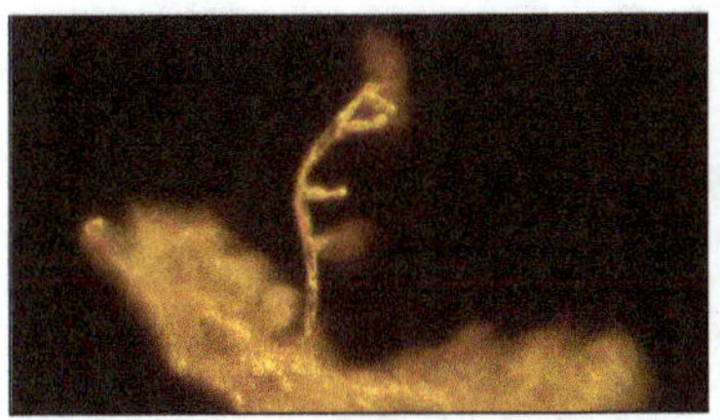

Figure 1.23: Mycelia of *H. thompsonii* Emerging from Infected Mite. (p. 29)

Figure 1.24: Grubs and Adult of *L.coneophora*. (p. 30)

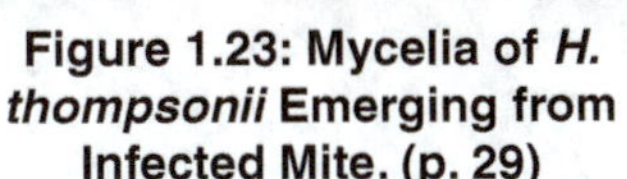

Figure 1.25: Life Stages and Damage Symptoms of Coreid Bug.
(a) Eggs, (b, c) Nymphs, (d) Adult, (e) Drying of inflorescence, (f, h) Infested buttons, (g) Infested mature nuts, (i) Barren nuts. (p. 32)

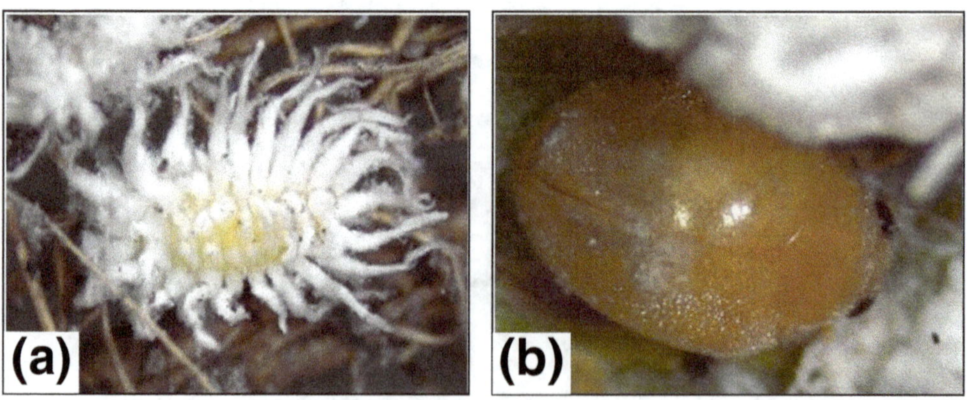

Figure 1.26: Scale Insects
(a) *Aspidiotus destructor;* (b) *Vinsonia stellifera,* (c) *Lepidosaphes megregori,*
(d) *Ceroplastes floridensis.* (p. 33)

Figure 1.27: Predators of Scale Insect–*Sesajiscymnus dwapikalpa*
(a) Grub and (b) Beetle. (p. 34)

Figure 1.28: Mealybugs.
(a-b) *Palmicultor palmarum,* **(c)** *Dysmicoccus finitimus,* **(d)** *Pseudococcus cryptus,* **(e)** *Nipaecoccus nipae.* (p. 35)

          Pests of Plantation Crops

Figure 1.29. Whiteflies.
(a, d) *Aleurocanthus arecae*, **(b, e)** *Aleurodicus dispersus* **and (c, f)** *Aleurodicus* sp.
rugioperculatus. **(p. 36)**

Contheyla rotunda Grub and Adult Moth   **Latoia lepida**

Figure 1.30: *Darna nararia* **Infestation on Coconut. (p. 38)**

Figure 1.31: Tender Coconuts Damaged by Rats. (p. 39)

Figure 1.32: *Brontispa longissima* and Infested Palm. (p. 40)

Figure 1.33: (a) *Wallacea* sp. infested coconut seedling, (b) Feeding damage lesions, (c) Grub, (d) Pupa and (e) Adult. (p. 41)

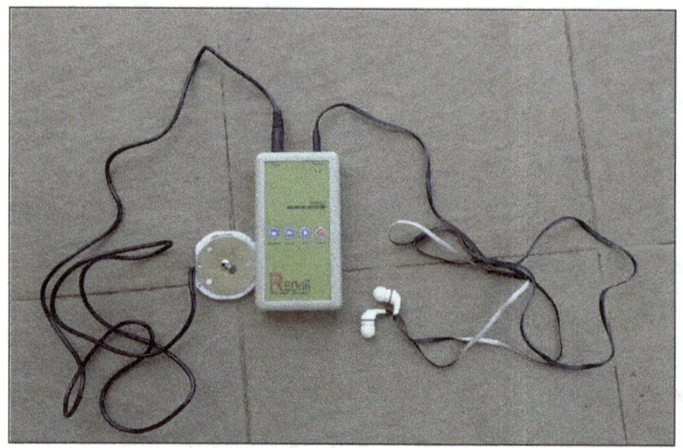

Figure 2.1: Portable Acoustic Device ReDrin™ for Early Detection of Red Palm Weevil Damage in Coconut Palms. (p. 57)

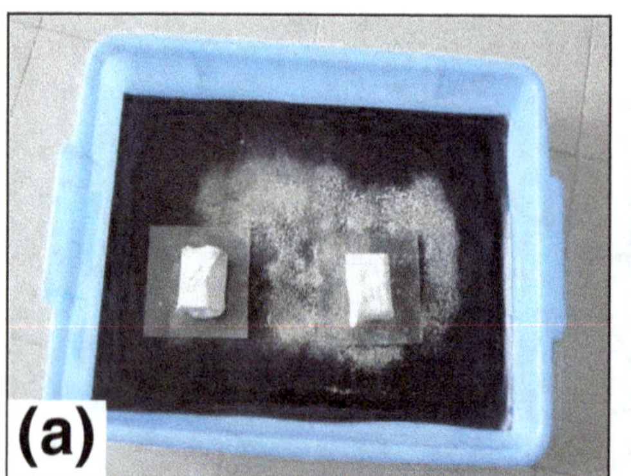

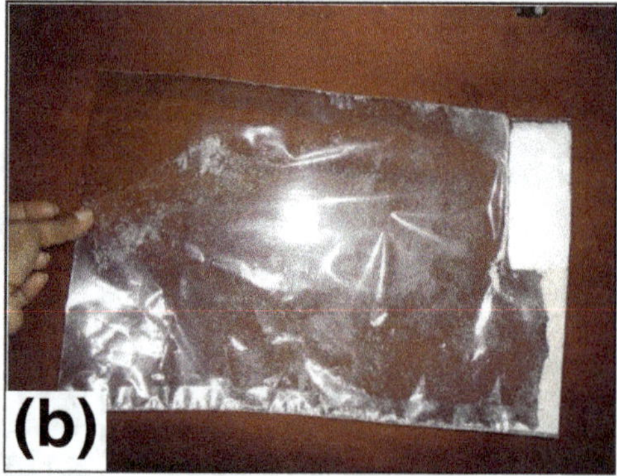

Figure 2.2: (a) Tray-type arena for the mass production of *N. baraki*, (b) Sachet-type rearing method of *N.baraki*. (p. 64)

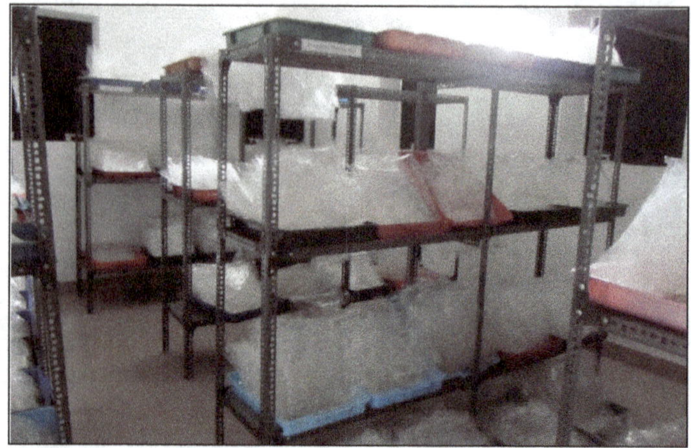

Figure 2.3: Mass Rearing Insectary Room of *N. baraki*. (p. 65)

Figure 3.1: Root Grub.

(a) Yellowing of fronds, (b) Stem tapering, (c) Root damage, (d) Reduction in crown size, (e) Root grub, *L. burmeisteri.* (p. 82)

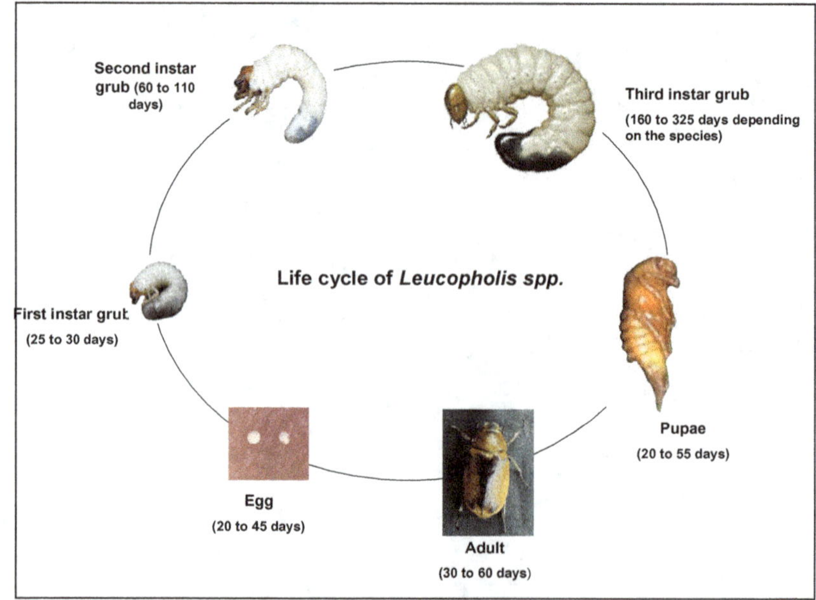

Figure 3.2: Life Cycle of White Grub. (p. 83)

Figure 3.3:

(a) Spindle bug Nymph, (b) Adult,
(c) Necrotic linear lesions on arecanut leaf caused by spindle bug. (p. 85)

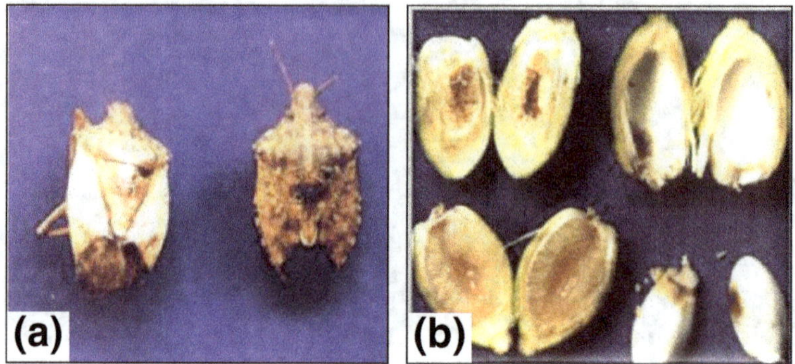

Figure 3.4

(a) *Halyomorpha marmorea*, (b) Damaged caused by pentatomid bug. (p. 88)

Figure 3.5: White Mite,
Oligonychus indicus. (p. 89)

Figure 3.6
(a) Mite–induced damage symptom on areca palms,
(b) Scarlet mite *Raoiella indica.* (p. 90)

Figure 3.7. Scale Infested
Arecanut. (p. 91)

Figure 3.8: Inflorescene Caterpillar and Infested Areca
Inflorescence. (p. 92)

Figure 4.1: Tea Mosquito Bug Damage. (p. 100)

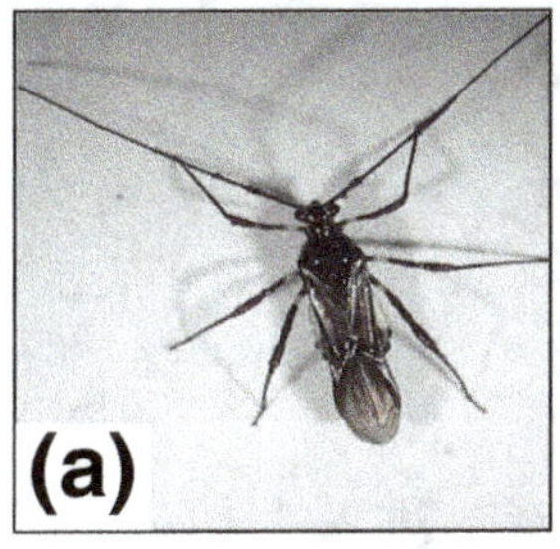

Figure 4.2
(a) Adult of *H. bradyi,* (b) Shoot with die back, (c) TMB infested cherelles. (p. 100)

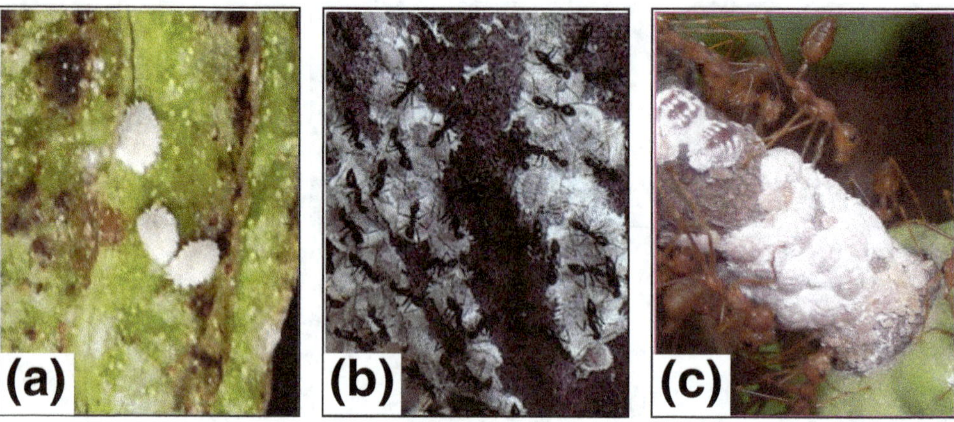

Figure 4.3: Mealybugs.
(a,b) *Crisicoccus hirsutus*, (c) *Planococcus lilacinus.* (p. 101)

Figure 4.4: Colonies of *T. aurantii* on Cocoa Leaves and Flower Bud. (p. 103)

Figure 4.5: Adults and Nymphs of Redbanded Thrips. (p. 104)

Figure 4.6

(a) Galleries on Stem made by *Z. coffeae*, (b) Larva boring into cocoa stem,
(c) Adult of *Z. coffea*. (p. 105)

Figure 4.7. Capsule Borer.

(a) Larva of *C. punctiferalis*, (b) Damaged cocoa pods with larva, (c and d) Rotten
cocoa beans due to *C. punctiferalis* damage. (p. 106)

Figure 4.8: Cocoa Pod Borer.
a) Larvae, b) Adult moth, c) Entry point, d) Tunnel made by CPB. (p. 108)

Figure 4.9: Vertebrate Pests.
(a) Rat damage, (b) Squirrel damage. (p. 112)

Figure 5.1: Shoot Borer, *Sesamia inferens* Walker Infested Oilpalm Seedling. (p. 122)

Figure 5.2
(a) Rat damage on oil palm seedlings, (b) Bamboo rat (Boi) *Cannomys badius.*
(p. 124)

Figure 5.3: Rat Damage on Juvenile Palms and Oil Palm Bunch. (p. 125)

Figure 5.4: Black Slug, *Laevicaulis alte* on Seed Sprouts. (p. 125)

Figure 5.5: Rhinoceros Beetle and its Damage Symptoms on Oilpalm. (p. 127)

Figure 5.6: *Metarhizium anisopliae* Infection on Grubs and Adults of Rhinoceros Beetle. (p. 128)

 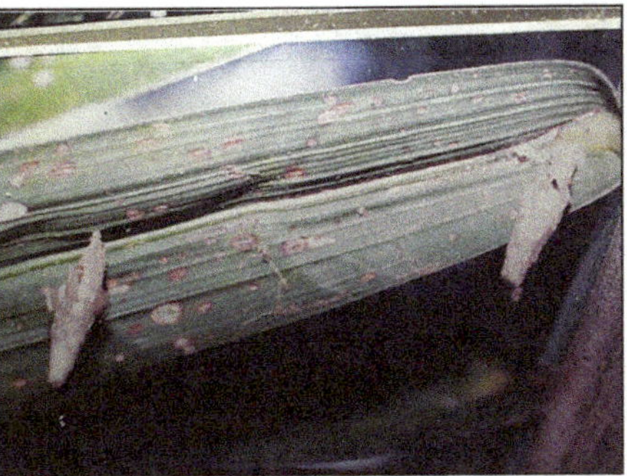

Figure 5.7: Psychid Damage on Oil Palm Leaves. (p. 130)

Figure 5.8: Leaf Webworm, *Acria meyerkii* and Infestation Symptoms on Oil Palm. (p. 131)

Figure 5.9: Slug Caterpillar, *Darna* spp. and Infestation Symptom on Oil Palm. (p. 133)

Figure 5.10: Termite Incidence on Adult Oil Palms. (p. 136)

Figure 5.11: Management Practices for Avian Pest Menace.
(a) Bio acoustic play, (b) Use of aterror, (c) Use of wire mesh and (d) Use of fishnet. (p. 138)

Figure 6.1: *Pollu* Beetle Infested Leaf and Berries of Pepper. (p. 144)

Figure 6.2: Pepper Leaf Infested by Coconut Scale and Mussel Scale. (p. 146)

Figure 6.3: Root Mealybugs of Blackpepper. (p. 147)

Figure 6.4: Top Shoot Borer. (p. 148)

Figure 6.5: Leaf Gall Thrips. (p. 149)

Figure 6.6: Thrips Damage on Cardamom. (p. 150)

Figure 6.7: Shoot and Capsule Borer. (p. 151)

Figure 6.8: Root Grub Infestation. (p. 152)

Figure 6.9: Shoot Borer. (p. 155)

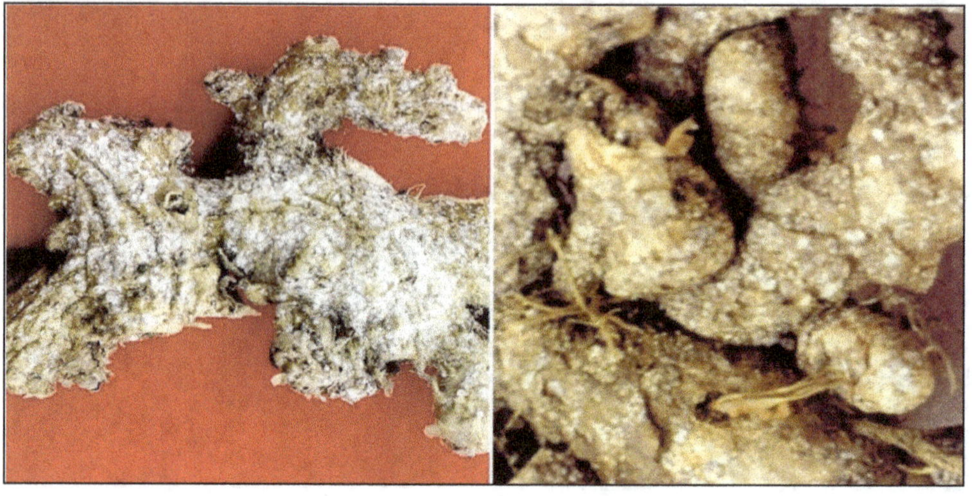

Figure 6.10: Scale Infested Rhizome. (p. 156)

2018, **Pests of Plantation Crops**   *Pages* **163–176**
Editors: **P. Chowdappa, Chandrika Mohan & A. Josephrajkumar**
Published by: **ASTRAL INTERNATIONAL PVT. LTD., NEW DELHI**

Chapter 7

Tea

☆ *B. Radhakrishnan*

1. Introduction

Tea (*Camellia* sp.) is one of the major foreign exchange earning commodities in India and is grown as a mono culture in an area of nearly, 5,20,000 ha. About 300 different insects are known to have been recorded on tea from all over the world. Of these, at least 200 have been found in south India. The major pest belongs to the order Acarina, Thysanoptera, Coleoptera, Lepidoptera and Hemiptera. The distribution and abundance of tea pests are largely influenced by weather, altitude, crop variety and the cultural operations such as pruning, manuring, regulation of shade, use of pesticides, natural enemies of pests and economics of tea production (Muraleedeharan, 1991). All parts of the tea plant, leaf, stem, root, flower and seed are fed upon by at least one pest species. IPM is an evolutionary stage in pest management rather than a revolutionary one and is continuing to absorb new ideas, new techniques that are practical, effective, economical and protective for crop and environment. It is based on ecological principles and integrates multidisciplinary methodologies in developing agro-ecosystem management. The present article deals with the recent trends in the integrated management of tea pests.

2. Crop Loss

There have been several assessments of crop loss caused by different tea pests. Crop loss caused by these major pests has been estimated to an extent of 15 to 20 per cent (Muraleedharan, 1987). In Vandiperiyar and Peermade areas of Kerala, the tea mosquito' was responsible for 4-46 per cent crop loss during the early 1950s. Thrips, is another important group of sucking insects, causing considerable yield reduction. Efficient control of this pest resulted in crop increase. Rao and Subramanian (1958) showed that satisfactory mite control could achieve 8 per cent crop increase. But, according to studies conducted in the Anamallais, the loss due to eriophyid mites was about 5 per cent (Muraleedharan and Radhakrishnan, 1990). Severe infestation of red spider mites resulted in a crop loss of 18 per cent (Radhakrishnan, 2004). Severe damage by termites and shot-hole borer leads not only the crop loss but also to capital loss resulting from the death of bushes. The average crop loss and the economic threshold level (ETL) for the major pests of tea are given in Table 7.1.

Table 7.1: Crop Loss Due to Major Pests and their Economic Threshold Levels (ETL)

Pest	Mean Crop Loss (Per cent)	ETL
Thrips	16	3 thrips/shoot
Pink and purple Mites	8	5 mites/leaf
Tea spider mite	18	4 mites/leaf
Tea mosquito	17	5 per cent
Shot hole borer	8	15 per cent

In addition to causing direct crop and plant losses, pest damage adversely affects the quality of made tea. Sucking pests like mites and thrips change the appearance of tea to 'dull'. Damage by flush worms results in the lowering of extractable solids and an increase in crude fibre content (Murthy and Chandrasekaran, 1979). Damage by tea mosquito bug results in reduction in polyphenols, catechins, amino acids, total soluble solids, nitrogen and phosphorus in made tea. The attack also increases the peroxidase activity and decreases the polyphenol oxidase activity at initial stages. In severe cases, reduction in photosynthetic rate up to 50 per cent was also reported. Decrease in chlorophylls, carotenoids, Theaflavin, Thearubigin, Total Liquor Colour and Highly Polymerised Substances was also noticed. Tea made out of Tea mosquito bug infested shoots scored significantly lower values on organoleptic evaluation (Sudhakaran and Muraleedharan, 2006).

3. Methods of Pest Control

The discovery of synthetic insecticides during the 1940's was a major breakthrough in pest control and for crop protection. However, the development of resistance to pesticides, resurgence of pests, secondary pest outbreaks and undesirable side effects of pesticides on human health and environment have compelled the plant protection scientists to review the pest situation in a new perspective and to integrate different control strategies into a pest management system. It is now realized that cultural and biological means of control are to be given utmost importance, whatever possible. At the same time, we cannot ignore the insecticides since they remain our most reliable immediate solution to a pest control at hand (Muraleedharan, 1991).

During the last few decades, several changes have taken place in the agronomic practices which have greatly contributed to the increase in tea productivity, but these have also magnified our pest problems. Obviously, the single line approach to pest control has to give way to a multi-faceted action.

3.1. Cultural Alternatives

Certain routine cultural operations such as plucking, pruning, shade regulation and weed control can be manipulated to reduce the incidence of pests and the intensity of their attack. Populations of leaf folding caterpillars can be suppressed to a considerable extent by their manual removal during plucking. Pruning removes a large part of the foliage and stems along with the pests. However, the newly emerging foliage is nutritionally more attractive to certain insects such as aphids, thrips, flush worms and leaf rollers. Shade regulation is not only essential for increasing productivity but also for reducing the intensity of pest damage. While fields devoid of shade trees suffer more from the attack of thrips and mites, heavily shaded and moist fields will be damaged more by the tea mosquito. Application of higher levels of potassium fertilisers is known to reduce the incidence of pests and disease in several crops. Application of higher rates of muriate of potash to the soil in the first year of the pruning cycle significantly reduces the infestation by shot-hole borer.

3.2. Host Plant Resistance

Host plant resistance is perhaps one of the least explored areas of non-insecticidal pest control in plantation crops, in general and tea in particular. Nevertheless, attempts have been made to screen different cultivars of tea in the laboratory and field conditions against pests such as mites, tea mosquito, shot hole borer and flush worm. A study on the susceptibility of clones to flush worm indicated that UPASI-17 was highly preferred by these caterpillars while UPASI-1 was the least infested. Certain Sri Lanka tea selections such as TRI-2024 and TRI-2025, which are planted in South India as well, are highly susceptible to the attack of shot hole borer. Murthy and Rao (1979) considered UPASI-10, UPASI-12 and UPASI-20 tolerant to the attack of these beetles. More recent surveys indicated that none of the clones released by UPASI could be classified as 'susceptible' to shot hole borer (Table 7.2).

The mechanisms of resistance are based on antixenosis (non preference), antibiosis and tolerance in terms of hosts' reaction to plant feeding insects. Antixenosis type of resistance may be due to the presence of certain chemicals in the plant which act as feeding or ovipositional deterrents, while a few others may act as insect repellents. Besides chemicals, morphological features of plants, especially the presence of trichomes, epicuticular waxes, silica content, colour and posture can impart antixensis. Toxic chemicals like alkabiosis. Many of these chemicals inhabit oviposition and adversely affect embryonic and post-embryonic development in insects (Muraleedharan, 1991). As in the case of resistance mechanisms, plants' susceptibility to insect attack is also highly influenced by the presence of chemicals which act as insect attractants and feeding and ovipositional stimulants. It is often amazing that the presence of specific chemicals may offer protection from one insect pest, but may increase the risk of invasion by others.

Table 7.2: Incidence of Shot Hole Borer on Certain Clones

Clone	Maximum Per cent of Attack Noticed in Clonal Blocks	Remarks
UPASI – 1	7.0	
UPASI – 2	38.3	
UPASI – 3	14.6	
UPASI - 6	14.0	
UPASI – 8	5.0	
UPASI – 9	28.8	
UPASI – 10	11.9	
UPASI – 12	*	Low level of attack noticed in field
UPASI – 13	*	Low level of attack noticed in field
UPASI – 15	10.0	
UPASI – 16	*	Low level of attack noticed in field
UPASI – 17	*	Low level of attack noticed in field
UPASI - 19	*	Low level of attack noticed in field
ATK – 1	3.9	More serious damage noticed in field
TRI - 2025	90.0	Considered highly susceptible to borer attack in Sri Lanka
TRI - 2024	**	Considered highly susceptible to borer attack in Sri Lanka
W-35	33.5	

* Percentage of attack not recorded; ** No data available from India.

Different cultivars of tea, having differing growth habits, vary in their susceptibility, resistance or tolerance to pests. "Chineric" varieties are more susceptible to the attack of scarlet and red spider mites whereas "Assam" varieties are more prone to infestation by eriophyids. The rhodoxanthin content of tea leaves has been shown to act as a feeding and reproductive stimulate for *Oligonychus coffeae* (Fernando, 1967). Soft wooded clones are easily damaged by termites. Similarly, clones with a high content of spinasterol are more susceptible to the attack of shot hole borer. The levels of this sterol is determined by the presence of several others like calcium saponins, theanine, arginine and chebulagic acid. Saponins could bind sterols and become a determinant of host resistance to shot-hole (Wickremasinghe *et al.*, 1976).

3.3. Plant Products

Neem formulation was effective against eriophid mites and flushworm of tea. Currently, neem formulations are recommended and used in tea plantations for the control of these pests (Muraleedharan and Selvasundaram, 1996). Experiments conducted on the effect of neem formulations on the development of tea flushworm *Cydia leucostoma* Meyr. have shown possibility for extensive use of neem formulations in tea pets management programme (Muraleedharan, 1993).

Several plants have shown potential in pest control. Formulations containing azadiractins obtained from neem seed kernels have been found effective against pink and purple mites and caterpillar pests such as flushworms and leaf rollers. But they have not been found effective against thrips. The commonly available weeds such as, *Ageratum houstonianum, Allamanda catharitica, Bidenspilosa, Casuarina equisetifolia, Conyza bonariensis, Crassocephalum crepidoides, Gliricidia sepium, Lantana camara, Ocimum basilicum* and *Tithonia diversifoila* found in tea plantations were evaluated for their efficacy against RSM under laboratory conditions. Among the plants, the aqueous extracts of *A. catharitica* and *C. bonariensis* showed 100.0 per cent and 80.0 per cent adult mortality (Radhakrishnan and Prabhakaran, 2014).

3.4. Cultural Control

The control of pests through timely manipulation of cultural practices is called cultural control. This, apparently, is the most economical and widely applicable method of pest control. In many crops, different methods like crop rotation, minimum tillage, multiple cropping *etc.*, have been developed traditionally to reduce the incidence of pests. In tea crop, cultural operations like plucking, pruning, manuring, regulation of shade and the use of agro chemicals influence the distribution and abundance of pests. Certain caterpillar pests such as flushworms, tortrix and tea leaf rollers can be manually controlled by removing the infested shoots during harvesting. Pruning and the developmental stage of the bush in the pruning cycle affect the distribution of pest species and the intensity of their attack. Pruning removes a large part of the foliage and stem, and also the pests feeding on these plant parts. Dieback and rot following pruning are significant factors facilitating the invasion of bushes by termites. Minimizing new access points by removal of dead wood, cankers and snags will help in reducing fresh infestation by termites. Removal of very badly affected stems will be useful in reducing the intensity of shot-hole borer attack. Leaf folding caterpillars such as flushworms and leaf rollers as well as the sucking insects like aphids and thrips are comparatively more abundant in fields recovering from pruning. A case in contrast is that of the shot-hole borer. In the first year of the pruning cycle, attack by beetles on bush frames will be extremely low. The decline in borer population in pruned bushes is related to the low levels of spinasterol and an increase in saponin content.

Intensity of shade plays a predominant role in determining the population density of certain sucking pests. Unshaded tea fields harbor large number of mites and thrips whereas *Helopeltis* damage is observed more in moist, shaded areas. Certain shade trees like *Indigofera* and *Albizzia* are alternate hosts for several caterpillar pests. *Grevillea robusta* A. Cunn., the common shade tree of south Indian tea fields, is comparatively free from pest attack. Shade trees function effectively as diversionary hosts for termites and help to reduce their incidence in fields (Sivapalan *et al.*, 1977). 'Dadap' trees, *Erythrina indica* Lam. and *E.lithosperma* Miq., non Blume are good hosts of root knot nematodes (*Meloidogyne* spp.). These trees should neither be grown in tea fields nor the soil from such areas used in nurseries without proper nematode control treatments.

The relative rate of major nutrients, especially nitrogen, in the fertilizer schedule influences the incidence of pests on several crops. When nitrogen is abundant relative to other macronutrients, plant succulence increase because carbohydrates are used for proteins rather than for cell walls. Such an increase in succulence and decrease in the toughness of plants have been correlated to the abundance of phytophagous insects and mites on crops. In tea, increased use of nitrogenous fertilizers has increased the number of shot-hole borer beetles (Gadd, 1944). Reviewing the role of potassium in tea husbandly, Ranganathan and Natesan (1984) pointed out the usefulness of this element in preventing algal (*Cephaleuros parasiticus* Karst.) and fungal (*Gloeosporium theasinensis* Miyake and *Hypoxylon nummularium* Bull. ex. Fr.) disease. They also noted that enhanced rate of potassium application in the pruned year was helpful in reducing the infestation by shot-hole borer.

The pattern of usage of agrochemicals is yet another factor, determining the composition of pest species. For example, the continuous use of copper fungicides against the blister blight (*Exobasidium vexans* Massee) of tea has been proved to be responsible for the heavy bid up of mites. While copper salts may not be directly responsible for this increase, inert diluents or the physical nature of the spray deposits may have a role in this. The use of DDT and certain related compounds have a role in this. The use of DDT and certain related compounds have also led to the high incidence of mites. Intensive use of herbicides for the control of weeds is probably a reason for the increasing incidence of grasshopper damage in tea and again emphasizing the need for taking an overall view for the management of any noxious organism whether they are weeds, pests or pathogens.

3.5. Legal Control

Almost all countries, including India, have legal enactments, called quarantine laws, to prevent the entry of foreign pests and pathogens. Government of India has also empowered the states to enact such laws as are necessary to prevent the spread of dangerous pests and diseases within their jurisdiction. Quarantine measures are advocated for the introduction of tea germplasm either as seed material or vegetative propague in the form of scion and cuttings (Venkata Ram, 1983). In Sri Lanka, the transport of rooted cuttings from one estate to another is prohibited under law to avoid the introduction of the lesion nematode, *Pratylenchus loosi*.

3.6. Biological Control

Several bio-control agents are active in the tea ecosystem exerting natural regulation on the populations of many pests (Muraleedharan *et al.*, 1988). Intensive searches have been made in South India tea areas resulting in the discovery of many parasitoids and predators. The notable contributions on the natural enemies of tea pests are those of Rao *et al.* (1970), Das (1974, 1979) and Sarma (1976). Pink and purple mites are attacked by the predatory mite, *Amblyseious herbicolus* and the predatory thrips, *Scolothrips rhagebianus*. The tea thrips is attacked by two predatory thrips, *Aelothrips intermedius* and *Mymarothrips garuda*. The tea aphid is attacked by one dozen syrphid and coccinellid predators and three parasitoids (Radhakrishnan, 1989). The tea tortrix is mainly attacked by two parasitoids namely, *Phytodietus spinipes* and *Palexorista solennis*. Together they are able to decimate more than 90 per cent of the tortrix population. Similarly, the tea leaf roller is attacked by five parasitoids of which *Sympiesis dolichogaster*is the most important. This small wasp can cause more than 80 percent mortality of the leaf rollers. In the case of the flush worm 14 species of parasitoids have been recorded. Among them *Apanteles aristaeus* is the most predominant species. Parasitism by this species goes up to 26 per cent.

In Sri Lanka, great success has been achieved in the biological control of *H. coffearia* by introducing an exotic parasitoid, *Macrocentrus homonae* Nixon, from Indonesia. This is an excellent example of classical biological control in tea. In addition to the parasitic and predatory insects, there are many nematode species which attack insect pests. A well-known group of nematodes, the mermithids, are obligate parasites of insects. Amermithid, *Hexamermis* sp. attacks the flushworms in south India (Subbiah, 1986).Use of entomopathogentic fungi in pest management as one of the important substitutes for chemical control has gained greater importance in the recent years. Reason for the change in the control strategies has emerged from the necessity of avoiding certain adverse effects of synthetic chemicals which are often used for pest control.

Naturally occurring epizootics of insect populations have revealed the potential of certain micro-organism such as viruses, bacteria, fungi and protozoa for biological control of harmful insects. Some of these microbes, especially viruses and bacteria are used in certain countries for the control of caterpillars belonging to *Spodoptera* and *Heliothis*, damaging a variety of crops. Although considerable work has been carried out on the efficacy of pathogens of many important pests, the controversy over the use of disease causing organisms in a country like India where silk industry is important remains unsolved. Consequently, none of these microbial formulations is commercially available in our country. However, in China, scientists have been carrying out studies on the viral disease of certain caterpillar pests of tea (Zhen *et al.,* 1985). Similarly, viruses were found effective in controlling the caterpillars, *Homona magnanima* Diakonoff and *Adoxophyes* sp. in Japan (Sato *et al.,* 1986). The Bacterial insecticide, *Bacillus thuringiensis* is useful against *H.magnanima*, *Caloptilia theivora* and Adoxophyes sp. (Kariya, 1977), *Buzura suppressaria* Guen. (Borthakur and Ragunathan, 1987) and the flushworm, *Cydia leucostoma* Meyrick (Muraleedharan and Radhakrishnan, 1990). Recently, Scientists have successfully incorporated the toxin producing gene of *Bacillus thuringiensis* in to living plants. Application of *V.lecanii.P.fumosoroseus* and *H.thompsonii* @ 3500g formulation per ha significantly reduced the population density of red spider mites in the field. The efficacy of *P.fumosoroseus* formulation @ 2500 g/ha was almost comparable to that of *V.lecanii* and *H.thompsonii* applied @ 3500 g formulations/ha. Though the percentage of mite control achieved by the application these fungal formulations was less in comparison to that of conventional acaricides, these biocontrolagentsare recommended for use in tea fields in view of their role in reducing the pesticide load on tea.

Leiophron sp. (Hymenoptera: Braconidae) was recorded as important endoparasite on nymphs and adults of tea mosquito bug. The parasitism under field conditions varied from 10-44.0 per cent (Srikumar *et al.,* 2014). The green lacewing, *Mallada desjardinsi* Navas was recorded as an important predator of red spider mite,

Oligonychus coffeae and its Life history, life table and predatory efficacy were studied under laboratory conditions (Vasanthkumar and Babu, 2013).

3.7. Pheromonal and Hormonal Control

Pheromones of insects are volatile compounds used for intraspecific communication and serve a wide variety of behavioural functions. They are useful for mess trapping and communication disruption between males and females of insects. Pheromone baited taps are effectively utilized to reduce infestation of bushes by the smaller tea tortrix, *Adoxophyes* sp. (Hiyori *et al.*, 1986) and the tortrix, *Homona magnanima* (Noguchi *et al.*, 1981). Besides the original pheromones of insects, their analogues and mimics are also employed in pest management programmes. The isolation, identification and synthesis of pheromones require sophisticated technology and this will remain a major hurdle in their widespread use in developing countries.

The tea mosquito bug (*Helopeltis theivora*), pheromone was successfully isolated and identified. Lure baited sticky traps were used for the effective monitoring/ controlling of TMB. Large scale field trials were conducted with satisfactory results (Radhakrishan and Srikumar, 2015). Tea mosquito pheromone traps with the catchers are shown in Figure 7.1.

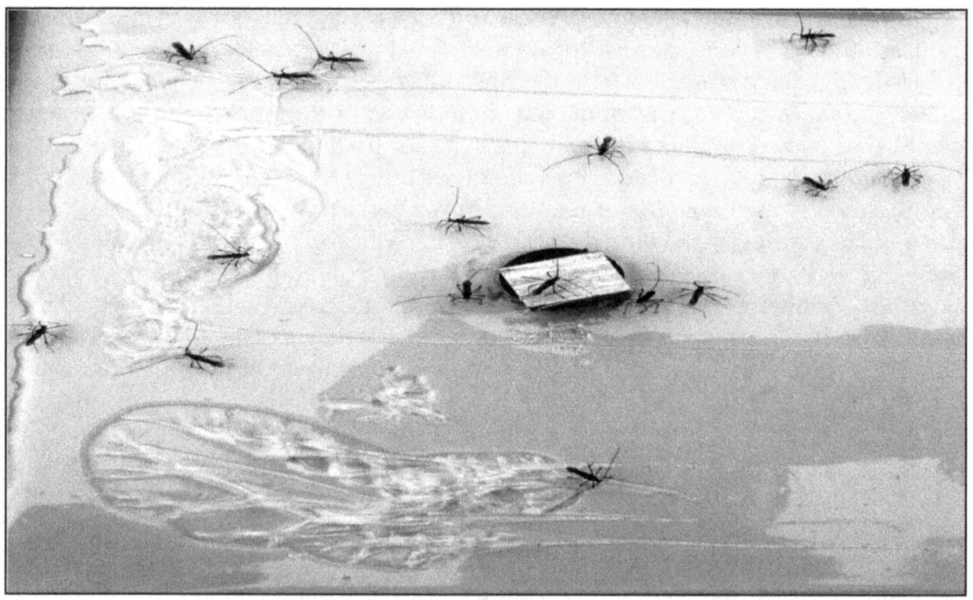

Figure 7.1: Tea Mosquito Bug Pheromone Trap.

Certain hormones of insects have shown potential as safer pesticides and these are called insect growth regulators (IGR). The Juvenile Hormone (JH) mimicking compound 'methoprene' is used as a larvicide in mosquito control. Similarly, the chitin inhibitor 'diflubenzuron' is presently used for the control of a number of caterpillar pests attacking many crops and has been tested against the flushworm of tea as well.

3.8. Chemical Control

The chemical method of pest control involves costly inputs like pesticides, fuel, labour and spraying equipment. The correct choice of pesticides, their dosage, timing and method of application are of paramount significance for the success of pest control. The misuse of pesticides and the improper execution of pest control technology may result in crop loss, pesticide resistance, secondary pest outbreaks, pest resurgence, health hazards and environmental pollution. Pesticides are classified in different ways. Based on the groups of organisms against which pesticides are primarily targeted, they are classified into (1) Fungicides (against fungi) (2) Herbicides (against weeds) (3) Insecticides (against insects) (4) Acaricides (against mites) (5) Nematicides (against nematodes) (6) Rodenticides (against rodents like rats and squirrels) (7) Molluscicides (against slugs and snails) (80) Bactericides (against bacteria).

On the basis of the mode of action, insecticides are categorized as (a) contact poison (b) stomach poison (c) fumigant. Certain pesticides move systemically within the plant and these are referred to as systemic.The synthetic organic insecticides are classified on the basis of their chemical structure as well. The familiar group are (a) Chlorinated hydrocarbons (DDT, BHC, endosulfan, heptachlor), (b) Organophosphates (ethion, phosalone, monocrotophos, malathion), (C) Carbamates (aldicarb, carbaryl, carbofuran), (d) Synthetic pyrethroids (permethrin, cypermethrin, fenvalerate,deltamethrin).

Most of the modern insecticides ranging from DDT to synthetic pyrethroids and insect growth regulators have been tested against many species of tea pests. However, for large scale and regular application on tea, our choice of insecticides and acaricides is limited. Tea being an important export commodity, it is essential that international regulations on pesticide residues are complied with. At present, only dicofol, endosulfan, ethion, phosalone, tetradifon, permethrin and cypermethrin can be readily applied on tea to conform to EPA (Environmental Protection Agency of USA) requirements. Sulphur is another chemical which can be readily applied on tea, but for its tainting properties. Food and Agricultural Organisation (FAO) has also recommended maximum residue limited for several pesticides on tea.

In all cases, it is advisable to spray insecticides and acaricides immediately following plucking so that the residues are at the lowest level. As already mentioned, the acaricide, sulphur imparts taint to made tea and care may be taken in its use. When chemicals not cleared by EPA or Codex Alimentarius commission are used on tea in plucking, the leaves harvesting in at least two plucking rounds after spraying are to be discarded.

The use of board spectrum chemicals is to restricted and for the control of mites, only acaricides are to be applied as far as possible. However, when a combined attack of mites and insect pests such as thrips or flushworms is notices, suitable chemicals with insecticidal and acaricidal properities may be preferred.Though pesticides remain our most important tools for pest control, their improper use can result in harmful side effects. The continuous use of one chemicals against a single pest species, over a long period, will ultimately lead to the development of resistance genes posses limitless persistence. The problem of resistant to an insecticide may be complicated by the development of cross resistance to related compounds or multiple resistances to a variety of compounds with differing modes of action and different detoxification mechanisms.

Further, indiscriminate application of insecticides leads to the outbreak of secondary pests. A familiar example from tea is the outbreak of several species of caterpillar pests after the application of heptachlor. This is mainly due to the disruption of food chain and elimination of natural enemies of secondary pests. Insecticides, more than any other pollutant, have been the subject of controversy because of their possible adverse effects on human health and environmental quality. Most of the widely used pesticides are general biocides and many of the chemicals which were once considered safe are being reviewed for their impact on environment. Some of the insecticides are known to induce cancer in laboratory animal. Such chemicals pose q risk to human health, but risk assessment is a separate area of research. Another problem is the delayed neurotoxic effects observed in people exposed to certain organ phosphorous compounds. However, there is no reliable worldwide data available on human morbidity and mortality caused by pesticides. Besides their harmful effects on human health, most of the pesticides cause damage to wild life, especially to bird populations by interfering with their reproduction. Many insecticides are high toxic to fish and accidental contamination of lakes and rivers has resulted in mass killing of fishes. A large number of chlorinated hydrocarbon insecticides are highly persistent in the soil. Their entry into the food chain. Despite their drawbacks, pesticides will continue to play a major role in pest control in the foreseeable future. Man must, therefore, learn to live with pesticides, though he should make every effort to reduce the risk through judicious and pragmatic use.

3.9. Integrated Pest Management

This is defined as the 'Pest management system that in the context of the associated environment and the population dynamics of the pest species, utilizes all suitable techniques and methods in as compatible a manner as possible and maintains the pest populations at levels below those causing economic injury'. In other words, the goal of IPM is to make pest control more efficient by co-ordination biological, chemical and culture means of pest control and by varying control practices with changing conditions in the field. IPM represents ecologically sound approach to pest control and it advocates marked changes in our present concept of pesticide usage. As per the IPM programme, pesticide applications are to be carried out only under unavoidable circumstances and the area of treatment also limited. Only selective chemicals are preferred and board spectrum insecticides with undue persistence are avoided. Special emphasis is always given to nonchemical control strategies. If strictly adhered to, these practices will considerably reduce the changes of pests developing resistance to pesticides.

Improved application methods, aimed at wastage reduction will heal to minimize the dosage rates. Rapid strides have been made in the field of pesticide application technology, but in tea plantations the reliance in mostly on conventional sprayers, requiring large volume of water, enormous manpower and incurring wastage of pesticides. Changes in nozzle design can produce sprays with a narrow droplet spectrum and the introduction of electrostatically charged sprays can provide greater control of droplet trajectories to increase deposition and reduce drip and draft (Matthews, 1979). The recent technologies on ultra low volume (ULV) and electrodyne spraying are directed towards this goal.

The damage potential of pests depends on weather, vigour and maturity of crop, time of pest infestation, sixe of beneficial insect populations and many other interacting factors. Development of mathematical models in pest management can help us to understand the ways in which the above factors interact with field populations of pests and their natural enemies in relation to need based usae of environmentally safe pesticides with limited persistence. Systems analysis can ultimate help us in decision making on the combination and timing of pest control practices to be adopted and some of these programmes come under the realm of artificial intelligence. The system approach has been widely adopted in North America to protect important crops such as cotton, alfalfa, soybean and citrus (Matthews, 1984).

It is expected that the widespread adoption of IPM programme will resurgence, health hazards to humans, livestock and birds and improve environmental Quality by lessening the pesticide burden. The components of IPM for pest control in tea are shown in Figure 7.2.

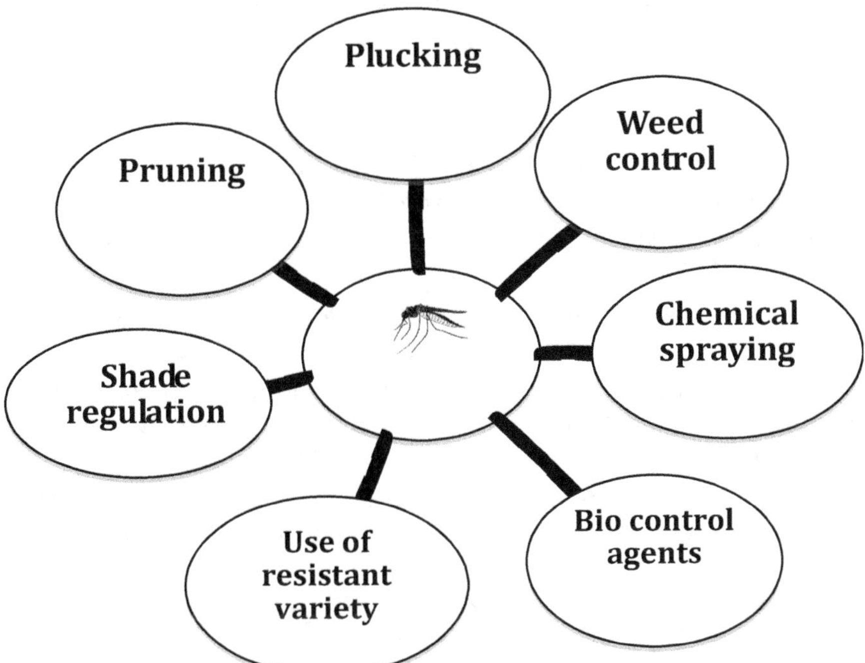

Figure 7.2: IPM Strategies for the Control of Tea Pests.

Acknowledgement

I have drawn liberally from several articles published by Dr.N.Muraleedharan, Former Director, UPASI and Tea Research Association, Tocklai for which I am extremely grateful to him.

References

Borthakur, M.C. and Raghunathan, A.N. (1987). Biological control of tea looper with *Bacillus thuringiensis. J. Coffee Res.* 17: 120-121.

Das, S.C. (1974). Parasites and predators of pests of tea, shade trees and ancillary crops in Jorhat circle. *Two and a Bud* 21: 17-21.

Das, S.C. (1979). Parasites and predators of pests of tea. *Two and a Bud* 26: 72-73.

Fernando, E.F.W.(1967). Studies on the aetiology and bio chemistry of *Oligonychus coffeae*Nietner (Acarina : Tetranychidae) the red spider mite of tea in Ceylon. M.S. Thesis, University of Paradeniya, Sri Lanka. pp. 146

Gadd, C.H. (1944). An unusual correlation between insect damage and crop harvested. *Ann. Appl. Biol.* 1: 47-51.

Hiyori, T., Kainoh, Y. and Nimomyia, Y. (1986). Wind tunnel tests on the disruption of pheromonal orientation of the male smaller tea tortrix moth, *Adxophyes* sp. (Lepidoptera : Tortricidae) I Disruptive effect of sex pheromone components. *Jap. J. Appl. Ent. Zool.*, 21: 153-158.

Kariya, A. (1970). Control of tea pests with *Bacillus thuringiensis. JARQ* 1: 173-178.

Matthews, G.A. (1984). *Pest management.* Longman, London. pp. 231.

Muraleedharan, N. (1987). Entomological Research in tea in southern India. *J. Coffee Res.*, 17: 80-83.

Muraleedharan.N. (1993) Entomology. 65-88. Annual Report for 1993, UPASI Scientific Department, UPASI Tea Research Institute, Valparai.

Muraleedharan, N. (1991). *Pest Management in Tea.* UPASI Tea Research Institute, Valparai.pp 130.

Muraleedharan, N and Radhakrishnan, B. (1990). Recent studies on tea pest management in south India. *UPASI Sci. Dept. Bull.* 43: 16-29.

Muraleedharan, N. and Selvasundaram, R. (1996). Control of tea pests in south India (Revised recommendations) In: *Handbook of Tea Culture,* Section 23 pp. 33. UPASI Tea Research Institute, Valparai 642 127, Coimbatore District, Tamil Nadu.

Muraleedharan, N., Selvasundaram, R. and Radhakrishnan, B. (1988). Natural enemies of certain tea pests occurring in southern India. *Insect Sci. Appl.* 9: 647-654.

Murthy, R.L.N. and Chanrasekaran, R. (1979). Effect of pest damage on quality of made tea. *Proc. PLACROSYM* 2: 275-283.

Murthy, R.L.N. and G.N. Rao (1979). Shot hole borer of tea: its present status and control. *Proc. PLACROSYM* 2: 243-257.

Noguchi, H., Tamaki, Y., Arai, S., Shimoda, M and Ishikawa, I. (1981). Field evaluation of synthetic sex pheromones of the oriental tea tortrix moth, *Homona magnanima* Diakonoff (Lepidoptera : Tortricidae). *Jap. J. Appl. Ent. Zool.* 25: 170-175.

Radhakrishnan, B. (1989) Studies on the aphid, *Toxoptera aurantii* (Boyer de Fonscolombe) (Hemiptera :Aphididae) and its natural enemies in southern Indian tea plantations. Ph.D. Thesis. Bharathiar University, Coimbatore, pp. 154.

Radhakrishnan, B. (2004). Seasonal abundance of red spider mite, *Oligonychus coffeae* (Nietner) and the crop loss caused by it in tea. *Journal of Plantation Crops* 32 (Suppl): 354-356.

Radhakrishnan, B. and Prabhakaran, P. (2014). Biocidal activity of certain indigenous plant extracts against red spider mite, *Oligonychus coffeae* (Nietner) infesting tea. *Journal of Biopesticides* 7(1): 29-34.

Radhakrishnan, B and Srikumar.K.K. (2015). Sex pheromone a new prospective in management of tea mosquito bug in tea. *Planters Chronicle* 111(7): 5-10.

Rao, G.N. (1970). Tea pests in southern India and their control. *Pesticides* 4: 71-79.

Rao, G.N. and Subramanian, H. (1968). Control of mites in tea results in increased yields. *Planters' Chronicle* 63: 412-413.

Sarma, P.V. (1976). Possibilities for integrated control of major pests of tea in India. *PANS* 25: 237-245.

Sato, R., Oho, N. and Kodomaro, S. (1986). Utilization of granulosis virus for controlling leaf rollers in tea fields. *JARQ* 19: 271-275.

Selvasundaram, R., and Muraleedharan, N. (1999). In: *Bio pesticides in Insect Pest Management* (Eds.,Ignacimuthu, S.J. and AlokSen), Phoenix Publ. House Pvt. Ltd., New Delhi, pp. 33-37.

Sivapalan, P. (1972). Nematode pests of tea, 285-311. In: *Economic Nematology*. John M.Webster (Ed.). Academic Press, London. pp.563.

Sivapalan, P., Senaratne, K.A.D.W and Karunaratne, A.A.C. (1977). Observations on the occurrence and behavior of live wood termites (*Glyptotermes dilatatus*) in low country tea fields. *PANS* 23: 5-8.

Srikumar.K.K, Radhakrishnan, B. and Sureshkumar, B. (2014).New record of parasitoid, *Leiophron* sp. (Hymenoptera: Braconidae) on Tea Mosquito Bug (*Helopeltis theivora*Wat.) In Tea. *Newsletter UPASI Tea Res. Foundn.* 24(2): 3-4.

Subbiah, K. (1986). New record of *Hexamermis* sp. (Mermethidae:Nematoda) from the larve of *Cydia leucostoma* (Eucosmidae:Lepidoptera). *Curr. Sci.,* 55: 1047.

Sudhakaran, R., and Muraleedharan, N. (2006). Biology of *Helopeltis theivora* (Hemiptera : Miridae) infesting tea. *Entomon* 3: 165-180.

VasanthaKumar, D and Babu, A. (2013). Life table and efficacy of *Mallada desjardinsi* (Chrysopidae: Neuroptera), an important predator of tea red spider mite, *Oligonychus coffeae* (Acari: Tetranychidae). *Experimental and Applied Acarology.* DOI 10.1007/s10493-013-9664-z.

Venkata Ram, C.S. (1983). Pathogens and pests of tea, 117-144. In: *Exotic plant quarantine pests and procedures for introduction of plant materials*. K.G. Singh (Ed.). ASEAN Plant Quarantine Centre and Training Institute, Selangor, pp. 333.

Wickremasinghe, R.L. Perera, B.P.M. and Perera, K.P.W.C. (1976). α-Spinasterol, temperature and moisture content as determining factors in the infestation of *Camellia sinensis* by *Xyleborus fornicates*. *Biochem. Syst. Ecol.* 4: 103-110.

Zhen, M., Chen, L., Xia, S., Liang, D., Zhao, K., Cai, Y. and Sung, A. (1985). Viruses of tea pests discovered in Yunnan and Guizhov Provinces (In chinese). *J. Tea Sci.* 5: 39-48.

2018, Pests of Plantation Crops   *Pages* **167–198**
Editors: **P. Chowdappa, Chandrika Mohan & A. Josephrajkumar**
Published by: **ASTRAL INTERNATIONAL PVT. LTD., NEW DELHI**

Chapter 8

Cashew

☆ *P.S. Bhat, T.N. Raviprasad, K. Vanitha*
and K.K. Srikumar

1. Introduction

Cashew (*Anacardium occidentale* L.) is an economically tropical evergreen tree that produces the cashew nut and the apple. India was the first country in the world to exploit international trade in cashew kernels in the early part of 20th century it self. Globally, it occupies an area of 53.15 lakh ha with a production of 41.53 lakh tonnes and India occupies the largest area (16.8 per cent) followed by Ivory coast, Brazil, Indonesia, Benin, Tanzania, Nigeria, Guinea-Bissau, Kenya, Vietnam and Philippines. The highest production was from Vietnam followed by Nigeria, India, Ivory Coast, Benin, Philippines, Guinea-Bissau, Tanzania, Indonesia, Brazil and Kenya. Large numbers of species of arthropods are associated as pests with cashew as all parts of the plant are fed upon by at pest species. Depending on the climate, location and age of plantation, however, each geographic region may have its own distinctive pest complex. The production loss is estimated to be about 20-30 per cent by tea mosquito bug alone and death of 5- 10 per cent of productive trees every year by Cashew Stem and Root Borer (CSRB) alone (Rai, 1984). Hence, insect pests pose a severe constraint for cashew production. There are significant contribution in the field of cashew entomology by Ayyar (1942), Abraham (1958), Pillai *et al.* (1976), Ohler (1979), Stonedahl (1991), Sundararaju (1984 and 2000a) and Sundararaju *et al.* (2006). A detailed knowledge about the pests is the pre-requisite in evolving suitable management approach against them.

2. Tea Mosquito Bug (TMB) - *Helopeltis* spp.

2.1. Species Composition and Distribution

Tea mosquito bug is a low-density pest and it could cause 36-75 percent damage at its mean population level of 0.15-0.36 nymphs and adults per shoot/panicle. Information on distribution, nature and extent of damage, biology, seasonal abundance, natural enemies, alternate host plants and control measures have been reviewed by Devasahayam and Nair (1986) and Stonedahl (1991). The genus *Helopeltis* has a palaeotropical distribution extending from West Africa to New Guinea and northern Australia. Of the recognized species, 26 are restricted to Africa, four are broadly distributed in the Oriental Region, four are endemic to the Philippine Islands, two are distributed throughout the Malay Peninsula, Sumatra and Java, and one species is endemic to South India and Sri Lanka, Pulo Laut Island (south-east coast of Borneo), Sulawesi and New Guinea and northern Australia (Stonedahl, 1991). Stonedahl (1991) had raised a doubt that most of the earlier reports pertaining to *H. antonii* Sign. may also be of *H. bradyi* Waterhouse since, *H. bradyi* has close resemblance with *H. antonii* and its existence was established for the first time in South India by him. *H. antonii* existed on varied host plants *viz.*, neem, cashew, guava, ber, drumstick, Indian gooseberry, cotton, *Ailanthus exelsa* Roxb. and cow pea, whereas in the Western Ghats, *H. antonii* co-existed as a dominant species along with *H. bradyi* and *H. theivora* Waterhouse cashew and cocoa. *H. antonii* also co-exists as dominant species on guava along with *H. bradyi*; only *H. theivora* exists on *Chromalaena odorata* (L.) and tea (Sundararaju and Sundarababu, 1999a). Venkata (2009) recorded the activity of *Helopeltis* spp on *Annona* spp while Srikumar and Bhat (2013a) recorded its activity on Singapore cherry (*Muntingia calabura* L.).

In India, this pest is common in most of the cashew growing regions of Kerala, Karnataka, Maharashtra, Tamil Nadu, Goa, Andhra Pradesh, Orissa and Madhya Pradesh (Sundararaju *et al.*, 1999) with loss in nut yield of 25 to 50 per cent. Apart from *H. antonii, H. theivora. H. bradyi* and *Pachypeltis measarum* Kirk. were also recorded on cashew causing similar damage in certain areas. *Helopeltis* spp. have been recorded from a vast number of plants in the Oriental and Australasian regions. From an economic perspective the primary plants attacked are cashew, cinchona, cocoa and tea. Estimates of crop loss attributed to damage by *Helopeltis* are variable and depend on factors such as agricultural practices, locality, climate and the plant and insect species involved as reported by Stonedahl (1991). A wide variety of host plants such as mahogany, neem, cocoa, cinchona, red gram, apple, guava, grapevine, drumstick, mango, *Syzygium* sp., *Lawsonia alba*, black bepper and all spice (Devasahayam and Nair, 1986) *Annona reticulata, Thespesia populnea, Solanum nigrum* and ornamental *Acalypha* (Sundararaju and Bhakthavatsalam, 1994) harbour this pest during off season.

The egg and nymphal period last 6-12 and 10-14 days respectively. The developmental period of eggs and nymphs depends mainly on temperature. It was longer during winter (December-January) and shorter during summer (April-May). The adults survive for more than a month and the female bug can lay up to 259 eggs during its life time. Both nymphs and adults suck sap from tender shoots, petioles, midribs and laminae of tender leaves, panicles, flower buds and immature fruits. Their feeding causes the drying up of new flushes resulting in a scorched appearance to the trees, shrivelling and drying of immature nuts.

2.2. Nature of Damage

Tea mosquito bug *Helopeltis* antonii Sign (Heteroptera:Miridae) is the most serious foliage and flower pest of cashew in India, Indonesia, and other major cashew growing areas. It is also a major pest on cocoa, tea and neem (Stonedahl, 1991; Sundararaju, 1996). The pest is distributed in most of cashew growing tracts of Kerala, Karnataka, Goa, Maharashtra, Tamil Nadu, Andhra Pradesh and Odisha with yield loss of 30-50 per cent (Desai *et al.*, 1977; Abraham and Nair, 1981). It generally spreads from neem trees to guava, ber, drumstick and cashew. The eggs are inserted in the tender parts of shoots, petioles and midribs of leaves during flushing and on flower buds, flowering panicles, peduncle rachii and immature fruits and nuts resulting in severe yield losses. A pair of fine thread-like unequal chorionic processes of 0.4-0.6mm length projecting outside is indicative of the presence of each egg inside the plant tissues.

The typical feeding damage results in the formation of brownish or darker necrotic lesions (Figure 8.1). The lesions are roundish on fruits and longitudinal on shoots. This is a low-density pest and a very small population is sufficient to cause conspicuous injury. Under vulnerable stage, its population increases very rapidly within a month leading to severe loss in yield. In guava and ber, feeding lesions are roundish on fruits. Sometimes greenish eruptions appear followed by necrosis. Each nymph or adult can cause more than 100 lesions on fruit buds or immature fruits and feeding lesions coalesce and the tender shoots and the entire panicles having tender immature apples and nuts wilt subsequently resulting in burnt-up appearance.

The life table analysis carried out in Indonesia (Siswanto *et al.*, 2008) and in India (Sundararaju and Sundarababu, 1998; Srikumar and Bhat, 2013b) indicated that the

Figure 8.1: Tea Mosquito Bug Damage on Cashew.
(a) On shoot, (b) Severe damage on plantations.

survivorship of the *H. antonii* population fed shoots of cashew was a type II with a high hatchability and bulk death occurring during early nymphal stages followed with a relatively lower death throughout the older stages. Oviposition peaking towards the end of the female life span was not an advantage for the population growth because the adult life span was short. Hence managing this pest during very early stage of attack will be quite effective. *H. antonii* was found to multiply very rapidly under most favourable conditions on neem and cashew, and also to sustain under unfavourable conditions on neem and guava. This suggests that appropriate management measures should be taken especially during vulnerable stages of cashew crop (flowering and fruiting period) before the onset of population build-up of *H. antonii* (Sundararaju and Sundarababu, 1998).

2.3. Seasonal Incidence

Build up of *H. antonii* population and its damage commenced from October--November onwards synchronising with the emergence of new flushes/panicles after the cessation of monsoon rains. Maximum shoot damage of 49.5 per cent during November and high panicle damage of 72.1-73.9 per cent from December to February with a peak pest population during February was recorded by Sundararaju(1984). In young plantations the pest is noticed continuously with a higher intensity during February and March (Sathimma, 1978). When weather parameters were related with weekly mean TMB population, only minimum temperature has shown consistently negative relationship. Thus the existence of low minimum temperature (*i.e.* below 20.0° C) continuously for more than one week during the months of December-January also provides a clue for further surveillance to decide insecticide control option. For undertaking the management decision on insecticidal control, only farm based monitoring of TMB damage and population from first month of flushing to fruit set appears to be most appropriate (Sundararaju, 2007).

3. Cashew Stem and Root Borer (CSRB)

3.1. Species Composition and Distribution

Plocaederus ferrugineus L. (Coleoptera: Cerambycidae) is the most important species that infests cashew in most parts of cashew growing areas in India and two other species, *viz.*, *P. obesus* G. and *Batocera rufomaculata* DeG. were also reported in association with *P. ferrugineus* (Abraham, 1958 and Rai,1984). The pest is capable of causing death of 1 to 10 per cent of the productive trees annually if left unnoticed (Ramadevi and Krishna Murty, 1983; Jena *et al.*, 1985 a, Samiayyan *et al.*, 1991; Mohapatra and Mohapatra, 2004; Mohapatra *et al.*, 2007). *Plocaederus* spp. is a serious pest both in west coast and east coasts of India (Figure 8.2). Infestation of pest varying from 1.6 to 10.0 per cent has been reported in few tracts of Kerala, Karnataka, Tamil Nadu and Maharashtra (Pillai *et al.*, 1976). *P. ferrugineus* L. is the predominant species while *P. obesus* Gahan and *Batocera rufomaculata* De G. also co-occur and infests Cashew (Rai, 1984). The extent of damage in different cashew plantations of forest departments of Kerala and Tamil Nadu was found to be 7-20 per cent and 30- 45 per cent respectively (Misa and Basu Choudhuri, 1985). Management of this pest is a tough job, as the borer remains in a concealed condition in the interface of

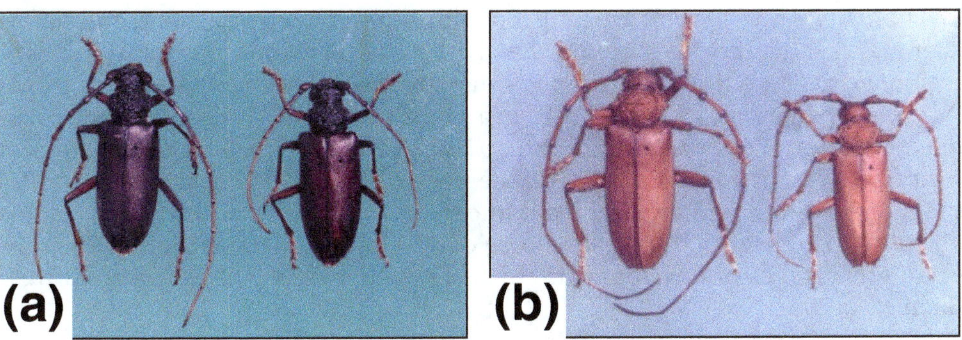

Figure 8.2: Stem and Root Borer.
(a) Adults of *Plocaederus ferrugineus*, (b) Adults of *P. obesus*.

bark and hard wood, normally it escapes from the attack of the natural enemies. Secondly, application of pesticides is not very effective as the grubs remain inside a thick protective layer.

Weevil borers of the genus *Marshallius* (Curculionidae) are found to be associated with the cashew tree *viz.*, *M. multisignatus* (Boheman) in Oara and Amazonas, in Brazil and in the French Guyana; *M. anacardi* Lima in states of Para, Piaui and Rio Grande do Norte, Rio de Janeiro, *M. bondari* Rosado-Neto in the states of Bahia and Piaui which are responsible for death of a high number of trees esp., dwarf types (Bleicher *et al.,* 2010).

3.2. Nature of Damage

Symptoms of damage include extrusion of frass, occurrence of gummosis, premature shedding of leaves, drying of twigs and, finally, death of the tree. The larval period continues for 6-7 months and during that time, it tunnels extensively the bark all round the collar region and completely damage stout lateral and taproots even up to a depth of 60 cm. Whenever, more than 50 per cent bark circumference damaged at the collar region, the tree will succumb with yellowing of leaves. Sometimes, the unexposed stout lateral and taproots will be extensively damaged without any external symptoms and such trees may look very healthy or with sickly appearance and suddenly die without any yellowing of leaves. A single grub can kill a two-year-old cashew tree during its development and it can tunnel around one square foot area of bark wood tissue. Infestation range varied widely recording 7-20 per cent and 30-35 per cent damaged trees from Kerala and Tamil Nadu respectively (Misra and Basu choudhuri, 1985). Its infestation is severe in unattended plantation and infested trees act as source of inoculum (Jena, 1990). For this reason, it is important to intercept early and prophylactic management to reduce intensity of infestation. The symptoms of damage include the presence of small holes in the collar region, gummosis and extrusion of frass through the holes, yellowing and shedding of leaves and drying of twigs. After hatching, the grubs bore into the fresh tissues of the bark and feed on the subsequent sub epidermal and sapwood tissues and make tunnels in irregular directions. As a result of the injury to the cells, resinous material oozes out which on exposure gets hardened.

The bore holes and tunnels are plugged with chewed fibers, frass and excreta. At later stages, attacked trees are not having desired softwood tissues in the bark region, extrusion of powdery frass is generally seen at the base of the tree trunk. As a result of damage by many grubs in a single tree, the bark all around the collar region withers away. Since the flow of sap had arrested completely through phloem region, leaves turn yellow and shed. The tree will be killed within a period of 1-3 years depending upon the pest load. In certain stages, even a two –year plant is killed by a single grub during its course of development and as high as 90 grubs of different stages of development are also seen damaging 15-20 years old trees (Bhat *et al.*, 2002).

3.3. Biology

Adults of *B. rufomaculata* are greyish, measuring 50 cm in length and has yellowish or orange spots on the forewings. The grubs of this species are apodus (legless) and pupate without forming any calcareous cocoon. The grub period lasts for about 6 months and the adults of *P. obesus* are chestnut coloured, longicorn beetles, measuring about 40 mm in length and with slight pubescence. Adults of *P. ferrugineus* are dark reddish brown, medium sized beetles (25 to 40 mm in length) and are sluggish on the day of emergence, mating starts on the second day, and repeated matings occur during the life time of the adults. Eggs are usually deposited in the crevices of the bark of main trunk up to one metre height from ground level and also on the exposed roots and also in soil close to collar region of the tree. Eggs are pale white, ovoid and smooth measuring about 4.5 x 2.0 mm. The nascent first instar grubs feed on the tissue near the site of oviposition and extrusion of fine dusty frass is noticed within few days of hatching. The larval period continues for 6-7 months. The fully grown grub measuring about 100 mm in length enters into heart wood for pupation and makes a circular exit hole of 1.5 cm width for adult emergence. The pupation takes place inside a calcareous cocoon. The adults form within 40 to 60 days but lie quiescent within the cocoon and emerge out after 45-60 days. Under laboratory conditions, pupation occasionally occurs without calcareous cocoon (Pillai, *et al.*, 1976; Godse *et al.*, 1990; Raviprasad and Bhat, 1998).

3.4. Seasonal Incidence

Even though, the occurrence of the pest is noticed throughout year in both East and West Coast, relatively large population of grubs and severe infestation could be seen in the coastal Karnataka and Andhra Pradesh during March-May and May-July, respectively (Abraham, 1958; Ramadevi and Krishnamurthy, 1983; Jena *et al.*, 1985a and b).

4. Shoot Tip Caterpillar

Two species of lepidopteran caterpillars are known to infest cashew shoot tips during flushing period and cause considerable damage. Gelechid caterpillar, *Anarsia epotias* Meyr causes damage to shoot tips. The pale yellowish green young caterpillars with black head bore in to the terminal shoots and tunnel inside up to 2-3 cm. A gummy substance oozes out from the infected tips and finally the attacked shoots dry up. The egg, larval and pupal period lasts for 3-4 days, 12-

16 days and 7-10 days respectively and the life cycle is completed in 25-29 days (Ramamony, 1965). Similarly, the tiny, yellowish to greenish-brown larvae of the moth *Hypotima* (=*Chelaria*) *haligramma* M. (Lepidoptera: Gelechidae) also damage shoot tips by folding the fresh leaves and feeding within. Tender shoot tips are bored occasionally up to 25-35 mm, leading to drying-up of shoot tips. This pest is regularly reported from the East Coast tracts (Mohapatra *et al.*, 1998). These larvae also damage inflorescences in the subsequent period. Extent of damage caused by the shoot tip caterpillar revealed 13.0 per cent incidence in newly emerging post-monsoon flushes (in October) and 10.5 per cent in pre-monsoon flushes (in May) and thus indicating the significance of this pest under the west coast condition (Pillai *et al.*, 1976; Mohapatra *et al.*, 1998).

5. Leaf Miner

The leaf miner, *Acrocercops syngramma* M. (Lepidoptera: Gracillaridae) is one of the serious pests of cashew during post monsoon period all over the country. Reports of occurrence of this pest are available for Goa (Sundararaju, 1984), Odisha (Jena and Satapathy, 1989) and also Andaman Islands (Jacob and Belvadi, 1990). The mining injury by caterpillars was noticed both in the tender leaf as well as in tender shoots. Young plants are observed to be more prone to attack by this pest. The caterpillars mine and feed below the epidermal layer of the tender leaves causing extensive leaf blisters which later dry up, causing distortion, browning and curling of the leaves. As the attacked leaf ages, the holes are formed due to drying out of the damaged portion. The adult is a silvery grey moth lays eggs on tender leaves. The freshly hatched larvae and younger larvae are pale white in colour, while full grown caterpillars measure about 5 mm in length, are reddish brown and feed by scraping the mesophyll below the epidermis. After full development, the larvae fall down to the soil where they pupate and the pupal period lasts for 7-9 days. Abraham (1959) estimated the leaf miner damage to be 26 per cent in severely infested leaves, while 70-80 per cent, 60 per cent, 6-20 per cent and 18-20 per cent leaf damage was reported by Basu Choudhuri (1962), Rai (1984), Ayyana *et al.* (1985) and Chatterjee (1997) from Kerala, Karnataka, Andhra Pradesh and West Bengal respectively. Though, negative relationship between rainfall and pest incidence was observed in Tamil Nadu, there was no significant influence of abiotic factors on this pest in Odisha region (Mohapatra and Barik, 1996). Besides cashew, jamun and mango serve as additional hosts for this pest (Butani, Sundararaju, 1984; Rai,1984; Sundararaju, 1984).

6. Hairy Caterpillars

The hairy caterpillars of *Euproctis* spp. (Lepidoptera: Lymantridae) feed in groups on the inflorescence and tender nuts of cashew. They scrape the green tissues on the inflorescence branches and feed on the shell of the nut in the tender green stage (listed 21 species of lepidopterans as leaf feeders of which only two species as *Metanastria hyrtaca* Cram (Lasiocampidae) and *Lymantria obfuscata* Wlk (Lymantridae) cause severe sporadic defoliation in cashew. Later, *L. ampla* was the correct name where, *L. obfuscata* was erroneously used. The caterpillars defoliate the cashew trees completely leaving only bare branches. The biology of *M. hyrtaca* was studied by Arjuna Rao *et al.* (1976) and early instar caterpillars were gregareous feeders on tender foliage and the full grown caterpillars fed voraciously on mature leaves. They congregate in large numbers on the ground under dry leaves near the base of the tree in crevices of bark or as lower parts of well shaded branches.

7. Leaf Thrips

Occurrence of foliage thrips *viz.*, *Selenothrips rubrocinctus*, *Rhipiphorothrips cruentatus* Hood and *Retithrips syriacus* (Mayet) have been reported on cashew (Ananthakrishnan, 1984; Ayyanna *et al.*, 1985; Jena *et al.*, 1985). The red banded thrips *S. rubrocinctus*, is a tropical-subtropical species thought to have originated in northern South America (Chin and Brown, 2008) and is found in parts of Asia, Africa, Australia and Pacific, North America, Central America, South America and West Indies. *S. rubrocinctus* and *R. cruentatus* cause severe damage to young plantations, particularly during summer in Tamil Nadu. The adults and immature stages of thrips colonise the lower surface of leaves. As a result of this rasping and sucking activity, the leaves become pale brown and slightly crinkled with roughening of the upper surface.

8. Leaf Beetles and Weevils

During rainy season (June–August), the chrysomelid leaf beetles and weevils defoliate cashew voraciously. Of the 14 coleopteran leaf beetles listed by Rai (1984), the chrysomelid beetles *Monolepta longitarsus* Jal. is an important regular pest in the west coast regions during the south west monsoon. These appear abundantly especially in young trees and skeletonise the leaves which gradually dry up. An ash coloured chrysomelid *Neculla pollinaria* also attacks the post harvest flushes and also the upcoming tender shoots and buds. *Microserica quadrinotata* Moscr (Melolonthidae) was recorded as a pest of cashew in Odisha from June–October with peak during September. The adults skeletonise the leaf by scrapping chlorophyll and up to 30 per cent leaf damage was reported by Jena *et al.* (1985b). In nursery, it appears in serious proportion in the middle of July. Damage of 87.3 per cent seedlings with 41.1 per cent leaf damage was recorded.

9. Pests of Cashew Apples and Nuts

Several Lepidopteran, coleopteran, dipteran and hemipteran pests are known to damage apples and nuts of cashew during different developmental stages. Under lepidopterans, two species, *Thylocoptila paurosema* M and *Hyalospila leuconeurella* R. (Lepidoptera: Pyralidae) are important. *T. paurosema* attack tender apples and nuts. Damaged nuts get deformed and dry away. Eggs are laid on the fruits. Dark pink larvae initially damage flowers by webbing the panicles and feed the unopened flower buds. Then they bore inside the tender nuts and developing apples resulting in shrivelling and premature fall. In the developed green nuts and apples, caterpillars tunnel near the junction of apple and nut and the boreholes are plugged with frass and excreta. Usually, these damaged apple and the nuts shrivel and fall prematurely. Damaged fruits can be easily located as they have frass. This pest was also reported as storage pests of cashew nuts (Rai, 1984). *Nephopteryx* sp. is common in Tamil Nadu and Andhra Pradesh (Ayyanna *et al.*, 1985; Dharmaraju *et al.*, 1976) and up to 60 per cent of nut damage was reported (Dharmaraju *et al.*, 1974). Besides leaves and shoot, *Orthaga exvinacea* also damages the apple (Sreeramulu *et al.*, 1976). Similarly, *Hyalospila leuconeurella* Ragnot (Pyralidae) and *Anarsia epotias* Meyr. (Gelechidae) were recorded as apple and nut borers in South India (Basu Choudhuri and Misra, 1973). The larvae of *H. leuconeurella* bore through the apple from one end to the other end and remain inside the apple till the fruit drops and when nuts are attacked they get deformed (Rai, 1984). However, *L. moncusalis*, *O. exvinacea* and *Euproctis* spp. are considered to be external feeders on tender fruits and apples.

Aphids (*Toxoptera aurantii*, mealybugs (*Planococcus citrii* and *Ferrisia virgata*) sucks saps of immature apples and nuts. Flower thrips such as *Rhynchothrips raoensis* G. and *Scirtothrips dorsalis* H., besides flowers, scraps on immature apples and nuts results in the malformation of nuts and immature fruit drop (Bhat *et al.*, 2002). A pentatomid bug, *Catacanthus incarnatus* also occur as apple pests occasionally (Bhat and Srikumar, 2013). One coreid bug, *Cletus rubidiventris* West. sucking sap from immature cashew apple, was also recorded as a minor pest (Sundararaju, 1984). *Drosophilla melanogaster* is the very serious apple feeding insect during fruiting stages followed by *Bactocera* spp. The incidence of TMB, along with flower thrips and fruit borers led to fruit drop of 1.0 to 9.0 per cent during the mustard stage, 6.4 per cent during the pea nut stage and 11.9 per cent fruit drop during later stages (Pillai and Pillai, 1975). Under coleopteran pests, *Carpophilus* sp. was recorded in India and *Macrodactylus pumilio* Burm. from Brazil feeding on ripe apples (Ohler, 1979).

10. Inflorescence Feeders

10.1. Leaf and Blossom Webber

Cashew shoots bearing fresh flushes and flowers are attacked by two species of leaf and shoot webbing caterpillars, *Lamida* (= *Macalla*) *moncusalis* Wlk. (Lepidoptera: Pyralidae) and *Orthaga exvinaceae* Hamps. (Lepidoptera: Noctuidae). Of these, *L. moncusalis* is a major pest in East Coast tracts. Symptoms of infestation are presence of webs on terminal portions, with clumped appearance, and drying of webbed shoot/inflorescences. Eggs are deposited ventrally on leaves and occasionally on tender shoots singly or in groups of six. The egg, larval, pre-pupal and adult stages last 4-7, 16-22, 9-15 and 3-6 days respectively (George *et al.*, 1984). This pest was found to be sporadic in certain pockets and maximum infestation of 26 per cent was noticed in one of the affected areas. During post monsoon period, the caterpillars feed on the terminal leaves of new shoots and blossoms after webbing them.

10.2. Flower Thrips

Flower thrips such as *Rhynchothrips raoensis* G., *Haplothrips ganglbaueri* (Schmutz), *Thrips hawaiensis* (Morgan), *H. ceylonicus* Schmutz, *Frankliniella schultzei* (Trybom) and *Scirtothrips dorsalis* H., cause premature shedding of flowers, scabs on floral branches, apples and nuts. The infestation on developing nuts results in the malformation of nuts and immature fruit drop (Bhat *et al.*, 2002). The occurrence of damage, extent of damage and seasonal incidence were reported for *R. roaensis* (Abraham, 1958; Ayyanna *et al.*, 1985; Patnaik *et al.*, 1986; Thirumalaraju *et al.*, 1990),

S. dorsalis, H. Ganglbeuer, T. hawaiiensis (Ayyanna *et al.*, 1985), *H. ceylonicus* and *F. schultzei* (Patnaik *et al.*, 1987).

11. Integrated Pest Management

11.1. Cultural Practices

11.1.1. Habitat Management

It is quite obligatory to keep proper surveillance at vulnerable habitats (Young cashew plantations or neem groves). In east coast, particularly in Tamil Nadu, the self sown neem trees exist in plenty on the border or fence side of fields where cashew and other horticultural crops (guava, drumstick *etc.*) are grown and these neem trees act as reservoir for this pest throughout the year (Sundararaju and Sundarababu, 1999b). Severely infested tree and dead tree should be uprooted before and after monsoon season as a main phytosanitory measure to manage CSRB. In the same pits of uprooted trees, new grafts should be planted for maintaining required optimum number of trees per hectare.

11.1.2. Monitoring

The population build up of TMB can be monitored by using single virgin (unmated) adult TMB female as bait insect. It is possible to detect the male population at 20m distance in few minutes during day time by this pheromone based technology (Sundararaju and Sundarababu, 1999b).

11.1.3. Phyto Sanitation–Weed Management

The weeds serve as host plants to important pests of cashew. For example, *Helopeltis theivora* is capable of completing its life cycle and multiply in a very common weed, *Chromolaena odorata* which is present in cashew plantations of West coast region (Srikumar and Bhat, 2013c). Fourteen weed species belonging to ten different families were found as alternate hosts of TMB during flushing period of cashew (September-October) for different species of TMB. Weeds observed to support TMB are *Terminalia paniculata, Getonia floribunda, Macaranga peltata, Chromolaena odorata, Melastoma malabathricum, Meremmia vitifolia, Solanum torvum, Cissus repanda, Strychnos nuxvomica, Ixora* sp., *Lantana camera* and *Leea* sp. Besides, ornamental plants *viz., Acalypha hispida* and *Acalypha wilkesiana* were also seen as hosts for TMB species. Maximum TMB infestation was seen during October on *C. odorata* (more than 30 per cent of weed population), *M. peltata* and *G. floribunda*. The number of weed species that support *H. antonii, H. bradyi, H. theivora* and *P. maesarum* are two, one, ten and seven species respectively. Hence, weed management is very important in this case. It contrast, there are some of the general predators like spiders, reduviids, praying mantises, coccinellids, green lace wings, syrphids, wasps that are feeding on specific weed pests also feed on pests of cashew. Hence, while insecticidal spraying, avoiding weeds may be important to sustain natural enemies and pollinators in the eco system. Removal of dead logs and pruned branches help in direct removal several pest stages of the important pest, cashew stem and root borer. A damage level of 6 - 10 per cent can be regarded as a control threshold for *H. pernicialis* in Australia (Peng *et al.*, 1997a). The pre-flowering flush appeared to be the most appropriate time to carry out the monitoring and spray programme to control *H. pernicialis*.

11.2. Mechanical Methods

Collection and destruction of various pest stages will be a very useful method of eliminating pests in the field especially larval caterpillars, pupae, cocoon, CSRB grubs and pupa, leaf webs. Yellow sticky traps and near-UV radiation reflective film have been used successfully in the tea plantations of Japan, China, and Malawai for mass trapping and repelling tea pests, which resulted in a 73 per cent and 79 per cent decrease of tea thrips and tea leafrollers, respectively. Similarly, possibility of usage of sticky traps, light traps, UV traps in cashew ecosystem has to be explored. Factors that could influence the trap catches, such as temperature, rain, moonlight, cloud-cover, shade trees, and landscape, also have to be studied in cashew plantations.

11.3. Host Plant Resistance

Histopathological investigations made in the tea mosquito bug infested tender cashew shoot revealed that cashew is inherently (genetically) provided with very active phenol-phenolase system (Sundararaju and Sundarababu, 1999b). Any feeding injury will result in rapid hypersensitive reactions leading to necroses, blighting and drying of affected parts especially tender shoot, panicle and fruits. The matured shoots of cashew irrespective of varieties exhibited highest oviposition and feeding deterrency. Every year, this type of resistant phenological stage brings down the population build up during non-flushing period (June-September) on older plantation. Mid season/late season flowering cashew varieties are able to escape from the severity of the pest infestation. One such accession, Goa 11/6, showed consistent performance with yield of 2 t/ha under unsprayed situations, under moderate level of pest incidence which was later released as variety 'Bhaskara' (Sundararaju *et al.*, 2006). Even though, incidence of shoot tip caterpillars and apple and nut borers was observed in all the recommended varieties of cashew, sometimes the fruit set was partially affected in the varieties which are having early mixed phase of flowering with male and hermaphrodite flowers, whereas in varieties which are having early male phase, the damage is severe resulting in poor fruit set.

11.4. Botanicals

Under laboratory conditions, neem extracts and their formulations are found effective for CSRB grubs and adults. It was found to have both repellent and oviposition deterrent effect against CSRB. Swabbing neem oil (10 per cent) and neem cake (1:1) twice during November and April followed by mud pasting of the base of the tree proved effective against CSRB (Chakraborthi, 2006).In a storage condition, use of plant products have several advantages over synthetic insectides and found to be most important approach in pest management. In cashew nuts, to manage *O. surinamensis*, use of extract of pomegranate, pig weed (*Chenopodium album*), sage tree (*Vitex negundo*) and Kankera (*Maytenus emerginata*) were found to have significant ovicidal effect (Neetu *et al.*, 2006).

11.5. Pheromones and Kairomones

The use of attractants in pest management systems can be precise, specific, and ecologically sound. The use of attractants and repellents against cashew pests is still in infancy which has to be explored since it is going to act a significant role in organic cashew production. Kairomonal effect of cashew bark and frass exuds towards CSRB adults is confirmed under laboratory study. The components responsible for their kairomonal activity were also studied. Similarly, in TMB, presence of sex pheromonal activity is confirmed for *H. antonii*. A confined virgin female TMB can attract more than 30 male insects in a single day under field condition. The whole body extract of virgin females of *H. antonii* was analysed through GC-MS and 17 components were identified including pinene, - hexadecenoic acid and 9-octadecenoic acid but none of them could be implicated as sex pheromone (Sundararaju and Sundarababu, 1999). The volatiles collected from virgin female and field collected female were analysed and methyl butyrate, a compound exhibiting pheromone activity in other insects of family miridae was one the compounds detected in the analyses (Bhat and Raviprasad, 2008).

11.6. Chemical Control

Spraying of insecticides during flowering season did not affect fruit set, even though cashew which is purely an insect pollinated crop. Therefore, in the endemic areas, it is appropriate to spray insecticides during most vulnerable period of the crop. Spraying coinciding with flushing, flowering and fruiting stages is quite adequate. Even though all groups of insecticides and several plant products (botanicals) were evaluated against this pest, none exhibited any ovicidal action (Raviprasad *et al.*, 2005). However, lambda-cyhalothrin (0.003 per cent) showed longest residual action against nymphs and adults (Sundararaju *et al.*, 1993). Although cashew is an insect-pollinated crop, spraying these insecticides during the flowering season did not influence fruit-set (Rai, 1984; Sundararaju *et al.*, 1993a)

Leaf and blossom webber and shoot tip caterpillars can be effectively controlled by sprays given to tea mosquito bug at the time of flushing (Patnaik *et al.*, 1987; Pillai *et al.*, 1976). Sprays given at flushing and flowering period are helpful in the management of foliage and flower thrips (Ganesh Kumar and Palaniswarny, 1983; Pillai *et al.*, 1976). For the management of apple and nut borers collections and burial of the damaged fruits and spraying with carbaryl (0.1 per cent). The injured portion and base of the tree, and any exposed root should be swabbed with/drenched in chlorpyriphos 0.2 per cent solution (Raviprasad *et al.*, 2009; Sundararaju and Bakthavatsalam, 1994) for the management of CSRB.

11.7. Biological Control

Natural enemy (NE) diversity in the cashew ecosystem has a significant role in biological control of various cashew pests. Several NEs have been recorded as parasitoids (various trichogrammatids, braconids, bethylids, eulophids, ichneumonids, tachinids), predators (coccinellids, syrphids, mirids, phytoseiids, and spiders), and pathogens [entomopathogenic fungi (EPF), entomopathogenic nematodes (EPN)] naturally.

11.7.1. Parasitoids

A parasitoid *Erythmelus helopeltidis* Gahan (Mymaridae : Hymenoptera) was recorded to parasitize the eggs of *H. antonii* (Devasahayam and Radhakrishnan Nair, 1986). Subsequently, two hymenopteran egg parasitoids namely, *Telenomus* sp. (Laricis group) (Scelionidae) and *Chaetostricha* sp. (Trichogrammatidae) were reported (Sundararaju, 1993a). *E. helopeltidis* Gahan (Hymenoptera: Mymaridae) *Telenomus* sp. (Laricis group) (Scelionidae) *Chaetostricha* sp. (Trichogrammatidae) and *Gonatocerus* sp. nr. *bialbifuniculatus* Subba Rao are the egg parasitoids reported on this pest from West Coast regions, while, *Ufens* sp. is an egg parasitoid reported from the East Coast (Vridhachalam). *E. helopeltidis* Gahan (Hymenoptera: Mymaridae), was recorded as an egg parasitoid from *Pachypeltis maesarum* (Heteroptera: Miridae) (Bhat and Srikumar, 2012). The build-up of TMB is naturally regulated through these egg-parasitoids (Devasahayam, 1989; Sundararaju, 1993a,1996). Two eulophid larval parasitoids, *viz.*, *Sympiesis* sp. and *Cirropilus* sp. (Hymenoptera: Eulophidae) were recorded on leaf miners in Goa, the former species is dominant one and highest parasitism up to 59 per cent parasitism was observed (Sundararaju, 1984). In Kerala, Karnataka and Orissa, the pest is parasitized by *Chelonus* sp., *S. purpurea* and *Sympesis* sp. respectively (Sundararaju and Bhakthavatsalam, 1994).

Panerotoma sp. (Braconidae) and *Trathala* sp. (Ichneumonidae), have been recorded as hymenopteran larval parasitoids of *T. paurosema*. Even though incidence of shoot tip caterpillars and apple and nut borers was observed in all the recommended varieties of cashew, fruit-set was only partially affected in varieties that showed early mixed-phase of flowering, with male and hermaphrodite flowers; whereas, in varieties with an early male-phase, damage was severe, resulting in poor fruit-set. The insecticidal sprays recommended against tea mosquito bug also manage all these minor foliage pests, if infestation levels are low to medium (Pillai *et al.*, 1984). However, indiscriminate spraying must be avoided as the above mentioned pests are parasitised by a number of larval parasitoids. Repeated spraying can also induce outbreak of secondary pests like mealybug (*Ferrisia virgata*) and aphids (*Toxoptera odinae*). *Perilampus microgastri* Ferr. (Perilampidae) was reared from a parasitised larvae collected from the field (Pillai *et al.*, 1976).

Two species of Braconid parasite (*Apanteles* spp.) and one Elasmid (*Elasmus johnstonii* Ferriere) were found to parasitise *L. moncusalis* to an extent of 10-50 per cent during July-August and, in Orissa, *Apanteles* sp., *Bracon brevicornis* W. (Braconidae), *E. johnstonii* and *Blepharella lateralis* T. were also reported (Jena *et al.*, 1985b).

11.7.2. Predators

An array of ants, spiders, mantids, reduviids, coccinellids and a few wasps have been identified as the main natural enemies of various cashew pests during different parts of the season. Many pests are preyed upon by a large assortment of natural predators such as spiders and mites, lacewings, predatory thrips, and predatory bugs, especially minute pirate bugs like red banded thrips (Chin and Brown, 2008), leaf beetles, TMB, aphids, thrips and flower pests (Sundararaju,1993b).

11.7.2.1 Spiders

Spiders are potential biological control agents in agroecosystems (Riechert 1990). Several species of spiders, *Hyllus* sp., *Oxyopes sehireta, Phidippus patch* and *Matidia* sp. *Sycanus collaris* (Fab), *Sphedanolestes signatus* Dist. and *Endochus inornatus* Stal., *Irantha armipes* Stal. and *Occamus typicus* Dist. been recorded as predators (Sundararaju,1993). The mean spider population varied from 0.22 to 0.31 per panicle and the spider population had influence on arthropod complex during fruiting season(Sundararaju, 2004). Bhat and Srikumar (2013a) reported spiders as indigenous natural enemies of TMB and the study revealed occurrence of 117 species of spiders belonging to 18 families *viz.*, araneidae, clubionidae, corinnidae, gnaphosidae, hersilidae, linyphiidae, lycosidae, miturgidae, nephilidae, oxiopidae, pholcidae, pisauridae, salticidae, sparassidae, tetragnathidae, theridiiae, thomisidae and uloboridae. Salticids were predominant (30 per cent) followed by araneidae (22 per cent). Field observation revealed that *Telamonia dimidiate* and *Oxyopes shweta* as the major predators of *Helopeltis* spp. The spiders *viz.*, *Agriope pulchella, Cyclosa fissicauda, Eriovixia laglazei, Neoscona mukerjeri, Nephila pilipes, Oxyopes sunandae, Bavia kairali, Carrihotus viduus, Epocilla aurantiaca, Hyllus semicupres, Achaeranea mundula, Camariacus formosus and Thomisus lobosus* were also superior with respect to its predatory activity. This rich diversity of spiders is indicative of overall biodiversity of cashew plantation since spiders are considered to be very useful.

11.7.2.2. Ants

The green ant, *Oecophylla smaragdina* (Fabricius), is an effective predator, and it can significantly reduce the numbers of over 30 important insect pest species of many tropical crops (Way and Khoo, 1992). The green ant can significantly reduce the damage levels of the main cashew insect pests, such as the tea mosquito bug, *Helopeltis pernicialis* (Stonedahl, Malipetil and Houston), the mango tip-borer, *Penicillaria jocosatrix* (Guenee), the fruit spotting bug, *Amblypelta lutescens* (Distant), the leaf-roller, *Anigraea ochrobasis* (Hampson), and the green vegetable bug, *Nezara viridula* (Fabricius) (Peng *et al.*, 1995; 1997a, b; 1998). However, its potential with respect to cashew yield, which is the major concern of cashew growers, has not been fully demonstrated. Under west coasts of India, in the cashew ecosystem, inter-colony rivalry and death of queen due to infection by broad spectrum mycopathogen (*Beauveria bassiana*) were commonly observed in the case of *O. smaragdina* and these might be possible reasons for low establishment (Sundararaju, 2004).

The role of *Doliclioderus thoracicus* to control *Helopeltis* spp. has been extensively studied and well understood (Way and Khoo, 1991; Khoo, 1992). The predatory ant, *O. smaragdina*, was also found in high numbers for each observation and no *H. antonii* was found on cashew plants occupied by this ant. In Northern Australia, *O. smaragdina* has been used to control *H. pernicialis* on cashews (Peng *et al.*, 1995, 1997a,b, 1999a,b). Other predators frequently found in quite high numbers were arachnids and to a lesser extent mantids and coccinellids. (Siswanto, 2008). Among the five species of ants *viz.*, *Camponotus* sp., *Anoplolepis longipes, Crematogaster* spp., *Paratrechina longicornis* and *O. smaragdina* observed under cashew trees, only *Oecophylla* could control TMB effectively, while other ants found to feed on extra floral nectaries and as scavengers (Sundararaju, 2000). *Crematogaster wroughtonii* Forel (Formicidae) has been recorded as a predator of nymphs of TMB (Ambika and Abraham, 1979).

Forty-nine species of ants (Fam: Formicidae) belonging to five subfamilies were recorded having multiple roles like predators, pollinators, scavengers, extra floral nectarine feeders *etc.* Ants belonging to Myrmicinae subfamily were dominant (22 species) followed by formicinae (13 species). Among the ant species, *Oecophylla smaragdina* (Fabricius) and *Anoplolepis gracillipes* Smith were most abundant, while *Camponotus compressus* and *C. sericeus* were found throughout the year. The activities of most ant species are predominant during flowering and fruiting period (November-April) and pre monsoon period (May), while during heavy rain *i.e.*, South West monsoon, activities of *Myrmicaria brunnea, C. sericeus, Prenolepis naoroji* and *C. angusticollis* were only seen. In a single tree, foraging activities of maximum of seven species were found at a time especially during flowering and initial fruiting season(Vanitha *et al.*, 2015).

11.7.2.3. Reduviids

Reduviids (Hemiptera: Reduviidae: Harpactorinae) are recorded as potential natural enemies of *Helopeltis* spp. (Stonedahl, 1991; Sundararaju, 1996). Five species of reduviids *viz., Sycanus collaris* Fab., *Sphedanolestes signatus* Dist., *Endochus inornatus* Stal, *Irantha armipes* Stal and *Occamus typicus* Dist., were reported as predators of *Helopeltis antonii* Sign. on cashew in India (Sundararaju, 1984). All these predators are capable of devouring 1-5 tea mosquito nymphs/adults per day. A total of 16 species of reduviids belonging to the subfamily Harpactorinae *viz., Alcmena* sp., *Biasticus* sp., *Cydnocoris gilvus* Burmeister, *Endochus albomaculatus* Stal, *Endochus* sp., *Endochus* sp., *Epidaus bicolor* Distant, *Euagoras plagiatus* Burmeister, *Irantha armipes* Stal, *Lanca* sp., *Panthous bimaculatus* Distant, *Rhynocoris fuscipes* Fabricius, *Rihirbus trochantericus* Stal var. *sanguineous, Rihirbus trochantericus* Stal var. *luteous, Sphedanolestes signatus* Distant and *Sycanus galbanus* Distant were recorded from cashew ecosystem. The damage to cashew trees by tea mosquito bug can be reduced by the introduction of assassin bugs (Sundararaju, 1984; Bhat *et al.*, 2013).

12. Conclusion

The perennial nature of cashew is ideal for assembly of vast number of arthropods. It is essential to have concerted efforts to popularize plant protection measures in cashew. Basically, cashew farmers have to be trained about the nature of the initial damage symptoms for correct identification of the pests. Need based and timely application of pesticides based on surveillance of pests is very essential, since the indiscriminate sprays may result in elimination of natural enemies.

References

Abraham, C. C. and Nair, G. M. (1981). Effective management of the tea mosquito bugs for breaking the yield barriers in cashew. *Cashew Causerie* 3(1): 6-7.

Abraham, E.V. (1958). Pests of cashew (*Anacardium occidentale*) in South India. Indian *J. Agric. Sci.* 28: 531-544.

Ananthakrishnan, T.N. (1984). Host relationships and damage potential of thrips infesting cashew. In: Cashew research and development (Eds., E.V.V.B. Rao and H.H. Khan) Indian Society for Plantation Crops, CPCRI, Kasargod, India, pp. 132-134.

Arjuna Rao., Dharmaraju, E.and Ayyanna, T. (1976). Biology of *Metanas triahyrtaca* Cram., a serious pest of cashew. *Andhra Agri.J.* 23: 61-68.

Ayyanna, T., Tejkumar, S. and Ramadevi, M. (1985). Distribution and status of pests of cashew in coastal districts of Andhra Pradesh. *Cashew Causerie* 7: 4-5.

Ayyar,T.V.R. (1942). Insect enemies of cashew plant (*Anacardium occidentale* L.) in South India. *Madras Agric. J.* 30: 223-226.

Barkade, D., Landge, S.A., Kadu, R.V. and Godase, S.K. (2010). Study on evaluation of different insecticides against leaf eating caterpillar (*Thalossades dissita* Walker) on mango and cashew. *International Journal of Plant Protection*, 3(2): 170-173.

Basu Choudhuri, J.C. (1962). Preliminary investigation on insect pests of cashew plantations in Kerala. *Indian Forster* 88: 516-522.

Basu Choudhuri, J.C. and Misra, M.P. (1973). *Hyalopsila leuconeurella* Ragnot (Lepidoptera: Pyralidae) and *Anarsia epotias* Meyrick (Lepidoptera: Gelechidae) two pests of cashew apple and nut in south India with a note on their control. *Journal of Plantation Crops* 1 (S): 168-170.

Bhat, P.S. and Raviprasad, T.N. (1996). Pathogenicity of entomopathogenic fungi against cashew stem and root borer *Plocaederus ferruginues* L. (Coleoptera: Cerambycidae). *Journal of Plantation Crops* 24: 265-271.

Bhat, P.S. and Raviprasad, T.N. (2008). Sex pheromone of tea mosquito bug *Helopeltis antonii* Signoret (Miridae:Heteroptera). *Journal of Plantation Crops* 36(3): 451-453.

Bhat, P.S. and Srikumar, K. K. (2012). Record of *Erythmelus helopeltidis* Gahan (Hymenoptera: Mymaridae), an egg parasitoid from *Pachypeltis maesarum* (Heteroptera: Miridae) infesting cashew. *Pest Management in Horticultural Ecosystems* 18(1): 103-104.

Bhat,P.S., Srikumar,K.K. and Raviprasad, T.N. (2013a). Seasonal diversity and status of spiders (Arachnida:Araneae) in cashew ecosystem. *World Applied Sciences Journal* 22(6): 763-770.

Bhat,P.S., Srikumar.K.K, Raviprasad, T.N. Vanitha.K, Rebijith.K.B. and Asokan. R. (2013). Biology, behaviour, functional response and molecular characterization of *Rihirbus trochantericus* Stal var.*luteous* (Hemiptera: Reduviidae: Harpactorinae) a potential predator of tea mosquito bug (*Helopeltis* spp.) *International Journal of Tropical Insect Science.*

Bhat, P.S., Sundararaju, D. and Ravi Prasad T.N. (2002). Integrated management of insects, pests and diseases in cashew. Indian Cashew Industry. (Eds.Singh, H.P., Balasubramanian, P.P. and Hubbali, V.N.). DCCD, Cochin, India.

Bleicher E., Rodrigues S.M.M., Melo Q.M.S. and Pinho J.H. (2010). Minimal effective dose of phosphine to control the cashew root borer, *Marshallius bondari* Rosado-

Neto (Coleoptera: Curculionidae). Rev. Ciênc. Agron. vol.41 no.2 Fortaleza Apr./June. http://dx.doi.org/10.1590/S1806-66902010000200021.

Butani, D.K. (1979). Insects and fruits. International Book Distributors, Dehradun, India.

Chatterjee, M.L. (1997). A study on leaf miner infestation and their control by insecticides.*Cashew Bull.* 4: 9-11.

Chakraborthi, S. (2006). Effect of some prophylactic management tactics for cashew stem and root borer. *J. Appl. Zool. Res.* 17(1): 77-79.

Chin, D. and Brown, H. (2008). Red-banded thrips on fruit Trees. Agnote. (http://www.nt.gov.au/dpifm/Content/File/p/Plant_Pest/719.pdf) (19 August 2008).

Desai, M.A., Kulkarni, S.V. and Rodrigues, A. (1977), Report on the marketing survey on cashew in Goa. Government of Goa, Deman and Diu, pp. 11-17.

Devasahayam, S. (1986). Some observation on insects visiting cashew inflorescence. *Cashew Bulletin.* 23(5): 1-5.

Devasahayam, S. (1989). *Erythmelus helopeltidis* Gahan(Hymenopera:Mymaridae) a new parasite of *Helopeltis antonii* Sign. on cashew. *J. Bombay Nat.Hist. Soc.* 86: 113.

Devasahayam, S. and Nair, C.P.R. (1986). The tea mosquito bug *Helopeltis antonii* Signoret on cashew in India. *Journal of Plantation Crops* 14: 1-10.

Dharmaraju, E., Ayyanna, T. and Sreeramulu, C. (1976). *Nephopteryx* sp. as an apple and nut borer on cashew. *Indian Cashew Journal* 10: 9-11.

Dharmaraju, E., Rao, P.A. and Ayyanna, T. (1974). A new record of *Nephopteryx* sp. as an apple and nut borer on cashew in *Andhra Pradesh. J. Res.* Andhra Pradesh Agric. Univ., 1 (4 and 5), 198.

George, M.V., Singh, Vijay and Thomas Peter, (1984). Aerial spraying against tea mosquito bug in cashew. In: Cashew Research and development.(eds.E. V. V.Bhaskar Rao and H.H.Khan) Indian Society for Plantation Crops, CPCRI, Kasaragod, India pp. 120-125.

Godse, S.K. Dumbre, R.B. and Karat, S.B. (1990). Bionomics of cashew stem and root borers, *Plocaederus ferrugineus* L. *Cashew Bull.* 27: 1-17.

Freitas, B.M. (1997). Number and distribution of cashew (*Anacardium occidentale*) pollen grains on the bodies of its pollinators, *Apis mellifera* and *Centris tarsata.* *Journal of Apicultural Research* 36: 15-22.

Ganesh Kumar, M. and Palaniswamy, K.P. (1983). The efficacy of certain synthetic pyrethroids against foliage thrips *Selenothrips rubrocinctus* Giard on cashew. *Cashew Causerie* 5: 11-13.

Jacob, T.K.and Belvadi.V.V.(1990).The cashew leaf miner, its pest status and larval size relationship with leaf area damage in Andamans. *The Cashew* 4: 10.

Jena, B.C. (1990a). Cashew stem-root and wood borers and their management. *The Cashew* 4: 5-7.

Jena, B.C, Patnaik, N.C. and Satapathy, C.R. (1985a). Insect pests of cashew. *Cashew Causerie*, 7(3): 10-11.

Jena, B.C, Patnaik, N.C. and Satapathy, C.R. (1985b). Insect fauna of cashewnut in Orissa. *Science and Culture* 51(11): 385-386.

Jena, B.C., Satpathy, C.R. (1989). Reaction of cashew varieties to the leaf miner (*Acrocercops syngramma* M.) incidence. *The Cashew* 3: 15-17.

Khoo. K.C. (1992). Manipulating predators for biological control with special reference to tropics. In Biological control Issues in the Tropics. Eds. Ooi, P.A.C., Lim, G.S. and Teng, P.S. Malaysian Plant Protection Society, Kuala Lumpur, Malaysia, pp. 15-22

Khoo, K. C. and Ho., C.T. (1992). The influence of *Dolichoderus thoracicus* (Hymenoptera: Formicidae) on losses due to *Helopeltis theivora* (Heteroptera: Miridae), black pod disease, and mammalian pests in cocoa in Malaysia. *Bulletin of Entomological Research*, 82: 485–491.

Lefroy, H.M. (1909). *Indian Insect Life. 4th Indian Reprint, 1984.* Jagmandir Book Agency, New Delhi, India, 786 pp.

Misra, M. P. and Basu Choudhuri., J. C. (1974). New Pest and Distributional Records of *Lymantria obfuscata* Walker (Lepidoptera: Lymantriidae) on Cashew from South India. *Indian Forester* 100(6): 391-393.

Misra, M. P. and Basu Choudhuri., J.C. (1985). Control of *Plocaederus ferrugineus* L.(Coleoptera:Cerambycidae) through field hygiene. *Indian J.Agric. Sci.* 55: 290-93.

Mohapatra, L.N. and Barik, H. (1996). Effect of abiotic factors on incidence of cashew leaf miner, *Acrocercops syngramma* Meyr. *Cashew Bull.* 3: , 3-4.

Mohapatra, L.N., Behara, A.K. and Satapthy, C.R. (1998). Influence of the environmental factors on the cashew nut shoot tip caterpillar, Hypotima haligramma Meyr. *Cashew Bull.*, 35: 17-18.

Mohapatra, L.N. and Mohapatra, R.N. (2004). Distribution, Intensity and damage of cashew stem and root borer, *Plocaederus ferrugineus* L. in Orissa. *Indian Journal of Entomology*, 66(1): 4-7.

Mohapatra, R.N., Jena,B.C. and Lenka, P.C. (2007). Effect of some IPM components in management of cashew stem and root borer. *J. Plant Prot. Environ.* 4(2): 21-25.

Neetu, K., Singhvi, P.M., Meenakshi, J. and Mathur, M. (2006). Effect of certain indigenous plant extracts on ovipositional behaviours of *Oryzaephilus surinamensis* (Linn.) in stored cashew nuts. *J. App. Zool. Res.* 17(2): 209-211.

Ohler, J.G. (1979). Cashew. Department of Agricultural Research. Royal Tropical Institute, Amsterdam, pp. 260.

Patnaik, H.P., Das, M.S., Sahoo, K.C. and Senapati, B. (1986). Preliminary studies on nut drying disease in cashew and its control. *Cashew Causerie* 8: 16-17.

Patnaik, H.P., Satapathy, C.R, Sontakke, B.K. and Senapathy, A.H.D. (1987). Flower thrips of cashew (Anacardium occidentale), their seasonal incidence and assessment of damage in coastal Orissa. *The Cashew* 1: 11-13.

Peng, R. K., K. Christian, and Gibb, K. (1995). The effect of the green ant, *Oecophylla smaragdina* (Hymenoptera: Formicidae), on insect pests of cashew trees in Australia. *B. Entomol. Res.* 85: 279–284.

Peng, R.K., Christian, K. and Gibb.K. (1997a). Control threshold analysis for the tea mosquito bug, *Helopeltis pernicialis* (Hemiptera: Miridae) and preliminary results concerning the efficiency of control by the green ant, *Oecophylla smaragdina* (Hymenoptera: Formicidae) in northern Australia. *International Journal of Pest Management* 43(3): 233-237. DOI: 10.1080/096708797228735.

Peng, R. K., Christian, K and Gibb, K. (1997a) Distribution of the green ant, *Oecophylla smaragdina* (F.) (Hymenoptera: Formicidae), in relation to native vegetation and the insect pests in cashew plantations in Australia. *International J.ournal of Pest Management* 43: 203-211.

Peng R. K., Christian, K. and Gibb, K. (1997b). Control threshold analysis for the tea mosquito bug, *Helopeltis* pernicialis (Hemiptera: Miridae) and preliminary results concerning the efficiency of control by the green ant, *Oecophylla smaragdina* (Hymenoptera: Formicidae) in northern Australia, *International Journal of Pest Management* 43: 233-237.

Peng, R.K., Christian, K. and Gibb, K. (1998). Management of *Oecophylla smaragdina* colonies in cashew plantations to control the main cashew insect pests. Sixth Australasian Applied Entomological Research Conference, Brisbane, Australia, 2: 332.

Peng, R.K., Christian, K. and Gibb, K. (1999a). The effect of colony isolation of the predacious ant, *Oecophylla smaragdina* (F.) (Hymenoptera: Formicidae), on protection of cashew plantations from insect pests. *International Journal of Pest Management* 45: 189-194.

Peng, R.K., Christian, K. and Gibb, K. (1999b). Utilisation of green ants, *Oecophylla smaragdina*, to control cashewe insect pests. pp 88, Rural Industries Research and Development Corporation, Canberra, Australia.

Pillai, G.B., Dubey, O.P. and Vijay Singh. (1976). Pests of cashew and their control in India-. A review of current status. *Journal of Plantation Crops* 4: 37-50.

Pillai, G.B., Singh, V., Dubey, O.P. and Abraham, V.A. (1979). Seasonal abundance of tea mosquito *Helopeltis antonii* on cashew tree in relation to meteorological factors. Proceeding of The International Cashew Symposium (pp. 103-110). Cochin.

Pillai, P.K.T.and Pillai, G.B. 1975. Note on shedding of immature fruits in cashew. *Indian J. of Agric. Sci.*, 45(5): 233-234.

Rai, P.S. (1984). Handbook on Cashew Pests. Research Co Publications, New Delhi.

Ramadevi, M. and Krishnamurthy, P.R. (1983). Schedule of pest occurrence on cashew (*Anacardium occidentale* L.) *Cashew Causerie* 5(1): 14-16.

Raviprasad, T.N. and Bhat, P.S. (1998). Laboratory rearing techniques for cashew stem and root borer, *Plocaederus ferrugineus* Linn. (Coleoptera:Cerambycidae). Abst. PLACROSYM XIII Coimbatore, 16-18 Dec. 1998. p.47.

Raviprasad, T.N., Bhat, P.S. and Sundararaju, D. (2009). Integrated pest management approaches to minimize incidence of cashew stem and root borers (*Plocaederus* spp.). *Journal of Plantation Crops* 37: 185-189.

Raviprasad, T.N., Sundararaju, D. and Bhat, P.S. (2005). Efficacy of botanicals against *Helopeltis* antonii Sig. infesting cashew. *The Cashew* 29: 9-14

Reddy, E. U. B. (1993). Pollination studies of cashew in India: an overview. In (Eds) G.K. Veeresh, R. Uma Shankar and K. N. Ganeshaiah. *Proc. Int. Symp. Polln. Trop.*, 321-234.

Remamony, K.S. (1965). Biology of *Anarsia epotias* Meyr. (Lepidoptera : Gelichidae) *Agric. Res. J. Kerala* 3: 46-47.

Riechert S. E. and Bishop, L. (1990). Prey control by an assemblage of generalist predators: Spiders in garden test systems. *Ecology* 71: 1441-1450.

Samiyyan, K, Palaniswamy, K.P., Ahmed Sah and Manivannan, K. (1991). Effect of prophylactic measures against cashew stem and root borer *Plocaederus ferrugineus* L. *The Cashew* 5(1): 16-17.

Sathiamma, B. (1978), Occurrence of insect pests on cashew. *Cashew Bull.*, 15(4): 9-10.

Sreeramulu, C., Ayyanna, T. and Dharmaraju, E. (1976). Cashew apple feeders. *Cashew Bulletin* 13: 7.

Srikumar, K. K. and Bhat, P.S. (2013a). Biology and feeding behaviour of *Helopeltis antonii* (Hemiptera: Miridae) on Singapore cherry (*Muntingia calabura*)-a refuge host. *J. Ent. Res.*, 37(1): 11-16.

Srikumar.K.K. and Bhat, P.S. (2013b) Demographic parameters of *Helopeltis antonii* Signoret (Heteroptera: Miridae) on neem, cocoa and henna. *African J. Agricultural Res.* 8(35): 4466-4473.

Srikumar.K.K. and Bhat, P.S. (2013c) Biology of the tea mosquito bug, *Helopeltis theivora* Waterhouse on *Chromolaena odorata* (L.) and notes on its egg parasitoids. *Chilean J. Agricultural Res.* 73(3): 309-314.

Stonedahl, G. M. (1991). The Oriental species of *Helopeltis* (Heteroptera: Miridae): a review of economic literature and guide to identification. *Bulletin of Entomological Research* 81: 465-490.

Subba Rao, V., Rajasekhar, P., Rama Subba Rao,V. And Shrinivasa Rao, V. (2006). Seasonal incidence and control of shoot tip and inflorescence caterpillar on cashew. *Ann. Pl. Protec. Sci.*, 14(1): 245-246.

Sundararaju, D. (1984). Studies on cashew pests and their natural enemies in Goa. *Journal of Plantation Crops* 12: 38-46.

Sundararaju, D. (1993a). Studies on the parasitoids of the mosquito bug, *Helopeltis* antonii Sign. (Hymenoptera: mymaridae) on cashew with special reference to *Telenomous* sp. (Hymenoptera: Scelionidae). *J. Biol. Control* 7: 6-8.

Sundararaju, D. (1993b). Compilation of recently recorded and some new pests of cashew in India. *The Cashew* 7: 15-19.

Sundararajau, D. (1996). Studies on *Helopeltis* spp. with special reference to *H.antonii* Sign in Tamil Nadu. Ph.D. The-sis, T.N.A.U. Coimbatore. 202 p.

Sundararaju, D. (2000). Foraging behaviour of pollinators on cashew. *The Cashew* 14(4).

Sundararaju, D. (2000a). Insects associated with extrafloral nectaries of cashew leaves. *J.Plant. Crops* 28(2): 175-178.

Sundararaju, (2004). Influence of spiders and insect predators on the incidence of tea mosquito bug in cashew. *The Cashew* 18 (1).

Sundararaju, D. (2002). Description of endoparasitism in nymphs and adults of *Helopeltis* spp. infesting cashew. *Journal of Plantation Crops* 30(3): 66-68.

Sundararaju, D. (2007). Pest and disease management in Cashew. Presented in 6th National Seminar Indian Cashew in the next decade – challenges and opportunities. 18-19 May, Raipur. 53-63p.

Sundararaju, D. and Bakthavatsalam, N. (1994). Pests of cashew. In: Advances in Horticulture, Vol.10 (Eds. K.L. Chadha and P. Rethinam), Malhotra Publishing House, New Delhi, India, pp. 759-785.

Sundararaju, D., T.N. Ravi Prasad and Bhat, P.S. (1999). Pests of cashew and their integrated management. IPM system in Agriculture. Rajeev, K., Upadhyay, K.G., Mukerji and O.P. Dubey. Aditya Books Pvt. Ltd., New Delhi. India. Vol. 6. Cash Crops, p. 525-544.

Sundararaju, D. and Sundara Babu, R.C. (1998). Life table studies of *Helopeltis antonii* Sign (Heteroptera: Miridae) on neem, guava and cashew. *J. Entomol. Res.* 22(3): 241-244.

Sundararaju, D. and Sundara Babu, R.C. 1999a. Species Composition of *Helopeltis* (Heteroptera : Miridae) in South India and Their Host Range. *Insect Environment* 4(4): 13-14.

Sundararaju, D. and Sundara Babu, R.C. (1999b). GC-MS analysis of virgin females of *Helopeltis antonii* Sign. (Heteroptera: Miridae). *Insect Environment* 4(4): 3-4.

Sundararaju, D. Yadukumar, N. Bhat, P.S., Raviprasad, T.N., Venkatakumar, R. and Sreenath Dixit. (2006). Yield performance of 'Bhaskara' cashew variety in coastal Karnataka. *J. Plantn. Crops*, 34: 216-219.

Thirumalaraju, G.T., Gowda, M.C., Krishnappa, K.S. and Narayana Reddy, M.A. (1990). Seasonal incidence of flower thrips under eastern dry zone of Karnataka. *The Cashew* 4: 3-4.

Vanitha, K., Bhat,P.S., Raviprasad, T.N. and Srikumar, K.K. (2015). Species composition of ants in cashew plantations and their inter-relationship with cashew. *Proc. Natl. Acad. Sci.,* India, Sect. B. Biol. Sci. DOI 10.1007/s40011-0150600-3.

Venkata, R. (2009). Record of *Helopeltis antonii* (Homoptera: Miridae) on the fruits of *Annona* spp. *Pest Management in Horticultural Ecosystems*, 15: 74-76.

Way, M.J. and Khoo, K. C. (1991). Colony dispersion and nesting habits of the ants, *Dolichoderus thoracicus* and *Oecophylla smaragdina* (Hymenoptera : Formicidae) in relation to their success as biological control agents on cocoa. *Bulletin of Entomological Research*, 81: 341-350.

Way M.J. and Khoo, K.C. (1992). Role of ants in pest management. *Annual Review of Entomology*, 37: 479-503.

2018, Pests of Plantation Crops   *Pages* **199–207**
Editors: **P. Chowdappa, Chandrika Mohan & A. Josephrajkumar**
Published by: **ASTRAL INTERNATIONAL PVT. LTD., NEW DELHI**

Chapter 9

Rubber

☆ *S. Thankamony*

1. Introduction

Rubber tree, *Hevea brasiliensis* Muell. Arg. (Euphorbiaceae) is a perennial, evergreen tree,which is the source of 99 per cent of the world's natural rubber production and is therefore an important economic plant. Because of the latex in the bark, rubber is not a preferred host plant for many common polyphagous insect pests and generally pest attacks are rather sporadic. and localised. However, there are few pests which become serious at times and cause considerable damage. The predominant pests of rubber include cock chafer grubs, bark feeding caterpillars, scale insects and mealybugs, termites, borer beetles, crickets, slugs and snails and rats. The cover crops raised in rubber plantations are also attacked by several pests. There are also few pests which cause health hazards and inconveniences to personnel working and residing in rubber plantations.

2. Cockchafer Grubs

Root grubs of the species *Holotrichia serrata* F., *H.rufoflava* Brenske, *H. fissa* and *Anomala varians* Ol. attack rubber plants and cause considerable loss in rubber nurseries adjoining virgin forests and in loose soils (Figure 9.1). Feeds on the roots of seedlings in the nursery and young plants (Nehru and Jayaratnam, 1993). Among the four species the damage caused by *H. serrata* was reported as severe in nursery plants and rendering them unfit for transplanting. Broadcasting phorate 10G @ 25kg/ha in the soil at the time of preparation of nursery beds or drenching the soil at the base of the affected plants with imidacloprid 0.005 per cent (confidor 1ml/L) or quinalphos 0.005 per cent were observed as effective for the control of attack of cockchafer grubs in rubber seedlings. Among entomopathogens, entomopathogenic nematodes (EPN) (272lakhs Ijs in 5 splits) and Beaumet (65x10^6) showed 71 to

Figure 9.1: Cockchafer Grubs, Adults and Infested Young Rubber Plants.

77 per cent control after two years. (RRII, 2009). Jayasinghe (1999) reported that *Lachnosterna (Holotrichia) bidentata, Holotrichia insularis, Psilopkolis vestita, Leucopholis rorida, Leucopholis nummicudens, Leucopholis tristis, Exopholis hypoleuca* and *Lepidiota stigma* are the major cockchafers infesting rubber tree in Sri Lanka.

3. Bark Feeding Caterpillar

Bark feeding caterpillar (Figure 9.2) was reported as the most serious endemic pest of mature rubber trees in India. The incidence of bark feeding caterpillar on rubber was sporadic up to 1968, but become extensive by 1980's (Nehru *et al.*, 1983; 1987; Nehru and Jayaratnam, 1987; Jayaratnam *et al.*, 1991). *Aetherastis circulata* Meyr. *Ptochoryctis rosaria* Meyr are the two species of bark feeding caterpillars reported in rubber plantations. These caterpillars build galleries with faecal matter and silk all over the trunk region and branches of trees. Generally it feeds on dead bark and occasionally on live bark causing exudation of latex (Figures 9.2 and 9.3). Among these, the incidence of *A. circulata* was found most severe and abundant. Deep scars are formed at the regions of feeding (Jose *et al.*, 2000; Rubber Board, 2013). Bark

Figure 9.2: Egg, Larva, Pupa and Adult of the Bark Feeding Caterpillar, *Aetherastis circulate* and Infested Rubber Tree.

feeding caterpillars *viz., Homodes brachteigutta* (*Euproctis subnotata*) *and Acanthopsyche snelleni* are also reported on rubber from Sri lanka (Jayasinghe, 1999).

Use of combination of insecticides are reported as more effective for the control of bark feeding caterpillars than single insecticides and biopesticides (Jose and Thankamony, 2010). Spraying the combination of fenvalarate 0.02 per cent and carbaryl 0.01 per cent recorded 94.78 per cent control of *A. circulata* followed by quinalphos 0.05 per cent and carbaryl 0.01per cent. Studies conducted on the occurrence of natural enemies revealed the presence of an entomopathogenic fungus and a natural parasitoid (Braconidae) from the pupae of bark feeding caterpillar. The fungus was isolated and identified as *Aspergillus flavus* (RRII,2013, 2014).

Figure 9.3: Bark Feeding Caterpillar, *Ptochoryctis rosaria* and Infested Rubber Tree.

4. Scale Insects and Mealybugs

Seen generally in young plantations and nurseries in almost all rubber growing areas. Occur on leaflets petioles and tender shoot portions and suck the sap. In heavy infestation the plants exhibit yellowing and shedding of leaves. Early signs of a mealybug infestation include drooping or dry-looking leaves and the appearance of cotton-like masses along leaf attachment sites and on the undersides of the leaves.

Severely affected portions dry up and die. Ants and sooty mould are associated with these pests (Figure 9.4). Species of scale insect and mealybug reported in rubber are *Saissetia nigra* Nietn and *Ferrisia virgata* Ckli. Spraying of organophosphorus insecticides like malathion at 0.1 per cent or quinalphos at 0.1 per cent are found effective for the control of scale insects and mealybugs. Application of chlorpyriphos 0.04 per cent or imidacloprid 0.005 per cent were also found effective for pest control. Natural enemies like insect parasites and entomogenous fungi keep this pest in check. Incidence of papaya mealybug, *Paracoccus marginatus* was recorded on mature tree branches, stem, foliage, inflorescence and seeds in Palakkad and Ernakulam districts during 2009-2011. Spraying of imidacloprid 0.005 per cent concentration was found effective for the control of *P. marginatus*. Efficacy of bioformulations *viz., Verticillium* and *Metarrhizium* combinations were evaluated against this pest but were found not effective. However, neem oil (0.1 per cent) reduced the pest population (12 per cent) after two weeks.

Figure 9.4: Scale Insect and Mealybug Infestation on Rubber Plant.

5. Termite (White Ant)

It occurs in dry regions of Central Kerala (Thrissur and Palakkad) and non-traditional areas like Dapchari in Orissa. It feeds on the dead bark of trees and young plants (Figure 9.5). Termites build covered passageways of soil on the tapping panel and collection cup. Sometimes young plants dry up due to attack. The most common species that infests rubber in India is *Odontotermes obesus* Rambar. Drenching the soil at the base of the affected plants with chlorpyriphos 0.1 per cent solution was observed as very effective for the control of termites. Arrowroot powder (*Curcuma zedoaria*) at 10 per cent was found effective in the management of termites (Rubber Board, 2013).

6. Borer Beetles

Being a non-durable wood, rubber is highly susceptible to fungi, wood borers, and termites (CIRAD, 2003; Wong *et al.*, 2005). Insect borers of families

Figure 9.5: Termite Infested Rubber Tree.

(Bostrichidae, Curculionidae: Platypodinae and Scolytinae) attack the wood at all stages from log to seasoned wood and finished products. In Malaysia, sixteen species of ambrosia beetles, in the subfamilies Scolytinae and Platypodinae, infesting felled trees and unseasoned rubber wood, nine powder post beetles (Bostrichidae) and one ambrosia beetle (Scolytinae) infesting seasoned sawn timber was reported by Browne (1961), Hussein (1981), and Ho and Hashim (1997). In India, Nair (2007) and Mathew (1982) reported six powder post beetles with predominance of Bostrichidae infesting stored rubber wood and sawn timber. Various borer beetles attack on partially dried bark of rubber and make tunnels inside wood and expel saw dust. The trees may fall due to trunk snap (Figure 9.6). The predominant borer beetles attacking rubber trees are *Heterobostrychu* spp., *Sinoxylon* sp., *Platypus* sp., *Minthea rugicollis*, *Dinoderus bifoveolatus and Platypus solidus*. A combination of carbaryl (0.25 per cent) + quinalphos (0.25 per cent) thrice at an interval of one week and carbaryl (0.50 per cent) + lambda cyhalothrin (0.02 per cent) twice at an interval of one week effective for in the control of borer beetles.

7. Crickets

The crickets (*Gryllacrys* sp.) are very destructive to the rain guards of tapping rubber trees. Because of the semi-circular cut inflicted by the cricket, water leaks in to the panel leading to panel rot. Application of malathion 0.01 per cent, neem oil 0.10 per cent and fenvalarate 0.02 per cent along the tapping panel once in a week reduce the attack. Direct application of used engine oil, maroti oil, cashew kernel oil on the interior periphery of the polythene were found more effective than neem oil.

Figure 9.6: Symptoms and Damage Caused by Borer Beetle Attack.

8. Slugs and Snails

The species of slugs and snails attacking rubber in India are *Mariaellae dussumieri* Grey (Slug) *Cryptozona (Xestina) bistralis* Beck (Snail). These mollusks climb on trees and feed on latex by lacerating the tender leaves and buds. Growth of affected buds is arrested and side shoots develop giving a bunchy appearance (Figure 9.7). Slugs drink latex from the tapping cut and collecting cup also (Figure 9.8). Snail kill, metaldehyde bait, when broadcast at the rate of 20 g per plant around the base of the affected plant gives better control of slugs. Application of 10 per cent Bordeaux paste around the stem above the bud union to a length of 30 cm was reported to repel slugs and snails for a period of 30 to 45 days.

9. Rat

Dominant rat pests of rubber reported in India are the Indian mole rat *Bandicota bengalensis* Gray, *Bandicota indica* Bech and *Rattus meltada* Gray. They destroy nursery plants and young plants in the field by eating the tap root just below the collar

Figure 9.7: Snail.

Figure 9.8: Slug.

region and often pulling the whole plants of up to two years growth down to the burrow (Figure 9.9). Poison baiting with two per cent zinc phosphide after two rounds of pre-baiting was found effective for the control of rats. Baits of single dose anticoagulant rodenticides like broadifacoum, bromadiolone (Roban) and flocoumafen (storm) at 0.005 per cent concentrations are observed as effective for the control of rats infesting rubber (RRII,1995).

10. Wild Animals

New clearings, nurseries and plantations adjacent to forests are found subjected to frequent invasions of wild animals. A large number of pests ranging from rats to elephants inflict severe damage to rubber plants from the nursery to mature plantations. Installation of a well maintained ordinary or electric fencing system was reported as best insurance against most of these pests.

11. Pests of Cover Crops

The cover crop *Pueraria phaseoloides* is found more susceptible to pest attack

Figure 9.9: Damage Caused by Rats.

compared to other cover crops such as *Calapagonium mucunoides, Centrosema pubescens, Mimosa invisa and Mucuna bracteata*. The most serious pests infesting *P. phaseoloides* are the stem borer *Eucomatocera vittata*, the leaf-lacerating flea beetle, *Pagria signata* and the flower and pod borer, *Maruca vitrata. Nacoleia vulgaris* is another common pest of *Pueraria* in India. *M. bracteata* is found attacked by leaf eating caterpillars. The damage was observed as highly at the time of establishment. Spray of carbaryl 0.1 per cent at 15 days intervals was noticed effective for the control of pests infesting cover crops.

12. Pests Affecting Plantation Workers

12.1. Mooply Beetles

The mooply beetle, *Luprops curticollis* Frm causes considerable nuisance because of its presence in large numbers and secretion of stain after invading dwellings in plantations (Figure 9.10). The beetles can be collected using light traps or killed by spraying insecticides like deltamethrin 0.0056 per cent or fenvalarate 0.02 per cent followed by chlorpyriphos 0.2 per cent and malathion 0.1 per cent. Evaluation of *Metarhizium anisopliae*(28×10^5) and *Beauveria bassiana* @50g/L water (10^8cfu/g) on the larvae of mooply beetles revealed 45 per cent mortality followed by 20 per cent by *M. anisopliae* and *B. bassiana*, respectively.

12.2. Leeches

The leeches (*Hirudinaria cochniana*), which inhabit swampy areas cause much menace by biting and sucking blood of persons involved in rubber plantation

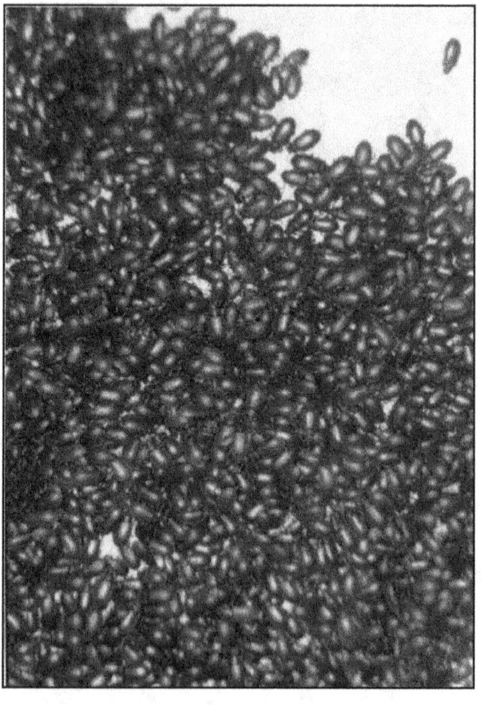

Figure 9.10: Mooply Beetles.

during rainy season. The extracts of *Cyclea peltata* (padakizhangu) in coconut oil, *C.peltata* in neem oil, odomos (direct application) mixture of neem oil and citronella oil (1:1) were reported as most effective repellents against this pest. Neem oil, maroti oil, citronella oil, dettol in coconut oil were the second effective repellents (RRII, 2009)

References

Browne, F.G. (1961). The Biology of Malayan Scolytidae and Platypodidae. *Malayan Forest Records* No.22, Forest Department, Kuala Lumpur, Malaysia.

CIRAD (2003). Rubber Wood General Properties. Tropix 5.0. French Agricultural Research Centre for International Development, France.

Ho, Y.F. and Hashim, S. (1997). Wood-boring beetles of rubber wood sawn timber. *Journal of Tropical Forest Products*. 3: 15-19.

Hussein, N.B. (1981) A preliminary assessment of the relative susceptibility of rubber wood to beetle infestation. *The Malaysian Forester*. 44: 482-487.

Jayarathnam, K, Nehru, C.R., and Jose, V.T. (1991). Field evaluation of some newer insecticides against bark feeding caterpillar *Aetherastis circulata* infesting rubber. *Indian Journal of Natural Rubber Research* 4(2): 131-133.

Jayasinghe, C.K. (1999) Pests and diseases of *Hevea* rubber and their geographical distribution. *Bulletin of the Rubber Research Institute of Sri Lanka*, 40: 1-8.

Jose, V.T., Thankamany, S., and Kothandaraman, R. (2000). Control of bark feeding caterpillar *Aetherastis circulata* (Meyr) (Ypnomutidae: lepidoptera) infesting rubber with insecticidal dust. *Advances in Plantation Crops Research*. pp. 352-354.

Mathew, G. (1982). A survey of beetles damaging commercially important stored timber in Kerala, Research Report, Kerala Forest Research Institute No. 10: 93.

Nair, K.S.S. (2007). Tropical Forest Insect pests: Ecology, Impact, and Management. Cambridge University Press, Cambridge, 404 pp.

Nehru C.R. and Jayarathnam, K. (1987). Field incidence of bark-feeding caterpillar *Aetherastis circulata* (Meyr.) on different alternative host plants and its control. *Pesticides* 21(7): 39-40.

Nehru C.R. and Jayarathnam, K. (1993). Evaluation of biological and chemical control strategies against the white grubs (*Holotrichia serrata*) infesting rubber seedlings. *Indian Journal of Natural Rubber Research* 6(1-2): 159-162.

Nehru C.R., Jayarathnam, K. and Pillai, P.N.R. (1983). Incidence of bark-feeding caterpillar *Aetherastis circulata* Myer on rubber (*Hevea brasiliensis* Muell. Arg.). *Indian Journal of Plant Protection* 11: 150.

Nehru, C.R., Jayarathnam, K. and Thankamany, S. (1987). Field incidence of bark-feeding caterpillar *Aetherastis circulata* (Meyr) on different alternative host plants and its control. *Pesticides* 21(7): 39-40.

RRII (2009). Annual report 2008-09. Rubber Research institute of India, Rubber Board, Kottayam, Kerala.

RRII (1995). Annual report 1994-95. Rubber Research institute of India, Rubber Board, Kottayam, kerala.

RRII (2013). Annual Report 2012-13 Rubber Research institute of India, Rubber Board, Kottayam, Kerala.

Rubber Board (2013). *http://rubberboard.org.in/Rubber Cultivation. asp. Rubber Board, India*. (assessed April 2013).

Wong, A.H.H., Kim, Y.S., Singh, A.P. and Ling, W.C. (2005). Natural Durability of Tropical Species with Emphasis on Malaysian Hardwoods - Variations and Prospects. The International Research Group on Wood Preservation, the 36th Annual Meeting, Bangalore, India, April 24-28, 2005, 33 pp.

2018, Pests of Plantation Crops
Editors: **P. Chowdappa, Chandrika Mohan & A. Josephrajkumar**
Published by: **ASTRAL INTERNATIONAL PVT. LTD., NEW DELHI**

Pages **209–219**

Chapter 10

Chemoecological Methods in Coconut Pest Management

☆ *Kesavan Subaharan*

1. Introduction

The need to produce inexpensive and abundant food supply for a growing population is a great challenge and warrants higher use of inputs like fertilizers and pesticides. Farmers resort to chemical control as their availability and application is done with ease. Though the use of pesticides keeps the pest at check, the experience gained during the recent past indicates that improper use of pesticide has caused undesirable side effects like buildup of residue in the marketable commodity, emergence of insecticide resistance in pests, negative impact on beneficial arthropods *etc.*

Increased dependence on pesticides have negative impact on farmers, consumers, non-target organism and environment. This situation prompts the search for safe alternative for pest control. Crop production and environmental protection can be balanced if emphasis is laid to adopt eco-friendly pest management strategies. In pest management alternative to chemical pesticides is the use of insect behavior modifying chemicals (IBMC). Semiochemicals are defined as chemicals that deliver behavioral messages between organisms (Dusenbury, 1992), they include pheromone (that aid in finding mates, food and habitat resources), allomone (favor the producer) and kairomone (favor the receiver).

In plantation crops the damage by insect pests have profound influence on crop production. Use of chemical pesticides leads to buildup of residues in the commodities that have high commercial value. Major pests that influence the crop yield in coconut are, rhinoceros beetle, *Oryctes rhinoceros*, red weevil, *Rhynchophorus ferrugineus*, black headed caterpillar, *Opisina arenosella*, eriophyid mite, *Aceria guerreronis* and white grub, *Leucopholis coneophora*. Safe alternate strategies in coconut pest management include use of biological agents and chemoecological approaches. Among the orders that cause damage to crops, coleopterans form a major share. Considering their long life cycle and their cryptic behavior using the conventional approaches becomes difficult to manage them. Though biocontrol agents are effective, their timely availability in quality and quantity is a question. This necessitates the need to depend on pheromones and kairamonesas they are an effective tool in monitoring and mass trapping.

Understanding the chemoecological approaches using cutting edge technologies involving chemical detectors (GCMS) and electrophysiological tools resulted in developing robust behavioural manipulations in coconut pest management. The compounds identified to be used in behavioral pest management will aid to decline the dependence on xenobiotics.

2. Methods Involved in Isolation and Identification of Semiochemicals

2.1. Sampling and Analysis of Semiochemicals

The increasing scientific interest in the biochemistry, physiology, ecology and atmospheric chemistry of VOCs has led to the development of a variety of systems for the collection and analysis of volatiles (Millar and Haynes, 1998). Volatile analysis has been improved by the design of relatively sensitive bench-top instruments for gas chromatography – mass spectrometry (GC-MS) (Dorothea *et al.*, 2006). Volatiles surrounding the airspace (headspace) around the insects are sampled and concentrated prior to analysis. Headspace sampling is a non-destructive method for collecting volatiles. When compared with solvent extractions of pheromone glands from insects, headspace analysis gives a more realistic picture of the volatile profile emitted by insects. It provides real time information for ecologically relevant applications. Materials like glass, metal and special plastics such as teflon that are inert are to be used for volatile trapping (Dorothea *et al.*, 2006).

An important advance in static headspace analysis is the development of solid phase microextraction (SPME) which is a fast and simple method for collecting volatiles at detection limits in the ppbv (parts per billion by volume) range. Solid phase microextraction is based on ad/absorption and desorption of volatiles from an inert fiber coated with different types of ad/absorbents. The fiber is attached within the needle of a modified syringe and volatiles can be sampled by inserting the needle through a septum of a headspace collection container and pushing the plunger to expose the fiber. Following equilibration between the fiber and the volatile sample, the fiber is retracted into the needle and can be transferred to a gas chromatograph for direct thermal desorption. Solid phase microextraction fibers can be re-used approximately 100 times. Thermal desorption of VOCs from the fiber eliminates the need for solvents that may contain impurities which will interfere with sample analysis (Dorothea *et al.*, 2006).

2.2. Gas Chromatographic Separation and Detection of Plant VOCs

Insect semiochemical adsorbing matrices are analyzed by the standard technique of GC (Handley and Adlard, 2005; Lockwood, 2001; Merfort, 2002). For GC analysis of semiochemicals, samples are either injected as solvent extracts into the heated injector in a split or splitless mode or desorbed from the adsorbent by placing it directly in a thermal desorption tube, heated to 250–300°C. In a two stage thermal desorber, the thermally released volatiles are concentrated by a cold trap (or cryotrap) prior to their injection into the GC column (Dorothea *et al.*, 2006).

For analytical purposes, volatiles are commonly separated on fused silica capillary columns with different stationary phases, such as the non-polar dimethyl polysiloxanes (*e.g.* DB-1, DB-5, CPSil 5), and the more polar polyethylene glycol polymers, including Carbowax-20M, DB-Wax, and HP-20M.

Following separation on a GC column, semiochemicals are analyzed by a variety of detectors. Flame ionization detector (FID) is commonly used for quantitative analysis because of their wide linear dynamic range, their very stable response and their high sensitivity with detection limits of the order of picograms to nanograms per compound. Mass spectrometry (MS) detectors are the most popular type of detector for routine semiochemical GC analysis. In the mass spectrometers of most standard GC-MS benchtop instruments, compounds exiting the GC column are ionized by electron impact (EI) and the resulting positively charged molecules and molecule fragments are selected according to their mass-to-charge (m/z) ratio by entering a quadrupole ion trap or a quadrupole mass filter. Total ion chromatograms are obtained, which provide information on the retention time of each compound and its mass spectrum consisting of a characteristic ion fragmentation pattern. Detection limits of highly sensitive mass spectrometers are in the picogram range for the full scan mode (scanning ions over a wide molecular range) and may be as low as in the femtogram range (in quadrupole mass filters) in the selected ion monitoring (SIM) mode scanning selected ions that are representative of a compound. Identification of compounds in GC-MS analysis is done by using the popular mass spectral libraries such as Wiley and NIST MS databases (Dorothea *et al.*, 2006).The orientation response of an insect to semiochemicals is evaluated by olfactometry and electrophysiological assays.

2.3. Electroantennography to Assess the Physiological Response to Volatile Organic Compounds

Electrophysiological techniques have been employed to study the nervous system of insects since the late 1950s. In 1957, Schneider for the first time announced that it was possible to measure the electrophysiological responses from the antennae of an insect, *Bombyx mori*, using an electroantennogram (EAG). The EAG technique has since been developed for several insect species of various insect orders (Millar and Haynes, 1998). Despite its usefulness, the method is limited in its use. For instance, the sensitivity of the technique is low, a shortcoming often observed in insects whose antennae have *e.g.* a limited number of sensilla (Millar and Haynes, 1998). To overcome this problem, a technique called single sensillum recordings

(SSRs) was developed, which can measure responses of single ORNs (Millar and Haynes, 1998). Single sensillum recordings allow for the analyses of specific types of sensilla and for the determination of the mechanisms by which insects code for different odors (Majid, 2007).

In SSRs, two sharpened tungsten electrodes are used: a ground electrode is in contact with the haemolymph while the recording electrode is inserted near the base or in the shaft of a single sensillum. Voltage differences generated between the electrodes, when amplified, can be viewed on an oscilloscope when placed within the circuit (de Bruyne*et al.*, 2001).

The introduction of gas chromatography (GC) coupled SSRs, first used in moths and aphids allowed for the first detection of active components within a complex blend of compounds. This technique has later been used to identify novel ligands of ORNs in a large number of insect species. At the GC part an extract likely to contain odorants detected by the ORNs is injected onto the GC-column. The column is located in an oven where it is possible to regulate the column temperature. As the temperature of the column is increased the components of the extract are separated while traveling down the column and exit the GC set-up. The separated components of the extracts encounter the single sensillum from which a stable electrical contact is established. Responses of the ORNs housed in a single sensillum to the extract components are recorded. The chemical identity of the response eliciting component(s) can be further identified using mass spectrometry (MS).The olfactory receptors respond to compounds that are attractive and repulsive. Hence the behavioural experiments have to be carried out to ascertain the nature of compounds using olafctometers and wind tunnels.

3. Chemoecological Approach in Coconut Pest Management

3.1. Coconut Red Palm Weevil, *Rhynchophorus ferrugineus* (Oliver)

The red palm weevil, *Rhynchophorus ferrugineus* (Oliver) is distributed in Asia and Europe. The host range of the weevil includes coconut, oil palm, date palm and sago (Wattanapongsiri, 1966). The adult female oviposit on young, damaged, stressed and healthy palms (Kalshoven, 1950). On hatching the larvae bore into the palm and develop into adult in two months (Giblin Davis *et al.*, 1996). Trapping by baiting with coconut stem tissue was done to reduce the weevil populations in India (Abraham and Kurian, 1975). Evidence of male produced aggregation pheromone was established in laboratory bio assay by Abraham (1987). The aggregation pheromone 4 methyl 5 nonanol (Ferrugineol) was synthesized by Hallet *et al.* (1993). Coconut logs treated with toddy, yeast and acetic acid were effective in trapping *R. ferrugineus* (Kurian *et al.*, 1984).

R. ferrugineus weevils are opportunistic oligophages to early fermentation volatiles like ethanol emanating from wounded host (Gunatilake and Gunawardane,1986), on feeding the palm tissue they produce an aggregation pheromone (4 methyl 5 nonanol) that attract their conspecifics (Giblin-Davis *et al.*, 1996). Food volatiles strongly enhance the attraction of *Rhynchophorus* species to aggregation pheromone (Rochat *et al.*, 1993). The palm esters, ethyl acetate, ethyl propionate, ethyl butyrate and ethyl isobutyrate were identified as kairomones for *R. phoenicis, R. palmarum, R.cruentatus, R. ferrugineus* and *R. vulneratus* (Gries *et al.*, 1994). The American palm weevil, *R. palmarum* adults were attracted to odors of variety of plant tissues (pineapple, banana and coconut) that were used as baits in traps (Jaffe *et al.*, 1993; Oehlschlager *et al.*, 1993). Among the complex pattern of odorants emitted by host plants only a few key compounds were used to locate the host. The identification of natural volatiles emitted by host plants aided in development of synthetic blends (rhyncophorol + host volatiles) to trap *R. palmarum* (Rochat and Avand-Faghih, 2000). Synergy between pheromone and plant volatiles (PVs) was confirmed by analyzing the locomotory responses of *R. palmarum* in a four choice olfactometer (Said *et al.*, 2003).

Racemic nonanoic lactone and 4 hydroxy 3 methoxy styrene from the steam volatiles of coconut bark caused electrophysiological response of *R. ferrugineus* antennae (Gunawardane *et al.*, 1998). Ethyl acetate, ethyl propionate, ethyl butyrate and ethyl isobutyrate synergized attraction of *R. cruentatus* to cruentol. None of the palm esters tested in combination with the pheromone were as attractive as palm or sugarcane tissue (Giblin-Davis *et al.*, 1994). Coconut petioles in traps attracted maximum weevils when they were 2-5 days old, the catch declined thereafter as volatile profile changed due to fermentation (Hallet *et al.*, 1993). The proportional changes in volatile from fermenting palm are attributed to abiotic conditions and microflora present (Nagnan *et al.*, 1992; Samarajeeva *et al.*, 1981). Fermented plant tissues produce a spectrum of odorants that are significantly different from those released by healthy plants (Giblin-Davis *et al.*, 1994; Rochat and Avand-Faghih, 2000). Fermented sap exuding from dead or wounded palms was highly attractive to *R. cruentatus* (Giblin-Davis *et al.*, 1996). Moist fermenting tissue from various palm species, fruits, sugarcane, pineapple and molasses are similarly attractive to palm weevils (Giblin-Davis *et al.*, 1994). The efficiency of the kairomones in attracting insects depends on the odour quality and/or the amount released. Timing of volatile trapping to identify the compounds and their ratios in their matrix is essential to formulate an effective pherosynergestic blend.

Acetoin a volatile product of anaerobic fermentation with ethyl acetate was an effective synergist with aggregation pheromone of *R. palmarum*. Stimulation by plant volatile blends excited three olfactory receptor neurons. Pheromone baited traps containing plant volatile blends attracted twice as many *R. palmarum*as control (Said *et al.*, 2005). Odors from plant tissue along with the aggregation pheromone that attract and orient the weevil from longer distances needs to be explored. No such studies are available for red palm weevil. Indigenously formulated CPCRI lure containing ferrugineol resulted in a weevil catch of 2– 6 weevils/month and it was 50 per cent efficient as compared to imported lures but was cost effective as compared to imported lures.

The electroantennogram assay on the adult *R. ferrugineus* showed higher antennal response to ethyl acetate (a major component in pineapple odor). A comparative study of the odorant binding protein genes and isolation of the partial OBP gene form *R. ferrugineus* was attempted. Amplification of putative OBP genes of *R. ferrugineus* yielded two genomic fragments. On sequencing they had homology with OBP of other insects (Rajesh *et al.*, 2008; Sree Smitha *et al.*, 2008). Commercial lures are loaded in polymembrane dispensers and they have a release rate ranging from 2-20 mg/day. In an attempt to reduce the loss of the chemistry smart delivery devices were developed at CPCRI. The smart delivery aids in controlled release of pheromone for over six months as compared to commercial lures loaded in polymer membrane that are exhausted in 3 months.

3.2. Coconut Rhinoceros Beetle, *Oryctes rhinoceros*

Thirty-nine species of *Oryctes* have been registered but only few of them have an impact on coconut production. They are distributed throughout South East Asia and South Pacific Islands. Apart from coconut, they also infest Palmyrah, date palm, wild date, areca, sago palm, pandanus, pineapple, colocasia, banana, oil palm and sugar cane. The pest occurs round the year with a spike in their population occurring during June to September, the period when the adults visit the crowns. The black colored beetle bores holes and feeds on the unopened fronds and spathe. On opening, the damaged leaves show geometric cuts (V shaped) on leaf lets. If the damage is severe, several cuts can be seen one above the other. The beetles cause damage to seedlings, young and adult palms. Damage when done to the leaves, reduces the photosynthetic area and renders them unsuitable for thatching purpose, but when damage is done to spathe it causes direct crop loss upto 10 per cent (Nair, 1986). Ramachandran *et al.* (1963) reported an yield loss of 5.5 – 9.1 per cent. The beetles also cause death of the seedling/young palms by destroying the growing points. Apart from feeding damage they serve as predisposers for red weevil attack and bud rot. Repeated damage done to the meristem may be lethal.

Male coconut rhinoceros beetles, *Oryctes rhinoceros* (L.), produce sex-specific compounds, ethyl 4-methylheptanoate, and 4-methyloctanoic acid, the first of which is an aggregation pheromone (Hallet *et al.*, 1995). Ethyl 4-methyloctanoate was synthesized by ChemTica and marketed as Oryctalure in the India. Buckets of 18 lit. capacity with black painted metal vanes were used as trap. The pheromone sachet was hung on the diamond shaped hole made in the vane. The trap was hung on a pole at 3-4 cm above the ground level. Coconut petioles were placed in the bucket trap for better attraction. A nonomatrix has been developed at CPCRI for the delivery of rhinoceros beetle pheromone

3.3. Chemoreception in White Grubs

Holotrichia is a genus of the melolonthine scarabs that occurs across the Indian subcontinent and through southeast and east Asia (Ward *et al.*, 2002). The chafer beetle, *Holotrichia serrata* F. (Coleoptera :Scarabaeidae) in its larval stage is a serious pest of coconut, (*Cocos nucifera*L.), sugarcane, (*Saccharum officinarum* L.), groundnut, (*Arachis hypogaea* L.) and vegetables in parts of Western and peninsular India (Ganeshaiah and Kumar,1993).The grubs cause damage by feeding on the roots and the adults emerge with the arrival of monsoon or heavy pre - monsoon showers (Yadav and Sharma, 1995). On emergence at dusk, they aggregate on plants like neem, (*Azadirachta indica* A. Juss), gulmohar, (*Delonix regia* L.), tamarind, (*Tamarindus indica* L.), mahagony (*Swietenia mahagony* L.), drumstick, (*Moringa oleifera* Lam.)

and subabul, (*Leucaena leucocephala* Lam.) for feeding and mating (A.R.V. Kumar Personal communication and Yadav and Sharma, 1995). Attraction of adult males by females has been documented in *H. serrata* (Ganeshaiah and Kumar, 1993). The pheromone of *H. consanguinea* and *H. reynaudi*, has been isolated and identified as anisole (Leal *et al.*, 1996; Ward *et al.*, 2002). Chemical insecticides are widely used by farmers to manage the grubs of *Holotrichia* (Anitha *et al.*, 2006) and they result in varying degree of success in grub management. Indiscriminate use of insecticides results in buildup of residues and cause negative impact on non target organisms. Hence, it is imperative to search for alternative pest control methods. One such option is to exploit the behavioral features of the insect. Ethological control has been successfully applied in the management of scarabs (Leal *et al.*, 1996; Ruther *et al.*, 2000).Though *H. serrata* is an important pest, there are no reports on its olfactory response till date. As a primary step to identify physiologically active compounds causing antennal responses, the electroantennography was used.

The non-aromatic esters, ethyl acetate and propyl acetate elicited a higher response in both male (-1.40 and -1.35 mV respectively) and female (-0.81 and -1.31 mV respectively) beetles while the hydrocarbon alkanes *viz.*, tridecane and nonane elicited the lowest responses. Non-aromatic esters – saturated open chain compounds were found to elicit significant EAG response from both males and females of *H. serrata*. The amplitude of the responses varied with the carbon chain length of the compound. Ethyl acetate a C4 elicited higher response in males. Female antenna was more responsive to propyl acetate. In either case the response decreased with increase in carbon chain length. In general the antennae of both male and female beetles were more responsive to esters followed by alcohol and hydrocarbon alkane.

Among the sex, the response of males to all the compounds tested was higher as compared to the females. Extract of pheromone glands in diethyl ether (-4.43 mV) when exposed to male antennae elicited a significantly higher response as compared to the extracts made in dichloromethane or hexane which were at par (-3.40 and -3.31 mV respectively). Among the solvents used to obtain extracts from the pheromone gland, diethyl ether, dichloromethane and hexane caused maximum EAG amplitude in the male antennae of *H. serrata*. Extraction of abdominal glands of *H. consanguinea* beetles in dichloromethane and ether yielded a clean profile of anisole and indole; the compounds responsible for attraction of *H. consanguinea* males (Leal *et al.*, 1996).

Both adult male and female antennae responded to host volatiles. But there was a sexual dimorphism in the olfactory perception of the host volatiles by *H. serrata*. The male antenna is more sensitive to host extracts. Perhaps, this indicates the necessity of finding the host first for their mate location, apart from using the female produced pheromone as the cue. The use of green leaf volatiles as a sexual kairamone has been observed in *M. melolantha* and *H. consanguinea* males (Reinecke *et al.*, 2005; Yadav and Yadav, 2004). A combination of the synthetic sex attractant (R,Z)-5-(Idecenyl) dihydro-2(3H)-furanone with a 3:7 mixture of phenethyl propionate (PEP) and eugenol caught significantly more *Popillia japonica* (Klein *et al.*, 1981).

Semiochemicals are a vital tool for monitoring and mass trapping in plantation crops. Being perennial crops the pest occurs round the year. To depend on pesticides is a difficult task in terms of application in the target area and also the frequency at which they have to be taken up. Hence exploiting the ethology is an effective method to monitor and trap the pest population in plantation crops. Though the adoption rate of semiochemicals in pest management is at a lower level as compared to pesticides, considering the shift in the policy by Government to scale down the use of pesticides due to health and environmental concerns there is a scope to increase the adoption rate. A note of caution is that semiochemicals may also face the same fate related issues raised over toxicity issues, as very little effort has been made to test the chemicals that are used in behaviour manipulation.

Another problem is that natural enemies use the cues used by the insect pest to identify its host and if the cues from the host plant to altered to bring in desirable pest control it would have a negative impact on natural enemies as its devoid of the cue to orient itself to its host. Hence a proper understanding of the volatiles role in various trophic levels have to be understood prior to attempting a pest management method. The demand for environmentally safe alternatives to broad-spectrum is on rise. The adoption of behavioral manipulation techniques can help to meet this demand, since the amount of chemicals released into the environment by behavioral manipulation is relatively small and are relatively nontoxic to vertebrates and are selective to the target pest species.

References

Abraham,V.A. (1987). Final report of the research project on Study of sex pheromone and other attractants for the major pest of coconut, Central Plantation Crops Research Institute, Regional Station, Kayangulam, pp. 1-18.

Abraham, V.A. and Kurian,C. (1975). An integrated approach to the control *Rhychophorus ferrugineus* the red weevil of coconut palm. In Proc of 4th session of the FAO technical work party on coconut production protection processing. Kingston, Jamaica, September 14 – 25.

Anitha, V., Rogers, D. J., Wightman, J. and Ward, A. (2006). Distribution and abundance of white grubs (Coleoptera :Scarabaeidae) on groundnut in Southern India. *Crop Prot.* 25: 732-740.

de Bruyne, M., Foster, K. and Carlson, J.R. (2001). Odor coding in the *Drosophila* antenna. *Neuron* 30: 537-552.

Dorothea Tholl, Wilhelm Boland, Armin Hansel, Francesco Loreto, Ursula S.R. Rose and Jorg-Peter Schnitzler. (2006). Practical approaches to plant volatile analysis. *The Plant Journal*: 45, 540-560.

Dusenbury, D.B.(1992). Sensory ecology. W.H.Freeman, New York.

Ganeshaiah, K. N. and Kumar, A. R. V. (1993). Self-organization and chemically mediated aggregation of adults in the white grub, *Holotichia serrata*.In T.N.Ananthakrishnan and A. Raman (Ed.) *Chemical Ecology of Phytophagous Insects*. New Delhi: Oxford and IBH Publishing Co. pp167 – 178.

Gblin Davis, R. M., Weissling, T. J., Oehlschlager, A. C. and Gonzales, L. M. (1994). Field response of *Rhynchophorus cruentatus* F. (Coleoptera: Curculionidae) to its aggregation pheromone and fermenting plant volatiles. *Florida Entomologist* 77: 164-177.

Giblin-Davis, R.M., Oehlschlager, A.C., Perez, A.L., Gries, G., Gries,R., Weissling, T.J., Chinchilla, C.M., Pen~a, J.E., Hallett, R.H., Pierce Jr, H.D. and Gonzalez, L.M. (1996). Chemical and behavioral ecology of palm weevils (Curculionidae: Rhynchophorinae). *Florida Entomologist* 79: 153–167.

Gries, G., Gries, R., Perez, A.L., Gonzales, L.M., Pierce, H.D. Jr, Oehlschlager, A.C., Rhainds, M., Zebeyou, M. and Kouame, B. (1994). Ethyl propionate: synergistic kairomone for African palm weevil, *Rhynchophorus phoenicis* L. (Coleoptera: Curculionidae). *Journal of Chemical Ecology* 20: 889–897.

Gunatilake, R. and Gunawardena, N. E. (1986). Ethyl alcohol: A major attractant of red weevil (*Rhynchophorus ferrugineus*). In Proc Sri Lanka Association for the Advancement of Science. 42nd Annual Sessions, p. 70.

Gunawardena, N. E., Kern, F., Janssen, E., Meegoda, C., Schäfer, D., Vostrowski, O., and Bestmann, H.J. (1998). Host attractants for red weevil, *Rhynchophorus ferrugineus*: Identification, electrophysiological activity, and laboratory bioassay. *J. Chem. Ecol.* 24: 425–437.

Hallet, R. H., Gries, G., Gries, R., Borden, J. H., Czyzewska, E., Oehlschlager, A.C., Pierce, H.D., JR., Angerilli, N.P.D. and Rauf, A. (1993). Aggregation pheromones of two Asian palm weevils, *Rhynchophorus ferrugineus* and *R. vulneratus*. *Naturwissenschaften* 80: 328-331.

Hallet, R.H., Perez, A.L., Gries, G., Gries, R., Pierce, H.D. Jr. Yue-Junming, Oehlschlager, A.C., Gonzalez, L.M., Borden, J.H., and Yue, J.M. (1995). Aggregation pheromone on coconut rhinoceros beetle, *Oryctes rhinoceros* L. (Coleoptera : Scaraebidae). *J. Chem. Ecol.* 21(10): 1549–1570.

Handley, A.J. and Adlard, E.R. (2005). Gas Chromatographic Techniques and Applications. Boca Raton, FL: CRC Press.

Jaffe´, K., Sanchez, P., Cerda, H., Hernandez, J.V., Jaffe´, R., Urdaneta, N., Guerra, G., Martinez, R., Miras, B. (1993). Chemical ecology of the palm weevil *Rhynchophorus palmarum* (L.) (Coleoptera: Curculionidae): Attraction to host plants and a male produced aggregation pheromone. *J. Chem. Ecol.* 19: 1703–1720.

Kalshoven, L.G.E.(1950). Pests of crops in Indonesia. Translated by van der Laan (1981). 701 pp.

Klein, M.G., Tumlinson, J.H., Ladd, T.U. and Doolittle, R.E. (1981). Japanese beetle (Coleoptera :Scarabaeidea) response to synthetic sex attractant plus Phenethyl propionate: Eugenol. *J. Chem. Ecol.* 7, 1–7.

Kurian, C., Abraham, V.A. and Ponnama, K.A. (1984). Attractants: An aid in red palm weevil management. In Proc : PLACROSYM V, Dec. 15 -18, Kasaragod, India.

Leal, W. S., Yadava, C. P. S. and Vijayvergia, J. N. (1996). Aggregation of scarab beetle *Holotrichia consanguinea* in response to female released pheromone suggests secondary function hypothesis for semiochemical. *J. Chem. Ecol.* 22: 1557-1566.

Lockwood, G.B. (2001). Techniques for gas chromatography of volatile terpenoids from a range of matrices. *J. Chromatogr. A*, 936: 23–31.

Majid Ghanina. (2007). Olfaction in Musquitoes. Ph.D Thesis, Swedish University of Agricultural Sciences, Alnarp.

Merfort, I. (2002). Review of the analytical techniques for sesquiterpenes and sesquiterpene lactones. *J. Chromatogr. A*, 967: 115– 130.

Millar, J.G. and Haynes, K.F. (1998). Methods in chemical ecology, chemical methods (Ed). Kluwer Academic Publishers. 390pp.

Nagnan, P., A. H. Cain, and D. Rochat. (1992). Extraction and identification of volatile compounds of fermented oil palm sap, candidate attractants for the palm weevil. *Oléagineaux* 47: 135-142.

Nair, M.R.G.K. (1986). Insects and mites on crops of India. ICAR, New Delhi, India 83p.

Oehlschlager, A. C., C. M. Chinchilla, L. M. Gonzalez, L. F. Jiron, R. G. Mexzon, and Morgan, B (1993). Development of a pheromone-based trapping system for *Rhynchophorus palmarum* (Coleoptera: Curculionidae). *Journal of Economic Entomology* 86: 1381-1392.

Rajesh, M.K., Subaharan, K., SatishKumar, R., Ritto Paul, Bobby Paul, SreeSimitha and George V. Thomas. (2008). A comparative study of insect odor binding protein genes and isolation of a partial OBP gene from red palm weevil, *Rhynchophorus ferrugineus*. *Journal of Plantation Crops* 36: 418-424.

Ramachandran, C.P., Kurien, C. and Mathew, J. (1963). Assessment of damage to coconut due to *Oryctes rhinoceros*. Nature and damage caused by beetle and factor involved in the estimation of loss. *Indian Coconut Journal* 17: 3-12.

Reinecke, A.; Ruther, J. and Hilker, M. (2005). Electrophysiological and behavioural responses of *Melolantha melolantha* to saturated and unsaturated aliphatic alcohols. *Entomol Exp Appl.* 115: 33-40.

Rochat, D. and Avand-Faghih, A. (2000). Trapping of red Palm weevil (*Rhynchophorus ferrugineus*) in Iran with selective attractants. In: Kleeberg, H., Zebitz, C.P.W. (Eds.), Practice oriented results on use and production of neem-ingredients and pheromones VI. Germany, pp. 219–224.

Rochat, D., Nagnan-Le Meillour, P., Esteban-Duran, J.R., Malosse, C., Perthuis, B. and Morin, J.P. (2000). Identification of pheromone synergists in American palm weevil, *Rhynchophorus palmarum*, and attraction of related *Dynamis borassi*. *J. Chem. Ecol.* 26: 155-187.

Rochat, D., Ramirez-Lucas, P., Malosse, C., Zagatti, P. and Mori, K. (1993). Pheromones and pheromone-related volatiles of four Rhynchophorinae weevils (Coleoptera: Curculionidae). *Bulletin of the International Organization for Biological Control/West Palearctic Regional Section* 16: 178–184.

Ruther, J., Reinecke, A., Tolasch,T., Hilker, M. (2000). Mate finding in the forest cocokchafer, *Melolantha hippocastani* mediated by volatiles from plants and females. *Physiol Entomol* 25: 172-179.

Said I., Tauban D., Rrnou M., Mori K.and Rochat D. (2003). Structure and function of the antennal sensilla of the palm weevil *Rhynchophorus palmarum* (Coleoptera, Curculionidae). *Journal of Insect Physiology*, 49: 857-872.

Samarajeewa, U., Adams, M.R and Robinson, J.M. (1981). Major volatiles in Sri Lanka arrack, a palm wine distillate. *Journal of Food Technol*ogy 16: 437-444.

SreeSmitha, Rajesh, M.K., Subaharan, K., Satish Kumar and George V. Thomas. (2008). Homology, modeling and docking studies in an odor binding protein from red palm weevil, *Rhynchophorus ferrugineus*. *Journal of Plantation Crops*, 36: 430-434.

Ward, A., Moore, C., Anitha, V., Wightman, J. and Rogers, D. J. (2002). Identification of sex pheromone of *Holotrichia reynaudi*. *J. Chem. Ecol.* 28 (3): 515-522.

Wattanapongsiri.A. (1966). A revision of the genera *Rhynchophorus* and *Dynamis* (Coleoptera: Curculionidae). *Dep. Agr. Sci. Bull., Bangkok* 1: 1-328.

Yadav, C. P. S. and Sharma, G. K. (1995). Indian white grubs and their management. All India Coordinated Research Project on White grubs, Technical Bulletin No.2. Indian Council of Agricultural Research, New Delhi.

Yadav, V.K. and Yadav, N. (2004). Identification of the chemical mediating attraction of *Holotrichia consanguinea* beetles to its most preferred host tree. Paper presented in IV International Crop Science Congress.

2018, Pests of Plantation Crops   *Pages* **221–231**
Editors: **P. Chowdappa, Chandrika Mohan & A. Josephrajkumar**
Published by: **ASTRAL INTERNATIONAL PVT. LTD., NEW DELHI**

Chapter 11

Insect Neuropeptides and Application in Pest Management

☆ *A. Josephrajkumar, M.K. Rajesh and P. Chowdappa*

1. Introduction

Neuropeptides are biologically active peptides that are mainly produced in neurosecretory cells of insects constituting a diverse widespread class of signaling substances in the nervous system. Insect neuropeptides function as neurotransmitters, neuromodulators and neurohormones and are therefore called 'master regulators' of metabolic, homeostatic, developmental, reproductive and behavioural events during an insect life (Holman *et al.*, 1990). Neurotransmitters are chemicals responsible for transmitting impulses between nerve cells by transiently altering the electrical excitability of cell membranes. Neuromodulators exert slow modulatory effects and control the level of excitability of whole group of nerve cells. Neurohormones tend to influence slow onset of events and act at a distance from the release site. In insects, a large number of neuropeptides are true neurohormones that regulate the above mentioned processes in a precise and controlled way. To interfere with these developmental processes in insects, and use the neuropeptides in safe and rational manner, it is logically important to characterize various neuropeptides and understand their functioning (Gäde and Goldsworthy, 2003).

In insects, neuropeptides have been extensively studied with respect to their roles as circulating hormones. Although role of neuropeptides in the insect central nervous system (CNS) is less understood, it is commonly considered that they act as neuromodulators or co-transmitters rather than as neurotransmitter (Nässel and Homberg, 2006). The term co-transmitter can be applied to a neuropeptide that is co-localised with a classical neurotransmitter that acts on an ion-channel type of receptor (Burnstock, 2004). When the neuropeptide is coreleased with the classical neurotransmitter, the activation of the peptide G-Protein Coupled Receptor (GPCR) leads to the modulation of the ion-channel mediated signaling. Neuropeptides can play a multitude of functional roles in the brain and even single neuropeptides are likely to be multifunctional (Nässel and Homberg, 2006).

Neuropeptides have been identified in insect neurons by immunocytochemistry and in several cases by *in situ* hybridization histochemistry. Most of the known insect neuropeptides are present in interneurons and in neurosecretory or endocrine cells (Nässel, 2002; Homberg, 2002). Great diversity exists in the patterns of distribution of the various neuropeptides, and the different peptidergic circuits display various degrees of complexity. In *Drosophila melanogaster*, certain peptidergic systems have been explored with molecular genetics approaches, such as cell specific interference with peptide signaling by using the binary GAL4-UAS system (Duffy, 2002) and in other insects more traditional pharmacological and physiological experiments have been performed. Remarkable advances in the field of neuropeptide research and the comparative approach paved way for the discovery of many novel peptide structures. The total number of peptides isolated from insect nervous system exceeded 200 and are classified into 20 families (Nässel, 1995). The first insect neuropeptide identified is proctolin.

2. Proctolin

Proctolin is the first pentapeptide isolated from the gut of American cockroach, *Periplaneta americana* and was proposed to function as a neurotransmitter with myotropic properties. It produces slow graded contraction of longitudinal muscles of proctodeum and modulates muscle excitability (Starrat & Brown, 1975). Using immunohistochemistry, it was found that proctolin-like immunoreactive neurons and processes are widely distributed throughout the central nervous system, stomatogastric nervous system and peripheral tissues such as oviducts and alimentary canal. Of late, proctolin-like immunoreactive lateral neurosecretory cells in the brain project processes to corpus cardiacum and corpora allata. Proctolin was now designated as a releasing factor capable of stimulating the release of adipokinetic hormone from the corpus cardiacum and of stimulating juvenile hormone production from the corpora allata.

3. Nomenclature

Insect neuropeptides are named after the first three and two letters of the genus and species name of the insect, respectively. Adipokinetic hormone from *Locusta migratoria* is therefore named as *Locmi*-AKH (Raina and Gäde, 1988).

This mini review focuses on few neuropeptides of potential application into a number of promising insect model system.

4. Adipokinetic Hormone (AKH)

AKH produced by corpora cardiac, mobilizes carbohydrates and lipids from insect fat body during extreme physical activities such as flight and locomotion.

Energy demand is profusely accelerated during high intensity muscular work like flight, where 100-fold enhancement is documented. Hence, flight muscles are depended on aerobic energy metabolism and oxidation of lipids, carbohydrates and amino acids (Gade and Auerswald, 1998). Intermediary metabolism is orchestrated by endocrine glands releasing short peptides from paired neurohemal glands, corpora cardiaca (Gade, 1996). Such metabolic regulators are termed adipokinetic, leading to higher levels in lipids, trehalose and proline in insect haemolymph as and when the demand is on (Gade and Goldsworthy, 2003).

4.1 AKH-Structure

AKH was purified and sequenced from the migratory locust, *Locusta migratoria* with 8, 9, or 10 amino acid residues having pQ group at N-terminus, aliphatic or amino acid residue at position 2 and amide-group at C-terminus position. AKH was identified in all insect orders, certain annelids, mollusks and nematodes and nearly thirty-six different isoforms of insect AKH are reported (Gade *et al.*, 1997). Recently, "AKH-like" and "proto-AKH" with more than 10 amino acid residues are also characterized (Li *et* al., 2016). In addition to the mobilization of energy substrates during high demand phase of flight and stress-shooting mechanism, AKH is also involved in inhibition of RNA synthesis of lipids and proteins in fat body, activities linked to cardio-excitatory properties in *Pyrrhocoris apterous*, accelerated long-term locomotion as well as insect immunity engulfing entomopathogens (Gade, 1996; Gade *et al.*, 1997).

5. Locomotion

It was well known and demonstrated that the availability of energy substrates are crucial for sustained flights in locusts rather than extracts from corpora cardiaca. Substantial availability of lipids in synergy with the extracts from corpora cardiac induced higher flight speeds in locusts, and that the flight speed was tremendously diminished under reduced titre of lipids even with the injection of corpora cardiaca extracts. Thus, the flight dynamics in locusts is influenced by the availability of lipids and regulated by AKH (Goldsworthy *et al.*, 1979). In *P. apterous*, AKH enhanced the mobilization of lipids and accelerated locomotion under its influence (Kodrik *et al.*, 2002). Topical application of con-specific AKH admixed with organic solvents penetrated the cuticle of the cricket, *Gryllus bimaculatus* and elicited flight-induced responses. Though structurally unrelated, AKH has similar function in insects as glucagon and adrenalin have in mammals.

6. Insect Immunity

A defensive response is exhibited by an insect when their tissues are invaded by entomopathogens. The immune response is elicited at both cellular and humoral level to counter the aggression but not akin to antibody production mechanism in vertebrates. Proteins in insect haemolymph recognize glucans in fungal cell walls and peptidoglycans in bacterial cell wall stimulating cellular mechanism involving phagocytosis and encapsulation. In certain cases, invading pathogens are engulfed by haemocytes to form nodules. Activation of prophenoloxidase cascade resulting in the synthesis of antimicrobial peptides to neutralize infection in heamolymph indicates the humoral defense mechanism (Lavine and Strand, 2002).

AKH has greater influence in signaling the activation of prophenoloxidase cascade in locusts. Mere injection of lipopolysaccharide and laminarin from infectious bacterium and fungal cells, respectively into the locusts could not trigger the prophenoloxidase cascade unless Locmi-AKH1 is co-injected. Nodule formation as well as increase in phenoloxidase in locusts could be elicited only in the simultaneous presence of AKH and lipopolysaccharide and laminarin (Goldsworthy *et al.*, 2003). Induction of apolipophorin III by the action of AKH was involved in the activation of prophenoloxidase cascade by lipopolysaccharide and laminarin (Dettloff *et al.*, 2001). Deciphering the interaction between the insect immunity and endocrine system has opened out to understand the role of certain eicosanoids on the activation of phenoloxidase and nodule formation in insects.

7. Homeostasis and Feeding Regulation by Neuropeptides

Water and ion balance in insects are well regulated and maintenance of sufficiently stable state of water balance has long been recognized. In the absence of significant blood pressure in the insect circulatory system, primary urine formation is mostly by secretion rather than by filtration. Malpighian tubules located at the junction of midgut and hindgut secreting KCl or NaCl rich solutions which are essentially isoosmotic to haemolymph and water and solutes follow by passive diffusion, with exception of some active transport involved for elimination of toxic substances. The excretory system is primarily responsible for homeostasis, following metabolic modification of toxic compounds to chemicals more readily excreted or could be safely stored (Ramsay, 1954).

7.1 Osmoregulation

Excretion and water balance are under neuroendocrine control. The hormone involved in the process of diuresis are called diuretic hormone (DH). They are responsible for the secretion of urine by the Malpighian tubules. Although, a number of diuretic hormones have been identified in insects, 5-hydroxytryptamine is the major diuretic hormone of *Rhodnius prolixus* which could consume 10-20 folds their body weight in a single meal. There are three primary diuretic hormones found in insects: Corticotropin Releasing Factor (CRF) and a family of smaller kinins, and serotonin. CRF-like DH and kinins are classified as neuropeptides and the serotonin, 5-hydroxytryptamine is classified as a neurotransmitter (Coast *et al.*, 2002). Diuretic hormones act in synergism to enhance the process of diuresis.

The most studied CRF-like DH are Manse-DH, Locmi-DH and Achdo-DH. CRF-like DH is similar to vertebrate-CRF that mediates the secretion of urine through the production of cyclic-AMP. Active transport of Na^+ and K^+ into the Malpighian tubules could be accomplished by the CRF-like DH through cAMP. In general, transport of ions from the haemolymph into the Malpighian tubules leads to the secretion of urine. Insect CRF-like diuretic peptides adopt a folded helix-loop conformation bringing C-terminal amide close to N-terminus whereas human CRF assumed rod-like ?-helical conformation (Coast *et al.*, 2002).

7.2 Feeding Regulation by DH

Sensitivity of a given species of an insect herbivore is dependent upon the quantity and chemical structure of the feeding deterrent. Food as well as non-food stimuli both from within and outside the animal is involved in regulation of feeding, amply modulated by feedback mechanism from stretch receptors of gut wall, hormones and composition of blood (Simpson and Bernays, 1983). Tobacco leaf painted with Manse-DH-11 and fed to first-instar larvae of *Manduca sexta* reduced food consumption and caused high mortality. Injection of diuretic peptides induced antifeedant activity for Manse-DH in *Heliothis viresencs* larvae and for Locmi-DH in *L. migratoria* (Coast *et al.*, 2002). In locusts, filling of foregut during the feeding process caused closure of the pores at the tip of the taste sensilla on the mouth by release of a factor from corpora cardiaca into haemolymph. It was therefore indicated that the normal release of endogenous CRF-like DH signals the end of the meal in locusts and thus altered the feeding behavior in *L. migratoria*. Development of analogues for CRF-like DH could interrupt the feeding behaviour in insects.

7.3 Feeding Regulation by Sulfakinins

Sulfakinins are myostimulatory peptides partially similar to gastrin in mammals. Sulfakinins cause muscle contraction on insect gut and occasionally releases amylase that is involved in digestion. By inducing muscle contraction, sulfakinins cause irregular peristaltic movement of gut wall and may interrupt the movement of food ingested from crop to midgut. Injection of sulfakinin into locust and cockroaches diminished the food intake in these insects by altering muscular contraction in gut (Wei *et al.*, 2000).

8. Disrupting Sex Attraction

Sex pheromones are intra-specific semiochemicals that attract conspecific males/females from over great distance. Basic chemistry of these pheromones is well understood in certain of the Lepidopteran and Coleoptera insects; however, research on regulation of pheromone biosynthesis has been initiated. In corn earworm moth, a unique factor from brain was found responsible for the biosynthesis of pheromones (Raina and Klun, 1984). This factor was later identified as a peptide containing 33 amino acid residues that stimulates the biosynthesis of moth's sex pheromone termed as Pheromone Biosynthesis Activating Neuropeptide (PBAN), which was later characterized by a common amino acid sequence FXPRLamide motif in the C-terminus (Raina *et al.*, 1989). In flesh flies, these peptides are pyrokinins and myotropins that stimulate the hind gut and oviduct *in vitro* accelerating puparium formation. PBAN is released into the hemolymph of females during the scotophase and is drastically reduced after mating, contributing to the loss in female receptivity. Pheromone production is age-dependent and Juvenile Hormone is involved in its regulation. It was determined that substitution of T at X-position of PBAN was more biologically active in pheromone bioassay than analogues containing V, S or G (Abernathy *et al.*, 1995).

8.1 Antagonists of PBAN

A linear lead antagonist that imposed conformational modulation on Helze-PBAN was designed by Alstein *et al.* (2000). It was observed that in Helze-PBAN, the hexapeptide $PBAN_{28-33}$ is as active in stimulating pheromone biosynthesis of *H. zea* as the complete $PBAN_{1-33}$. Replacement of L-amino acids (Ser^{30} or Arg^{32}) by the D-hydrophobic amino acid D-Phe was found to be an effective PBAN antagonist capable of inhibiting sex pheromone biosynthesis in the female moth, *H. zea*. Compounds with antagonistic activity up to 96 per cent could be obtained and the most active peptide had an alkyl chain length of two and a spacer chain length of three carbons (Alstein *et al.*, 2000). Development of non-peptide analogues with insecticidal properties is being attempted.

8.2 Pseudopeptide Analoges of PBAN

Neuropeptides, which are polar in nature could not penetrate the non-polar lipid layer of insect cuticle and these peptides are broken down by peptidases in the gut and haemolymph as well. Nachman *et al.* (2001) developed paseudopeptides that were able to penetrate the hydrophobic cuticle and had enhanced peptidase resistance. Addition of various hydrophobic groups to the N-terminus of the C-terminal pentapeptide active core, which in conjunction with the polar Arg side chain, confer an amphiphilic property. Hydrophobic groups appended to N-terminus included fatty acids of various chain length, cholic acid, aromatic acids and carboranylpropionic acid. Amphiphilic analogues are more resistant to digestion by aminopeptidases and have higher binding affinity for the receptor. Topical application of native PBAN and the pentapetide fragment had no pheromonotropic activity but the amphiphilic analogues had significant pheromonotropic activity. Higher number of phenyl rings conferred higher resistance to aminopeptidases.

9. Growth and Development

The steroid hormone, ecdysteroids and the terpenoid, juvenile hormone regulate growth and reproduction in insects. Ecdysteroid triggers moulting events and the characteristic of the moult is orchestrated by juvenile hormone. Both the hormones are produced by endocrine glands and their synthesis is regulated by neuropeptides (Gade *et al.*, 1997). The neuropeptide, Prothoracicotropic hormone (PTTH) stimulates the synthesis and release of ecdysteroids. PTTH was first identified in *B. mori* as 109 amino acid residue long, with seven Cys residues forming disulfide bridges and a N-glycosylation site (Gade, 1997). Juvenile hormone is a sesquiterpene produced by the corpora allata. Several slightly different forms known as JH0, JH1, JHII and JHIII containing 19, 18, 17 and 16 carbon atoms, respectively have been isolated. JHIII is the ubiquitous form occurring in most insects. Titre of juvenile hormone in the haemolymph is determined mainly by the rate of biosynthesis from corpora allata (Tobe and Stay, 1985). It was now understood that two types of neuropeptides regulate the JH production *in vitro*, allototropins by stimulation of JH biosynthesis and allotostatins by its inhibition.

9.1 Moth Allatostatin or C-type

The Moth Allatostatin is a 15-residue peptide reported first from horn worm, *Manduca sexta* which is characterized by an N-terminal pGlu residue, a free C-terminus and two cys residues forming an intramolecular disulfide bridge (Kramer *et al.*, 1991). This peptide inhibits JH biosynthesis *in vitro* in several moths but is found cardioinhibitory in fruit fly. It also inhibits the feeding of tomato moth, *Lacanobia oleracea* L. *in vitro* (Audsley *et al.*, 2001).

9.2 Snow Drop Lectin Fusion Protein

Topical sprays of neuropeptides were not successful on account of its polar nature and as such peptides are prone to degradation in the environment. In this context, a novel delivery of neuropeptides was attempted. It was observed that the allotostatin identified in *L oleraceae* is identical with that of *M. sexta*. Injection of allotostatin did not inhibit JH biosynthesis in larvae of *L. oleracea* but feeding was reduced and growth was retarded inducing up to 80 per cent mortality. Dietary incorporation of allotostatin was ably inactivated by the proteases in the digestive system. In a classical study it was observed that the mannose-binding plant lectin from snow drop (*Galanthus nivalis* agglutinin, GNA) was not digested by the gut proteases and could be detected in the haemoloymph (Fitches *et al.*, 2001). GNA was found to be potent carrier of peptides into haemolymph as it could cross the gut epithelium without being disintegrated. In order to make the delivery successful, a fusion protein of GNA and Manse-AST was made and expressed in *Escherichia coli*. Feeding of fusion protein by the fifth-instar *M. sexta* larvae drastically affected the feeding and weight gain. Fusion of GNA to Manse-AST protected from proteolytic breakdown in the haemolymph and induced antifeedant effect, which free allatostatin could not exert. It was now observed that the fusion of neuropeptide to GNA could effectively release the peptide into the haemolymph and cause detrimental effect. Fusion protein could be genetically engineered into plants and the bio-engineered product would combat pest invasion (Gade and Goldworthy, 2003).

10. Reproduction

Insect reproduction is well regulated and the hormones ecdysteroids and juvenile hormone along with neuropeptides control adult reproduction with great precision. Juvenile hormone and ovary maturating parsin regulate oocyte maturation in locusts. Peptide hormone that regulates the expression of serine proteases, the key luminal proteases in insects, was identified (Borovsky *et al.*, 1993). These peptides form the core technology upon which Insect Biotechnolgy Inc., California is developing a variety of products.

Digestion of blood meal by mosquitoes is very crucial for ovary maturation and egg development. Gut proteases especially trypsin-like serine proteases are activated after the blood meal and are responsible for the digestion of blood. Oocytes in mosquitoes are developed after the transport of free amino acid from intestine into the fat body leading to the biosynthesis of vitellogenin. Borovsky *et al.* (1993) identified a decapeptide synthesized from the mosquito ovary, 18-24 h after the blood meal and termed it as Trypsin Modulating Oostatic factor (TMOF). Once TMOF is released into the haemolymph, there is complete stoppage of trypsin biosynthesis and no more egg maturation occurred. Injection of TMOF indicated inhibition of trypsin biosynthesis (translation), however, transcription of mRNA of trypsin is not affected (Borovsky *et al.*, 1993).

It was observed that TMOF was not inactivated in the gut of mosquitoes and subsequently stopped trypsin biosynthesis and egg maturation, when fed. Higher dosages of TMOF @ 1 ng per larva absorbed on yeast and fed to wrigglers caused mortality of mosquitoes in laboratory. It was further proved that TMOF gene could be fused with coat protein gene of tobacco mosaic virus at the restriction site for trypsin and this construct @ 140 pg inhibited proteases and killed mosquito wrigglers in a period of five days (Borovsky *et al.*, 1998). Protease inhibitors through trypsin inhibition are a very important tool employed in pest management in a wide array of crops. Most of lepidopteran insects have protease activity in the mid gut and the key pest of coconut *viz.*, red palm weevil was also found to have trypsin-like protease for protein metabolism. In *H. virescens*, a factor closely associated with Aedae-TMOF was unraveled, which has an important role in biosynthesis of trypsin. This could be reality in coming years of advancement in science in pest management.

Some of the insect neuropeptides have been discussed for a possible utilization as one of the tools in pest management programme. Refinement is further required for commercial exploitation especially against major pests of our tropical zone. Insect neuropeptides isolated on the basis of potential inhibitory control will emerge as a likely antifeedant lead molecule. The disruption of any step leading to biosynthesis of neuropeptides, their modifications during storage, their release into the haemolymph as well as their interaction with the target-cell membrane-bound receptors offer multiple modes of action for a novel-neuropeptide based pest management programme. Indeed, not all biochemical mechanisms will be worth exploiting nor will all neuropeptide be of equal importance with regard to pest control. Science has gone so deep that even evolutionary relationship are being unraveled with the study on neuropeptides in invertebrates and some of them are likely to be potential markers in the near future. Insect pest management is so dynamic and no one tool is likely to eliminate them and we have to learn to live with insects, of course not allowing to cause economic damage to our livelihood. Such novel tools will be very useful in the long run.

References

Abernathy RL, Nachman RJ, Teal PEA, Yamashita O, Tumlinson JH (1995). Pheromonotropic activity of naturally occurring pyrokinin insect neuropeptide (FXPRLamide) in *Helicoverpa zea*. *Peptides* **16**: 215-219.

Alstein M, Ben-Aziz O, Schefler I, Zeltser I, Gilon C. (2000). Advances in the application of neuropeptides in insect control. *Crop Prot* **19**: 547-555.

Audsley N, Weaver RJ, Edwards JP. (2001). *In vivo* effects of *Manduca sexta* allatostatin and allatotropin on larva of the tomato moth, *Lacanobia oleracea*. *Physiol Entomol* **26**: 181-188.

Borovsky D, Carlson DA, Griffin PR, Shabanowitz J, Hunt DF. (1993). Mass spectrometry and characterization of *Aedes aegypti* trypsin modulating oostatic factor (TMOF) and its analogs. *Insect Biochem Mol Biol* **23**: 703-712

Borovsky D, Janssen I, VandenBroeck J, Huybrechts R, Verhaert P, DeBondt HL, Bylemans D and DeLoof A. (1996). Molecular sequencing and modeling of *Neobellieria bullata* trypsin. Evidence for translational control by *Neobellieria* trypsin – modulating oostatic factor. *Eur J Biochem* **237**: 279-287.

Borovsky D, Powell CR, Dawson WO, Shivprasad S, Lewandowski DJ, DeBondt HL, De Ranter C, De Loof A. (1998). Trypsin modulating oostatic factor (TMOF): a new biorational insecticide against mosquitoes, In *Insects: Chemical, Physiological and Environmental Aspects 1997,* (Eds) Konopinska D, Goldsworthy DJ, Nachman RJ, Nawrot J, Orchard I and Rosinski G, University of Wroclaw, 131-140.

Burnstock G. (2004). Cotransmission. *Curr Opinion Pharmacol* **4**: 47-52.

Coast GM, Orchard I, Phillips JE, Schooley DA. (2002). Insect diuretic and antidiuretic hormones. *Adv Insect Physiol* **29**: 279-409.

Dettloff M, Wittwer D, Weise C, Wiesner A. (2001). Lipophorin of lower density is formed during immune responses in the lepidopteran insect *Galleria mellonella*. *Cell Tissue Res* **306**: 449-458.

Duffy JB. (2002). GAL4 system in *Drosophila*: A fly geneticist's Swiss army knife. *Genesis* **34**: 1-15.

Fitches E, Audsley N, Gatehouse JA, Edwards JP. (2002). Fusion proteins containing neuropeptides as novel insect control agents: snowdrop lectin delivers fused allatostatin to insect haemolymph following oral ingestion. *Insect Biochem Mol Biol* **32**: 1653-1661.

Fitches E, Ilett C, Gatehouse AMR, Gatehouse LN, Greene R, Edwards JP, Gatehouse JA. (2001). The effects of *Phaseolus vulgaris* erythro and leucoagglutinating isolectins (PHA-E and PHA-L) delivered via artificial diet and transgenic plants on the growth and development of tomato moth (*Lacanobia oleracea*) larvae; lectin binding to gut glycoproteins *in vitro* and *in vivo*. *J Insect Physiol* **47**: 1389-1398.

Gäde G, Auerswald L. (1998). Insect neuropeptides regulating substrate mobilization. *S Afr J Zool* **33**: 65-70.

Gäde G, Goldsworthy GJ. (2003). Insect peptide hormones: a selective review of their physiology and potential application for pest control. *Pest Manag Sci* **59**: 1063-1075.

Gäde G, Hoffman KH, Spring JH. (1997). Hormonal regulation in insects: facts, gaps, and future directions. *Physiol Rev* **77**: 963-1032.

Gäde G. (1996). The revolution in insect neuropeptides illustrated by the adipokinetic hormone/red pigment-concentrating hormone family of peptides. *Z Naturforch* **51c**: 607-617.

Gäde G. (1997). The explosion of structural information on insect neuropeptides, In: *Progress in the Chemistry of Organic Natural Products* Vol 71, (Eds) Herz W, Kilby GW, Moore RE, Steglich W, Tamm Ch, Springer-Verlag, New York, 1-128.

Goldsworthy GJ, Jutsum AR, Robinson NL. (1979). Substrate utilization and flight speed during tethered flight in the locust. *J Insect Physiol* **25**: 183-185.

Goldsworthy GJ, Mullen L, Opoku-Ware K, Chandrakant S. (2003). Interactions between the endocrine and immune systems in locusts. *Physiol Entomol* **28**: 54-61.

Holman GM, Nachman RJ, Wright MS. (1990). Insect neuropeptides. *Annu Rev Entomol* **35**: 201-217.

Homberg U. (2002). Neurotransmitters and neuropeptides in the brain of the locust. *Microsc Res Tech* **56**: 189-202.

Kodrik D, Socha R, Zemek R. (2002). Topical application of Pya-AKH stimulates lipid mobilization and locomotion in the flightless bug, *Pyrrhocoris apterus* (L). *Physiol Entomol* **27**:15-20.

Kramer SJ, Toschi A, Miller CA, Kataoka H, Quistad GB, Li JP, Carney RL, Schooley DA. (1991). Identification of an allatostatin from the tobacco hornworm *Manduca sexta*. *Proc Natl Acad Sci USA* **88**: 9458-9462.

Lavine MD, Strand MR. (2002). Insect haemocytes and their role in immunity. *Insect Biochem Mol Biol* **32**: 1295-1309.

Li, S., Hauser, F., Shadborg, S.K., Nielsen, S.V., Kirketerp-Moller, N., Cornelis, J. and Grimmikhuijzen, P. (2016). Adipokinetic formone and other G protein-coupled receptor emerged in Lophotorchozoa. *Scientific Reports* doi:10.1038/srep32789.

Nachman RJ, Teal PEA, Strey A. (2002). Enhanced oral availability/pheromonotropic activity of peptidase resistant topical amphiphilic analogs of pyrokinin/PBAN insect neuropeptides. *Peptides* **23**: 2035-2043.

Nachman RJ, Teal PEA, Ujvary I. (2001). Comparative topical pheromonotropic activity of insect pyrokinin/PBAN amphiphilic analogs incorporating different fatty and/or cholic acid components. *Peptides* **22**: 279-285.

Nässel DR, Homberg U. (2006). Neuropeptides in interneurons of the insect brain. *Cell Tissue Res* **326**: 1-24.

Nässel DR. (1995). Neuropeptide diversity in the insect nervous system. Facts and speculation, In: *Recent Advances in Insect Endocrine Research*, Muraleedharan D and Mariamma J (Eds) 1-38.

Nässel DR. (2002). Neuropeptides in the nervous system of *Drosophila* and other insects: multiple roles of neuromodulators and neurohormones. *Progr Neurobiol* **68**: 1-84.

Raina AK, Gäde G. (1988). Insect peptide nomenclature. *Insect Biochem* **18**: 785-787.

Raina AK, Klun JA. (1984). Brain factor control of sex hormone production in the female corn earworm moth. *Science* **225**: 531-533.

Raina AK, Jaffe H, Kempe TG, Keim P, Blacher RW, Fales HM, Riley CT, Klun JA, Ridgway RL, Hayes DK (1989). Identification of a neuropeptide hormone that regulates sex pheromone production in female moth. *Science* **244**: 796-798.

Ramsay JA. (1954). Active transport of water by the Malpighian tubules of the stick insect, *Dixippus morosus. J Exp Biol* **31**: 104-113.

Simpson SJ, Bernays EA. (1983). The regulation of feeding: locusts and blowflies are not so different from mammals. *Appetite* **4**: 313-346.

Starrat AN, Brown BE. (1975). Structure of the pentapeptide Proctolin, a proposed neurotransmitter in insects. *Life Sci* **17**: 1253-1256.

Tobe SS, Stay B. (1985). Structure and regulation of corpus allatum. *Adv Insect Physiol* **18**: 305-432.

Wei Z, Baggerman GJ, Nachman R, Verhaert P, De Loof A, Schoofs L. (2000). Sulfakinens reduce food intake in the desert locust, *Schistocerca gregaria. J Insect Physiol* **46**: 1259-1265.

2018, Pests of Plantation Crops
Editors: **P. Chowdappa, Chandrika Mohan & A. Josephrajkumar**
Published by: **ASTRAL INTERNATIONAL PVT. LTD., NEW DELHI**

Pages **233–250**

Chapter 12

Entomopathogenic Nematodes in Pest Management

☆ *Rajkumar and A. Josephrajkumar*

1. Introduction

Plantation crops are perennial in nature, cultivated extensively in tropics/ subtropics. The important plantation crops in India are coffee, tea, rubber, cashew, cocoa, coconut, arecanut, oil palm and intercrops like black pepper, cardamom, ginger and turmeric, which are not the exception to be damaged by insect pests at all life stages. Insect pests cause damage not only to standing crops but seedlings in nursery and seeds during storage. By virtue of quick knock down effect, chemical insecticides are the preferred options for these insect pests. But increased awareness on the side effects caused by indiscriminate use of chemical pesticides and global concern regarding possible traces of pesticides residues in export plantation commodities had made eco-friendly integrated pest management (IPM) the need of the present era. Among the various components of IPM, the most efficient tool for ecological sustainability is the biological pest suppression. Biological control agents broadly involve parasitoids, predators and entomopathogens like bacteria, fungi and nematodes which cause lethal infections. The nematodes which are not parasitic on plants but infect insects and used as biological control agents in insect pest management are known as entomopathogenic nematodes (EPN). Using EPN as a component of IPM will aid to decline the dependence on insecticides and promotes eco-friendly pest management techniques in plantation crops as they are safe for plants and animals. They recycle and persist in environment and highly host specific. EPN play a vital role in the bio-suppression of various agricultural pests in particularly soil pests (root grubs) of plantation crops have been investigated and effectively utilized in the case of coconut, arecanut, cashew and cardamom pests

(Sosamma 2003; Varadarasan *et al.*, 2006; Vasanthi, 2012). Some of the important plantation crops and their pests are given in Table 12.1.

Table 12.1: Pest Attacking some Important Plantation Crops

Name of the Crop	Important Pests
Coconut	Rhinoceros beetle (*Oryctes rhinoceros* L.)
	Red palm weevil (*Rhynchophorus ferrugineus* F.)
	Leaf eating caterpillar (*Opisina arenosella* Walker)
	White grub (*Leucopholis coneophora* Burm.)
Arecanut	White grub (*Leucopholis* sp.)
	Spindle bug (*Carvalhoia arecae* Miller and China)
Oil palm	Rhinoceros beetle (*Oryctes rhinoceros* L.)
	Red palm weevil (*Rhynchophorus ferrugineus* F.)
Cashew	Stem and root borer (*Plocaederus* sp. and *Batocera rufomaculata* De Geer)
	Leaf miner (*Conopomorpha syngramma* M.)
Cardamom	Root grub (*Basilepta fulvicorne* Jacoby)
	Capsule borer (*Conogethes punctiferalis* Guen.)
	Cardamom thrips (*Sciothrips cardamomi* Ramk.)
Tea	Pale mite (*Acaphyllisa parindiae*)
	Scarlet mite (*Brevipalpus australis*)
	Cut worm (*Spodoptera litura*)
Coffee	Coffee berry borer (CBB), *Hypothenemus hampei*

The interest in the use of EPNs as biological control agents for control of insect pests has increased exponentially over the past decades. A decade ago, the idea of using nematodes to control pest populations was vague promise held by handful of researchers working with these obscure insect parasites. Today, they are no longer a laboratory curiosity but have begun to gain acceptance as environmentally benign alternatives to chemical insecticides. However, their potential as IPM component has been realized recently. These nematodes have been widely used to treat insect pest problems in agriculture, horticulture, plantation and forestry crops in India and abroad (Hussaini *et al.*, 2003; Banu and Rajendran 2002; Grewal *et al.*, 2004). The ease of mass production and exemption from registration requirements are the two major reasons for early interest in the commercial developments of EPNs and are now commercially mass-produced in six of the seven continents.

2. Entomopathogenic Nematodes

Entomopathogenic nematodes (Families Heterorhabditidae and Steinernematidae) are soft bodied, non-segmented roundworms similar in morphology to plant parasitic nematodes that are obligate or sometimes facultative parasites causing death to insects known as entomopathogenic nematodes (EPNs). They occur naturally in soil environments and locate their host in response to carbon dioxide, vibration and other chemical cues (Kaya and Gaugler, 1993), have been described from 23 nematode families (Koppenhofer, 2007). Of all of the nematodes studied for biological control of insects, the steinernematids and heterorhabditids are produced commercially and used as bio-agents, because, they possess many of the attributes of effective biological control agents (Kaya and Gaugler, 1993; Grewal *et al.*, 2005a; Koppenhofer, 2007) and have been utilized as classical, conservational, and augmentative biological control agents. Their unique association with symbiotic bacteria (*Xenorhabdus* for Steinernematidae and *Photorhabdus* for Heterorhabditidae) make them more effective compared to chemical insecticides. Soil has been one of the most difficult environments in which to achieve biological control of insect pests. These nematodes are adapted to soil and have been especially effective as inundative biological control agents against a number of soil insect pests (Klein, 1990). They are also effective against a number of insect pests that occur in cryptic habitats (*e.g.* tree boring insects). Major EPN species being used for biological control presently are *Steinernema carpocapsae, S. feltiae, S. riobrave, S. scapterisci, Heterorhabditis bacteriophora, H. megidis* (Rabindra and Hussaini, 2003)

The use of EPN as biological control agents is challenging, and application techniques are still under development (Piggot and Wardlow, 2002). Effectiveness depends on the targeted host, and environmental conditions such as temperature and relative humidity (Hara *et al.*, 1993) as well as application technology, because EPN are susceptible to desiccation, temperature extremes, and ultraviolet radiation (Mason and Wright, 1997).

Despite logistical issues, the use of EPN has been successful in field and greenhouse environments to manage certain insect pests, including the black vine weevil *[Otiorhynchus sulcatus* (Coleoptera: Curculionidae)], cranberry girdler *[Chrysoteuchia topiaria*(Lepidoptera: Pyralidae)], mint root borer *[Fumibotys fumalis* (Lepidoptera: Pyralidae)], citrus weevil *[Pachnaeus litus* (Coleoptera: Curculionidae)], mole crickets *[Scapteriscus* spp. (Orthoptera: Gryllotalpidae)], billbugs *[Sphenophorus* spp. (Coleoptera: Curculionidae)], white grubs (Coleoptera: Scarabaeidae), fungus gnats *[Bradysia* spp. (Diptera: Sciaridae)] western flower thrips *[Frankliniella occidentalis* (Thysanoptera: Thripidae)], and serpentine leafminer *[Liriomyza trifolii* (Diptera: Agromyzidae)] (Hara *et al.*, 1993).

2.1. Attributes of EPNs

The EPNs have very impressive attributes as biological control agents at par with any other bioagents in use. These include, ability to kill hosts within short periods, they have a broad host range, are safe to vertebrates, plants and other non-target organisms, have no known negative effect on the environment, are easy to mass produce *in vivo* and *in vitro*, are easily applied using standard spray equipment, have the potential to recycle in the environment, high virulence, are compatible with many chemical and other biological pesticides, and wider genetic diversity, presence of chemoreceptors, are amenable to genetic selection for desirable traits, and are exempt from registration in many countries (Kaya and Gaugler, 1993). There is no need for personal protective equipment and re-entry restrictions. Insect resistance problems are unlikely. Negative attributes include their broad host range (although no negative effects on non-target hosts have been observed, this broad host range may include some beneficial insects), narrow tolerance to environmental conditions (*e.g.* moisture requirement), poor long-term storage, poor field persistence, and relatively high cost in comparison to chemical pesticides (Kaya, 1993).

2.2. Pathogenicity, Biology and Life Cycle of EPN

Entomopathogenic nematodes are obligate, soil dwelling rhabditids, capable of infecting broad range of insect pests. The only stage that can survive outside the host is the non-feeding third infective juvenile (IJs) stage. The IJs has adapted to survive in harsh soil environment, locate and infect suitable host through natural openings (mouth, spiracles and anus) or in some species through intersegmental membranes of the cuticle before developing into adult male, female or hermaphrodite. If the mode of entry is by mouth or anus, the nematode penetrates the gut wall to reach the hemocoel, and if by spiracles, it penetrates the tracheal wall. When the infective juvenile reaches the hemocoel of a host, it releases the bacteria, which multiply rapidly in the hemolymph. The infective juveniles are closely associated with their symbiotic bacterium, which kills their host within 1-4 days after infection and protect the cadaver from colonization by other microorganisms (Thomas and Poinar, 1979). Even though the bacterium is primarily responsible for the mortality of most insect hosts, the nematode also produces a toxin that is lethal to the insect. Inside host, infective juvenile becomes feeding third stage juveniles, feeds on *Xenorhabdus* for steinernematids, *Photorhabdus* for heterorhabditids bacteria and liquefying host tissue and molts to the fourth stage prior to maturing into a adults males and females of the first generation. After mating, the females lay eggs that hatch as first-stage juveniles that molt successively to second, third, and fourth-stage juveniles and then to males and females of the second generation. The adults mate and the eggs produced by these second-generation females hatch as first-stage juveniles that molt to the second stage. The late second-stage juveniles cease feeding, incorporate a pellet of bacteria in the bacterial chamber, and molt to the third stage (infective juvenile), retaining the cuticle of the second stage as a sheath, and leave the cadaver in search of new hosts as nutritive conditions are depleted in dead cadavers (Nguyen and Smart, 1992; Wouts, 1980; Johnigk and Ehlers, 1999). Steinernematid infective juveniles may become males or females, where as heterorhabditids develop all juveniles into self-fertilizing hermaphrodites in first generation. In the second generation, males, females, and hermaphrodites are produced. Thus, steinernematids require a male and a female infective juvenile to invade an insect host to produce progeny, whereas heterorhabditids need only one infective juvenile to penetrate into a host as the resulting hermaphroditic adult is self-fertile. The life cycle is completed in a 6 to 8 days in steiernematids and 12 to 14 days in heterorhaditids at 25°C in *Galleria mellonella* (Kaya and Koppenhofer, 1999). The nematode/bacterium association is highly specific. In the infective juvenile, the bacterial cells are housed in a vesicle in the anterior part of the intestine for steinernematids and in the intestinal tract for heterorhabditids.

The insect cadaver becomes red if the insects are killed by heterorhabditids and cream if killed by steinernematids (Kaya and Gaugler, 1993). The color of the host body is indicative of the pigments produced by the monoculture of mutualistic bacteria growing in the hosts. Nematode growth and reproduction depend upon conditions established in the host cadaver by the bacterium. The bacterium further contributes anti-immune proteins to assist the nematode in overcoming host defenses, and anti-microbials that suppress colonization of the cadaver by competing secondary invaders. Conversely, the bacterium lacks invasive powers and is dependent upon the nematode to locate and penetrate suitable hosts.

2.3. EPN Bacterium Complex

Xenorhabdus and *Photorhabdus* are motile, Gram negative, facultative, non-spore forming, anaerobic rods in the family Enterobacteriaceae. Major differences occur between the 2 bacterial genera (Boemare, 2002). For example, most *Photorhabdus* spp. are luminescent and catalase positive, whereas *Xenorhabdus* spp. have no luminescence and are catalase negative. Both bacterial genera produce phenotypic variant cell types called primary form (phase I) and secondary form (phase II) (Forst and Clarke, 2002). The primary form is the cell type naturally associated with the nematodes, whereas the secondary form can arise spontaneously when the bacterial cultures are in the stationary non-growth stage. The *Xenorhabdus* secondary form can revert to the primary form, but this phenomenon has not been documented for *Photorhabdus* spp. Differences between the primary and secondary forms occur. For instance, the primary form produces antibiotics, adsorbs certain dyes, and develops large intracellular inclusions composed of crystal proteins, whereas the secondary form does not or only weakly produces antibiotics, does not adsorb dyes, and produces intracellular inclusions inefficiently. The primary form is superior to the secondary form in its ability to support nematode propagation in vitro, although some evidence suggests that this is not always the case (Volgyi, 1998). The reason for the occurrence of the 2 forms is not known. The relationship between the nematode and bacterium is truly mutualistic for the following reasons: the nematode is dependent upon the bacterium for (1) quickly killing its insect host, (2) creating a suitable environment for its development by producing antibiotics that suppress competing microorganisms, (3) transforming the host tissues into a food source, and (4) serving as a food resource. The bacterium needs the nematode for (1) protection from the external environment, (2) penetration into the host's hemocoel, and (3) inhibition of the host's antibacterial proteins. Without the nematode the bacteria cannot survive well in the natural environment and are generally not pathogenic when ingested by a host (Akhurst, 1982; Akhurst and Boemare, 1990).

Entomopathogenic nematodes and their endosymbiotic bacteria are potent bioinsecticides that can control a wide variety of economically important agricultural pests (Shapiro - Ilan *et al.*, 2002). Due to their sensitivity to ultraviolet light and desiccation (Georgis and Gaugler, 1991), entomopathogenic nematodes have been most successful at suppressing populations of ground-dwelling pests or pests in other protected environments (*e.g.*, greenhouses). Successful pest control with nematodes requires a proper match of the nematode to the host species and favorable economics relative to the value of the commodity and the cost of competing pest control strategies. To be effective, entomopathogenic nematodes must generally be applied at rates of 2.5×10^9/ha or higher (Georgis and Hague, 1991; Georgis *et al.*, 1995). Some of the pests that have been targeted commercially with entomopathogenic nematodes are listed in Table 12.2. In addition to controlling harmful insect pests, new frontiers are opening by using entomopathogenic nematodes, and more so, their symbiotic bacteria or associated metabolites to suppress plant parasitic nematodes (Gouge *et al.*, 1994) and as antimicrobial agents in pesticide and pharmaceutical applications. Furthermore, toxins produced by the bacteria are being investigated for their suitability as alternatives to other orally active insecticides such as toxins produced by *Bacillus thuringiensis* (Bowen *et al.*, 1999).

Table 12.2: Commercial Use of EPNs,
Steinernema and Heterorhabditis as Bio-insecticides

EPN Species	Major Pest(s) Targeted - as Recommended by various Commercial Companies
Steinernema glaseri	White grubs (scarabs, especially Japanese beetle, *Popillia* sp.
Steinernema kraussei	Black vine weevil, *Otiorhynchus sulcatus*)
Steinernema carpocapsae	Turfgrass pests- billbugs, cutworms, armyworms, sod webworms, chinch bugs. Orchard, ornamental and vegetable pests - codling moth, cranberry girdler, dogwood borer and other clearwing borer species, black vine weevil, peachtree borer, shore flies (*Scatella* spp.)
Steinernema feltiae	Fungus gnats (*Bradysia* spp.), shore flies, western flower thrips
Steinernema scapterisci	Mole crickets (*Scapteriscus* spp.)
Steinernema riobrave	Citrus root weevils (*Diaprepes* spp.)
Heterorhabditis bacteriophora	White grubs (scarabs), cutworms, black vine weevil, flea beetles, corn root worm
Heterorhabditis megidis	Weevils
Heterorhabditis indica	Fungus gnats, root mealybug, grubs
Heterorhabditis marelatus	White grubs (scarabs), cutworms, black vine weevil

2.3. Searching Behaviour

Entomopathogenic nematodes use two search strategies: ambushers or cruisers (Grewal *et al.*, 1994a). Ambushers such as *S. carpocapsae* have an energy-conserving approach and lie-in-wait to attack mobile insects (nictitating) in the upper soil by direct contact (Campbell *et al.*, 1996). Cruisers like *S. glaseri* and *H. bacteriophora* are highly active and generally subterranean, moving significant distances using volatile cues and other methods to find their host underground. Some nematode species such as *S. feltiae* and *S. riobrave* use an intermediate foraging strategy (combination of ambush and cruiser type) to find their host.

2.4. Recycling of Nematodes

Recycling is desirable after an application of entomopathogenic nematodes because it can provide additional and prolonged control of a pest. The abiotic and biotic factors that affect persistence, infectivity, and motility of infective juveniles influence nematode recycling. Because they are obligate pathogens, the availability of suitable hosts is a key to recycling of the nematodes. Recycling is rather common (Klein, 1993) after nematode application but is probably not sufficient for prolonged host suppression, and the nematodes have to be re-applied to maintain adequate control of soil insect pests.

2.5. Dispersal of Juveniles

The juveniles of steinernematids and heterorhabditids disperse vertically and horizontally, both actively and passively (Epsky *et al.*, 1988; Parkman *et al.*, 1993). Passively, they may be dispersed by rain, wind, soil, humans, or insects. Active dispersal may be measured in centimetres, while passive dispersal by insects may be measured in kilometres (Smart and Nguyen, 1994).

2.6. Survival of Juveniles

In general, entomopathogenic nematodes do not have a long shelf life. Many microbial insecticides, including *Bacillus thuringiensis*, have a resting stage facilitating longterm storage. The infective juveniles do not feed but can live for weeks on stored reserves as active juveniles, and for months by entering a near-anhydrobiotic state. This is almost certainly the most important survival strategy for the nematode. The length of time that juveniles survive in the soil in the absence of a host depends upon such factors as temperature, humidity, natural enemies, and soil type. Generally, survival is measured in weeks to months, and is better in a sandy soil or sandy-loam soil at low moisture and with temperatures from about 15-25° C than in clay soils and lower or higher temperatures (Ames, 1990; Kaya, 1990; Kung, 1991). Extended exposure to temperature extremes (below 0°C or above 40°C) is lethal to most species of entomopathogenic nematodes. In the soil environment, infective juveniles are normally buffered from temperature extremes. For storage, the best longevity of infective juveniles is between 5 and 15°C. At higher temperatures, the infective juveniles have increased metabolic activity and deplete their energy reserves, shortening their life span (Brown and Gaugler, 1996).

UV can kill nematodes within minutes. Direct exposure to UV light (*i.e.* sunlight) can be minimized by applying infective juveniles early in the morning or evening, or using sufficient amounts of water to wash the infective juveniles into the soil. Infective juveniles can survive low moisture conditions by lowering their rate of metabolism. Gradual water removal from the infective juveniles gives them time to adapt to the desiccating conditions (Patel *et al.*, 1997; Solomon *et al.*, 1999).

Soil texture affects infective juvenile survival, with the poorest occurring in clay soils. The poor survival rate in clay soils is probably due to the lower oxygen levels in the smaller soil pores. Oxygen is also a limiting factor in water-saturated soils and soils with high organic matter content, but pH does not have a strong effect on infective juvenile survival.

3. Mass Production and Formulation of EPN

3.1. Mass Production

A key factor in the success of entomopathogenic nematodes as biopesticides is their amenability to mass production. These nematodes were first cultured more than 70 years ago (Glaser, 1940), and currently they are commercially produced using three culture methods: *in vivo* and *in vitro* solid and liquid culture (Friedman, 1990). Each approach has advantages and disadvantages relative to cost of production, capital outlay, technical expertise required, economy of scale, and product quality, and each approach has the potential to be improved. A variety of formulation options are available (Georgis *et al.*, 1995)

Entomopathogenic nematodes are easily cultured either *in vivo* or *in vitro* for laboratory tests or for commercial production (Friedman, 1990). *In vivo* culture is a two-dimensional system that relies on production in trays and shelves. The wax worm, *G. mellonella*, is the insect of choice for *in vivo* production because it is produced commercially in large numbers and well defined diet material is available. *In vivo* production is labor intensive, lacks economies of scale, and is costly, but it is also simple and reliable and results in high quality nematodes (Shapiro-Ilan, 2003). A system based on the White trap (White, 1927), which takes advantage of the infective juvenile's natural migration away from the host cadaver upon emergence is widely used. The methods described consist of inoculation, harvest, concentration, and (if necessary) decontamination. Insects are inoculated with nematodes on a dish or tray lined with absorbent paper (*e.g.*, filter paper) or another substrate conducive to nematode infection such as soil or plaster of Paris. After 2–5 days, infected insects are transferred to the White traps; if infections are allowed to progress too long before transfer, harm to nematode reproductive stages may occur, and the cadavers will be more likely to rupture (Shapiro – Ilan, 2001). White traps consist of a dish on which the cadavers rest surrounded by water, which is contained by a larger dish or tray. The central dish (containing the cadavers) provides a moist substrate for the nematodes to move upon, *e.g.*, an inverted petri dish lid lined with filter paper or filled with plaster of Paris. The progeny infective juveniles that emerge migrate to the surrounding water where they are trapped and subsequently harvested. The choice of host species and nematode for *in vivo* production should ultimately rest on nematode yield per cost of insect and the suitability of the nematode for the pest target. Nematode quality appears to be greater when cultured in hosts that are within the nematode's natural host range (Abu Hatab and Gaugler, 2001). Furthermore, nematodes can adapt to the host they are reared on (Stuart and Gaugler, 1996), which could reduce field efficacy if that host is not related to the target. Therefore, although *G. mellonella* may often be the most efficient host to use, it may not be the most appropriate "medium" for maximizing efficacy versus a particular target pest.

There are only a couple of entomopathogenic nematodes not amenable to culture in *G. mellonella* (due to extremes in host specificity): *Steinernema kushidai* is most amenable to culture in scarab beetle larvae (Coleoptera: Scarabaeidae) (Kaya and Stock, 1997) and *Steinernema scapterisci* is most amenable to mole crickets (*Scapteriscus* spp.) (Grewal *et al.*, 1999). Other hosts in which *in vivo* production has been studied include the navel orangeworm (*Amyelois transitella*), tobacco budworm (*Heliothis virescens*), cabbage looper (*Trichoplusia ni*), pink bollworm (*Pectinophora gossypiella*), beet armyworm (*Spodoptera exigua*), corn earworm (*Helicoverpa zea*), gypsy moth (*Lymantria dispar*), house cricket (*Acheta domesticus*) and various beetles (Coleoptera) including the yellow meal worm (*Tenebrio molitor*) (Blinova and Ivanova, 1987).

For large-scale production, *in vitro* methods using 3- dimensional solid media or liquid fermentation methods have been employed but it is not cost effective and high capital requirement is needed and the inability of the amphimictic adults to mate under liquid culture conditions (Gaugler and Han, 2002). Yang *et al.* (1997) reported reduced quality in *S. carpocapsae* produced in solid culture compared with *in vivo* culture. Without sophisticated mechanization (*e.g.*, bulk sterilization) solid culture may not offer substantial advantages in cost efficiency relative to *in vivo* production (a cost analysis is warranted). Yet large - scale mechanization for solid culture requires substantial capital. If *in vitro* solid culture is to be adopted on wider scale, efficiency will have to be increased by finding less capital - intensive methods of mechanization.

3.2. Formulation

Regardless of culture method, once entomopathogenic nematodes are commercially produced they must be formulated for delivery and application (Georgis, 1990). An effective formulation provides a suitable shelf life, stability of product from transport to application, and ease of handling. Shelf life, in most entomopathogenic nematode formulations, is obtained by reducing nematode metabolism and immobilization, which may be accomplished through refrigeration and partial desiccation. Optimum storage temperature for formulated nematodes varies according to species: generally, steinernematids tend to store best at temperatures near 4–8°C whereas heterorhabditids have longer shelf life at temperatures close to 10–15°C. The climate of origin is predictive of the optimum storage temperature, *e.g.*, *H. indica*, a nematode originating only in warm climates, stores better at 15–20 than at 10°C (Shapiro *et al.*, 1999).

Various formulations for entomopathogenic nematodes have been reported including activated charcoal, alginate and polyacrylamide gels, baits, clay, peat, polyurethane sponge, vermiculite, and water-dispersible granules (WDG) (Georgies *et al.*, 1995). Due to cost, *in vivo* producers tend to use low-technology formulations such as sponge and paste. The nematodes are not desiccated and tend to retain high viability. However, these formulations cannot be packaged at high densities and are therefore not appropriate for large - scale usage because of labor requirements in application. Formulations used by most *in vitro* producers include clay, gels, vermiculite, and WDG. For example, a successful non-desiccated formulation has been developed for *in vitro* produced nematodes based on vermiculite, which allows a shelf life of at least one month for *H. megidis* and 2–3 months for steinernematids.

4. Plantations Insect Pests are Promising Targets–Why ?

Plantations in India are major commercial crops grown varying from tropical evergreen Western ghats regions of Kerala, Karnataka and Tamil Nadu to dry regions of Andhra Pradesh, Maharasthra and Odisha. They are one of the higher foreign exchange earners besides contributing neutraceuticals supply. Insects, which are an integral part of the existing ecosystem are also the major biological determinants influencing the development and destruction of plantation crops *viz.*, seedlings damage in nurseries, young plants at establishment stage in main field, mature old plantation destruction, seeds in stand and storage. As against agriculture, which is man-made practice in the limited and controlled area, plantations some time cover the thick forests, incessant continuous rains and unapproachable areas, imposing serious restrictions on control operations. Biological control programmes for the management of insect pests in such areas with entomopathogenic nematodes holds good promise along with judicious use of chemicals, wherever feasible. In India, a number of insect pests belonging to the coleoptera, lepidoptera, thysonoptera, hemiptera *etc.* are serious pests of plantations from seedling stage in plantations nursery to the standing tree in plantation gardens, as root feeders, defoliators, stem borers or soil inhibiting borne pests. Some of these are difficult to manage by virtue of their long life cycle, sometimes extending to one to two year and peculiar feeding habit in hidden tunnels or galleries, which are otherwise, difficult to manage by any other known method of insect pest management.

In line with all the above, the EPNs, by virtue of the presence in the local environment, if explored and proved promising against the plantation insect pests, could become an important component of IPM programme against plantation insect pests. The strongest benefit is the ability of self propagation and establishment for many year in the ecosystem, offering continuation of population at the site of release Hussaini *et al.* (2003) have discussed that the characterization of traits related to the control of potential of species (strains) of EPN is the key for a successful biological pest control with insecticidal nematodes. This has to be done with wide range of insect pests including horticulture, agriculture and plantation importance, which has been missing till date in India.

5. Current Use of EPN as Bio-agents in IPM of Plantation Crops in India

As EPN are compatible with many control measures, numerous opportunities exist for including these successfully in IPM programmes with minimal reliance on chemical pesticides, and involving more and more of other natural enemies and pathogens.

In India work on steinernematids started in the nineteen sixties. Use of DD-136 for control of pests of rice, sugarcane and apple were discussed by Rao and Manjunath (1966). In the seventies Singh and Bardhan (1974) worked on mortality in laboratory and field trials, life cycle and compatibility of DD-136 with insecticides and fertilizers.

5.1. Coconut

5.1.1. Red Palm Weevil (*Rhynchophorus ferrugineus*)

In general, 5-7 per cent palms are infested by red palm weevil in the country and being a concealed borer it becomes fatal enemy of coconut on most occasions. Higher virulence of local entomopathogenic nematode (EPN) strain of *Heterorhabditis indica* (LC$_{50}$ =355.5 IJ) in the suppression of grubs as well as greater susceptibility (82.5 per cent) of pre-pupal stage than that of grubs stage was indicated (Figure 12.1). Synergistic effect of *H. indica* (1500 IJs) with imidacloprid (0.002 per cent) in field trials was indicated. Placement of three filter paper sachets containing 12-15 *H. indica* infected *Galleria mellonella* cadavers on the leaf axils after application of 0.002 per cent imidacloprid could recover 60 per cent infested palms (Josephrajkumar *et al.*, 2013).

Figure 12.1

(a) Infective juveniles of EPN, (b,c) EPN infected Red palm weevil grubs and pupa, (d) EPN infected cadavers in paper boats for leaf axil placement.

Talc based local strain of *H. indica* formulation fortified with chitosan 0.25 per cent and admixed with sand were filled in leaf axils of coconut crown as prophylactic treatment against infestation of red palm weevil on a susceptible host *viz.*, Chowghat Green Dwarf (CGD). It was found that *H. indica* formulation proved efficacious to the tune of 94.4 per cent in preventive treatment while 33.3 per cent untreated control palms (CGD) succumbed to red palm weevil infestation (CPCRI, 2010).

5.1.2. Rhinoceros Beetle (*Oryctes rhinoceros*)

The entomopathogenic nematode, *Steinernema abbasi* was pathogenic to third instar grubs of rhinoceros beetle at 350 IJS/cc of vermicompost in laboratory bioassay. *Steinernema carpocapsae* was oriented to over 7.3 cm in 72 hours of inoculation using volatile cues in vermicompost to find *O. rhinoceros* grubs. EPN, *S. carpocapsae* infected *G. mellonella* cadaver @ one/500 cm^3 was found effective in the bio-management of rhinoceros grubs (neonates) in vermicompost (CPCRI, 2014; Jagadish Patil *et al.*, 2012).

5.1.3. Root Grub (*Leuchopholis coneophora*)

S. carpocapsae (900 IJ and 1200 IJ) admixed with imidacloprid (0.250, 0.125, 0.125, 0.063, 0.031, 0.015 and 0.008 per cent) and exposed to white grub indicated a significantly higher mortality in all nematode-imidacloprid combinations after 7 days. The interaction between imidacloprid and nematodes was found to be synergistic in all combinations (Jagadish Patil *et al.*, 2013). The EPN, *Steinernema carpocapsae* at 8000 to 16000 IJs per grub in combination of 1 to 0.0001 per cent imidacloprid caused 72 per cent mortality in root grub compared to challenging of grubs with EPN alone under *in vitro* condition. Two rounds of root zone drenching of liquid EPN formulation, *S. carpocapsae* @0.5x10^6 IJs palm^{-1} during June-July and September-October as on trial resulted in 61 per cent reduction of root grub population in costal sandy soils of Kasaragod. The reduction of root grub population increased with increase in nematode density per palm and number of treatments (CPCRI, 2014).

5.2. Arecanut

5.2.1. Root Grub (*Leuchopholis* sp.)

One round root zone drenching of liquid EPN formulation, *S. carpocapsae* @0.5x10^7 IJs palm^{-1} during September-October as on trial resulted in 41 per cent reduction of root grub population in red soils of Sringeri in Chikkamangaluru. The reduction of root grub population increased with increase in nematode density per palm and number of treatments. Nematodes in combination of imidacloprid 17.8 SL (0.004 per cent), 1ml 5L^{-1} water palm^{-1} was found synergistic and reduced root grub population to the tune of 60 per cent. The nematode establishment was found in all treated plots (Rajkumar *et al.*, 2014)

5.3. Coffee

5.3.1. Coffee Berry Borer (CBB), *Hypothenemus hampei*

Potential of *S. carpocapsae* as a control for CBB in fallen coffee berries in Hawaii coffee fields was accomplished. All life stages of CBB are being killed by *S. carpocapsae*, with highest mortality in larvae (Jessical Manton *et al.*, 2012).

5.4. Cardamom

5.4.1. Root Grub [*Basilepta fulvicorne* (Jacoby)]

Root grub is a serious pest damaging the roots of cardamom. Nutrient uptake is reduced due to root damage leading to yellowing of leaves; the pest problem is severe in less shaded area. Beetles occur in March-April and August-September. Females lay about 124-393 eggs in batches of 12-63 on dry cardamom leaves or mulches. The minute creamy white grubs hatch out from eggs, fall on the ground, reach root zone and start feeding the roots. Grubs have two periods of occurrence, the first during April-July and the second during September to January. Grubs (larvae) feed on roots, become mature in 45-60 days (Varadarasan *et al.*, 2009). Soil application of *Metarhizium anisopliae* (@ 108 spores/gm) 25gms/plant mixed with compost. Native strain of EPN (*Heterorhabditis indica*) application @ 1,00,000 nematodes (IJs/plant) against early stage grubs during April/May and September/October.

On account of its versatality, persistence and specificity EPN is all likely to become an essential component of IPM strategies in sustainable pest mangment of plantation crops, as organic farming is widely promoted in this part of the country.

References

Abu Hatab M. and Gaugler R. (2001). Diet composition and lipids of *in vitro -* produced *Heterorhabditis bacteriophora*. *Biological Control* 20: 1–7.

Abu Hatab, M., Gaugler, R. and Ehlers, R. (1998). Influence of culture method on *Steinernema glaseri* lipids. *Journal of Parasitology* 84: 215–221.

Akhurst, R. J. (1982). Antibiotic activity of *Xenorhabdus* spp., bacteria symbiotically associated with insect pathogenic nematodes of the families Heterorhabditidae and Steinernematidae. *Journal of General Microbiology* 128: 3061–3065.

Akhurst, R. J. and Boemare, N. E. (1990). Biology and taxonomy of *Xenorhabdus*. In: Gaugler R and HK Kaya, (Eds.), Entomopathogenic Nematodes in Biological Control. CRC Press, Boca Raton, FL, pp. 75–87

Ames, L. M. (1990). The role of some abiotic soil factors in the survival of *Steinernema scapterisci*. M.S. thesis, University of Florida, Gainesville.

Banu, J. G. and Rajendran, G. (2002). Host record of an entomopathogenic nematode, *Heterorhabditis indica*. *Insect Environment*, 8: 61-62.

Blinova, S. L. and Ivanova, E. S. (1987). Culturing the nematode–bacterial complex of *Neoaplectana carpocapsae* in insects. In: Sonin MD (Ed.), Helminths of Insects. American Publishing, New Delhi, pp. 22–26.

Boemare, N. (2002). Biology, taxonomy and systematics of *Photorhabdus* and *Xenorhabdus*. In: Gaugler R. ed. Entomopathogenic Nematology. CABI Publishing. Wallingford, UK. pp. 35-56

Bowen, D., Blackburn, M., Rocheleau, T. A., Andreev, O., Golubeva, E. and French -Constant, R. H. (1999). Insecticidal toxins from the bacterium R and HK Kaya (Eds.), Entomopathogenic Nematodes in Biological Control. CRC Press, Boca Raton, FL, pp. 173–194.

Brown, I. M. and Gaugler, R. (1996). Cold tolerance of steinernematid and heterorhabditid nematodes. *Journal of Thermal Biology* 21: 115-121.

Campbell, J., Lewis, E., Yoder, F. and Gaugler, R. (1996). Entomopathogenic Nematode Spatial Distribution in Turfgrass. *Parasitology* 113: 473–482.

Chakravarthy, A. K., Ashok Kumar, Abraham Verghese and Thiagaraj, N. E. (2013). International Congress on Insect Science, 14th-17th February, Bengaluru, pp. 48-49.

CPCRI (2010). Annual report, 2009-2010, Central Plantation Crops Research Institute, Kasaragod, Kerala. India, 148p.

CPCRI (2012). Annual report, 2011-12, Central Plantation Crops Research Institute, Kasaragod, Kerala, India, pp. 35.

CPCRI (2014). Annual report, 2013-14, Central Plantation Crops Research Institute, Kasaragod, Kerala, India, 139p.

Epsky, N. D., Walter, D. E. and Capinera, J. L. (1988). Potential role of nematophagous microarthropods as biotic mortality factors of entomogenous nematodes (Rhabditida: Steinernematidae, Heterorhabditidae). *Journal of Economic Entomology* 81: 821-825.

Forst, S. and Clarke, D. (2002). Bacteria-nematode symbiosis. In: Gaugler R. ed. Entomopathogenic Nematology. CABI Publishing. Wallingford, UK; pp. 57-77.

Friedman, M. J. (1990). Commercial production and development. In: Gaugler R and HK Kaya (Eds.), Entomopathogenic Nematodes in Biological Control. CRC Press, Boca Raton, FL, pp. 153–172.

Gaugler, R. and Han, R. (2002). Production technology. In: Gaugler R, (Ed.), Entomopathogenic Nematology. CABI

Gaugler, R. and Kaya, H. K. (1990). Entomopathogenu: Nematodes in Biological Control. CRC Press, Boca Raton. FL.

Georgis, R. (1990). Formulation and application technology. In: Gaugler R and HK Kaya (Eds.), Entomopathogenic Nematodes in Biological Control. CRC Press, Boca Raton, FL, pp. 173–194.

Georgis, R. and Gaugler, R. (1991). Predictability in biological control using entomopathogenic nematodes. *Journal of Economic Entomology*, 84, 713–720.

Georgis, R. and Hague, N. G. M. (1991). Nematodes as biological insecticides. *Pesticide Outlook* 2: 29–32.

Georgis, R., Dunlop, D. B. and Grewal, P. S. (1995). Formulation of entomopathogenic nematodes. In: FR Hall and JW Barry (Eds.), Bio-rational Pest Control Agents: Formulation and Delivery. American Chemical Society, Washington, DC, pp. 197–205.

Glaser, R. W. (1940). The bacteria - free culture of a nematode parasite. *Proc Soc Exp Biol Med* 43: 512–514.

Gouge, D. H., Otto, A. A., Schirocki, A. and Hague, N. G. M. (1994). Effects of steinernematids on the root - knot nematode *Meloidogyne javanica*. *Annals of Applied Biology* 124: 134–135

Grewal, P. S., Converse, V. and Georgis, R. (1999). Influence of production and bioassay methods on infectivity of two ambush foragers (Nematoda: Steinernematidae). *Journal of Invertebrate Pathology* 73: 40–44.

Grewal, P. S., Ehlers, R. U. and Shapiro-Ilan, D. I. (eds.). (2005). Nematodes as biological control agents. Wallingford: CABI Publishing.

Grewal, P. S., Lewis, E, Gaugler, R and Campbell, J. (1994). Host finding behaviour as a predictor of foraging strategy in entomopathogenic nematodes. *Parasitology* 108: 207-215.

Grewal, P. S., Powar, K. T., Grewal, S. K., Suggars and Hauprucht, S. (2004). Enhanced consistency in biological control of white grubs (Coleoptera: Scarabaidae) with new strains of entomopathogenic nematodes. *Biological Control*, 30: 73-82

Hara, A.H., Kaya, H.K., Gaugler, R., Lebeck, L.M., Mello, C. L. (1993). Entomopathogenic nematodes for biological control of the leaf miner, *Liriomyza trifolii* (Dip.: Agromyzidae). *Entomophaga* 38: 359–369.

Hussaini, S. S., Rabindra, R. J. and Nagesh, M. (Eds.). (2003). Current status of research on entomopathogenic nematode in India, Project Directorate of Biological Control, Bangalore, India

Jagadeesh Patil, Rajkumar and Kesavan Subaharan (2012). Impact of entomopathogenic nematodes on rhinoceros beetle larvae, *Oryctes rhinoceros* (Coleoptera: Scarabaeidae). Paper presented in 22nd Swadeshi Science Congress, CPCRI, Kasargod from 6 to 8th November 2012. pp. 82-83.

Jagadish Patil, Rajkumar and Kesavan Subaharan (2013). Synergism of Entomopathogenic nematode and Imidacloprid: A Curative toll to white grub, *Leucopholis coniophora* (Coleoptera: Melolonthinae) control in plantation crops, 4th International congress on Insect Science, UAS, GKVK, Bangalore, February 14th–17th 2013, pp. 46.

Jessical Manton, Robert, G., Hollingsworth Roxana, Y. M. and Cabos (2012). Potential of *Steinernema carpocapsae* (Rhabditida: Steinernematidae)against *Hypothenemus Hampeie* (Coleoptera: Curculionidae) in HawaiiI. Bioone, pp. 1194-1197.

Johnigk, S. A. and Ehlers, R, U. (1999). Juvenile development and life cycle of *Heterorhabditis bacteriophora* and *H. indica* (Nematoda: Heterorhabditidae). *Nematology* 1: 251-260.

Josephrajkumar, A., Chandrika M. and Rajan, P. (2013). Evaluation of entomopathogenic nematodes against red palm weevil, *Rhynchophorus ferrugineus* (Olivier) and synergistic interaction with the neonicotinoid, imidacloprid. In: New Horizons in Insect Science (Eds.)

Kaya, H. K. (1990). Soil ecology. In; R. Gaugler and H. K. Kaya, eds. Entomopathogenic nematodes in biological control. Boca Raton, FL: CRC Press. pp. 93-115

Kaya, H. K. (1993). Entomogenous and entomopathogenic nematodes in biological control. In: Evans K, Trudgill DL and Webster JM. eds. Plant Parasitic Nematodes in Temperate Agriculture. CAB International, Wallingford, UK pp. 565-591.

Kaya, H. K. and Stock, S. P. (1997). Techniques in insect nematology. In: Lacey LA (Ed.), Manual of Techniques in Insect Pathology. Academic Press, San Diego, CA. pp. 281–324.

Kaya, H. K. and. Koppenhöfer, A. M. (1999). Biology and ecology of insecticidal nematodes. In Workshop Proceedings: Optimal Use of Insecticidal Nematodes in Pest Management Edited by S. Polavarapu, Rutgers University. pp. 1–8.

Kaya, H. K., and Gaugler, R. 1993. Entomopathogenic nematodes. *Annual Review of Entomology* 38: 181–206

Klein, M. G. (1990). Efficacy against soil-inhabiting insect pests. In: Gaugler, R. and Kaya, H. K. ed. Entomopathogenic Nematodes in Biological Control. CRC Press. Boca Raton, FL. pp. 195-214.

Klein, M. G. (1993). Biological control of scarabs with entomopathogenic nematodes. In: Bedding R, Akhurst R and Kaya H. K. eds. Nematodes and the Biological Control of Insects. CSRIO Publications. East Melbourne, Australia. pp. 49-57,

Koppenhofer, A. M. (2007). Nematodes. In: L. A. Lacey and H. K. Kaya, eds. Field manual of techniques in invertebrate pathology: Application and evaluation of pathogens for control of insects and other invertebrate pests, second ed. Dordrecht: Springer. pp. 249–264

Kung, S. P., Gaugler, R. and Kaya, H. K. (1991). Effects of soil temperature, moisture, and relative humidity on entomopathogenic nematode persistence. *Journal of Invertebrate Pathology* 57: 242-249.

Mason, J. M. and Wright, D. J. (1997). Potential for the control of *Plutella xylostella* larvae with entomopathogenic nematodes. *Journal of Invertebrate Pathology* 70: 234–242.

Molyneux, A. S. (1985). Survival of infective juveniles of *Heterorhabditis* spp., and *Steinernema* spp. (Nematoda: Rhabditida) at various temperatures and their subsequent infectivity for insects. *Revue de Nematologie* 8: 165-170.

Nguyen, K. B. and Smart, Jr. G. C. (1992). Life cycle of *Steinernema scapterisci* Nguyen and Smart, 1990. *Journal of Nematology* 24: 160-169.

Nguyen, K. B., and Smart, G. C. Jr. (1990). Vertical dispersal of *Steinernema scapterisci*. *Journal of Nematology* 22: 574-578.

Parkman, J. P., Frank, J. H., Nguyen, K. B. and Smart, G.C. Jr. (1993). Dispersal of *Steinernema scapterisci* (Rhabditida: Steinernematidae) after inoculative applications for mole cricket (Orthoptera: Gryllotalpidae) control in pastures. *Biological Control* 3: 226-232.

Patel, M. N., Perry, R. N. and Wright, D. J. (1997). Desiccation survival and water contents of entomopathogenic nematode *Steinernema* spp. (Rhabditida: Steinernematidae). *International Journal of Parasitology* 27: 61-70.

Piggot, S. and Wardlow, L. (2002). A fresh solution for the control of western flower thrips: Dramatic results in trials. Commercial Greenhouse Grower.

Rabindra, R. J. and Hussaini, S. S. (2003). Scope of biological control of crop pests using entomopathogenic nematodes in India. In: Current status of research on entomopathogenic nematode in India. (eds, Hussaini, S. S., Rabindra, R. J. and Nagesh, M.), Project Directorate of Biological Control, Bangalore, India. pp. 15-26.

Rajkumar, Jagadeesh Patil and Kesavan Subaharan (2014). Efficacy of entomopathogenic nematode in combination with imidacloprid against root grub (*Leucopholis burmesteri*) in arecanut. Paper presented in 'International conference on 'Changing scenario of pest problems in Agri-Horti ecosystem and their management', held at MPUA and T, Udaipur, Rajasthan during 27-29 November, 2014. pp. 199.

Rao, V. P. and Manjunath, T. M. (1966). DD-136 nematode that can kill many insect pests. *Indian Farming* 16: 43.

Shapiro - Ilan, D. I., Gouge, D. H. and Koppenhofer, A. M. (2002). Factors affecting commercial success: case studies in cotton, turf, and citrus. In Entomopathogenic Nematology Gaugler R, ed CABI, In press.

Shapiro - Ilan, D. I., Lewis, E. E., Behle, R. W. and McGuire, M. R. (2001). Formulation of entomopathogenic nematode - infected - cadavers. *Journal of Invertebrate Pathology* 78: 17–23.

Shapiro, D. I., Cate, J. R., Pena, J., Hunsberger, A. and McCoy, C. W. (1999). Effects of temperature and host range on suppression of *Diaprepes abbreviatus* (Coleoptera: Curculionidae) by entomopathogenic nematodes. *Journal of Economic Entomology* 92: 1086–1092.

Shapiro-Ilan D, Lewis, E. E. and Tedders, W. L. (2003). Superior efficacy observed in entomopathogenic nematodes applied in infected-host cadavers compared with application in aqueous suspension. *Journal of Invertebrate Pathology* 83: 270-272.

Singh, J. and Bardhan, V. (1974). Effectiveness of DD-136, an entomophilic nematode against insect pests of agricultural importance. *Current Science* 43: 662.

Smart, G.C., Jr. and Nguyen, K.B. (1994). Role of entomopathogenic nematodes in biological control. In: D. Rosen, F. D. Bennett, and J. L. Capinera, eds. Pest management in the subtropics: Biological control—A Florida perspective. Andover, UK: Intercept. pp. 231-252.

Solomon, A., Paperna, I. and Glazer, I. (1999). Desiccation survival of the entomopathogenic nematode *Steinernema feltiae*: Induction of anhydrobiosis. *Nematology* 1: 61-68.

Sosamma, V.K. (2003). Utilization of EPN in plantation crops. In: current status of research on entomopathogenic nematodes in India (Eds, Hussaini, S. S., Rabindra, R. J., Nagesh, M.) Project Directorate of Biological Control, Bangalore, India. pp. 109-112.

Stuart, R.J. and Gaugler, R. (1996). Genetic adaptation and founder effect in laboratory populations of the entomopathogenic nematode *Steinernema glaseri*. *Canadian Journal of Zoology* 74: 164–170.

Thomas, G.M. and Poinar, G.O. 1979). *Xenorhabdus* gen. nov., a genus, of entomopathogenic nematophilic bacteria of the family Enterobacteriacea. *International Journal of Systematic Bacteriology* 29: 352-360.

Varadarasan, S., Hafitha, N.M., Hussaini, S.S., Chandrasekar, S.S., Ansar Ali, M.A. and Thomas, J. (2009). Field evaluation of different formulation of Entomopathogenic nematodes (EPN) for the management of cardamom root grub. Paper presented at fifth International Conference on Biopesticides: Stakeholders perspective, 26–30 April 2009 by society for promotion and innovation of Biopesticides. Abstracts, pp. 195.

Varadarasan, S., Sooravan, T., Chandrasekar, S. S., Ansar Ali, M. A. and Thomas, J. (2006). Survey for entomopathogenic nematodes in cardamom growing areas of Kerala and Tamil Nadu. *Journal of Plantation Crops* 34: 392–400.

Vasanthi, V. and Raviprasad, T. N. (2012). Relative susceptibility of cashew stem and root borers (CSRB), *Plocaederus* spp. and *Batocera rufomaculata* (De Geer) (Coleoptera: Cerambycidae) to entomopathogenic nematodes. *Journal of Biological Control* 26: 23-28.

Volgyi, A., Fodor, A. and Szentirmai, A. (1998). Phase variation in *Xenorhabdus nematophilus*. *Application of Environment Microbiology* 64: 1188-1193.

White, G. F. (1927). A method for obtaining infective nematode larvae from cultures. *Science* 66: 302–303.

Wouts, W. M. (1980). The biology, life cycle, and redescription of *Neoaplectana bibionis* Bovien, 1937 (Nematoda: Steinernematidae). *Journal of Nematology* 12: 62-72.

Yang, H., Jian, H., Zhang, S. and Zhang, G. (1997). Quality of the entomopathogenic nematode *Steinernema carpocapsae* produced on different media. *Biological Control* 10: 193–198.

Index

Figure 8.1: Tea Mosquito Bug Damage on Cashew.
(a) On shoot, (b) Severe damage on plantations. (p. 179)

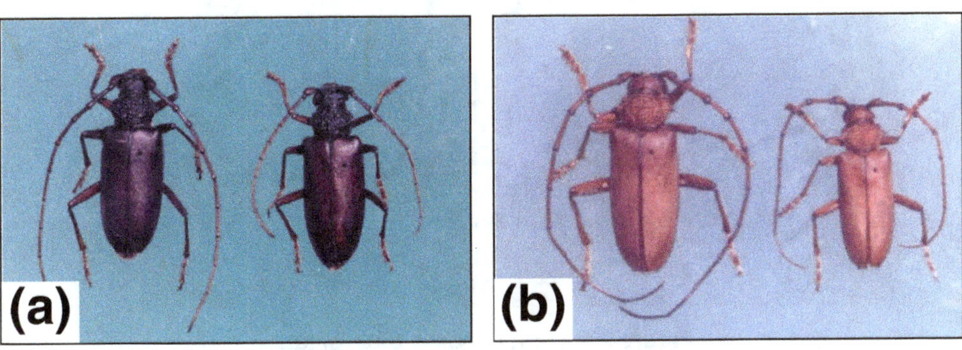

Figure 8.2: Stem and Root Borer.
(a) Adults of *Plocaederus ferrugineus*, (b) Adults of *P. obesus*. (p. 181)

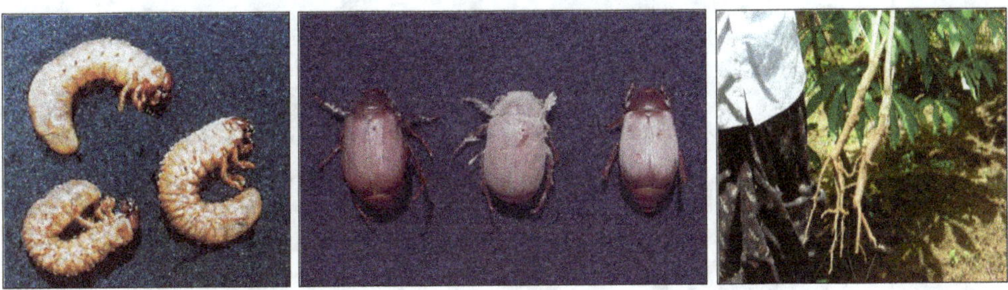

Figure 9.1: Cockchafer Grubs, Adults and Infested Young Rubber Plants. (p. 200)

Figure 9.2: Egg, Larva, Pupa and Adult of the Bark Feeding Caterpillar, *Aetherastis circulate* and Infested Rubber Tree. (p. 201)

Figure 9.3: Bark Feeding Caterpillar, *Ptochoryctis rosaria* and Infested Rubber Tree. (p. 201)

Figure 9.4: Scale Insect and Mealybug Infestation on Rubber Plant. (p. 202)

Figure 9.5: Termite Infested Rubber Tree. (p. 203)

Figure 9.6: Symptoms and Damage Caused by Borer Beetle Attack. (p. 204)

Figure 9.7: Snail. (p. 205)

Figure 9.8: Slug. (p. 205)

Figure 9.9: Damage Caused by Rats. (p. 205)

Figure 9.10: Mooply Beetles. (p. 206)

Figure 12.1

(a) Infective juveniles of EPN, (b,c) EPN infected Red palm weevil grubs and pupa, (d) EPN infected cadavers in paper boats for leaf axil placement. (p. 243)